D1552059

SOLUTION OF CONTINUOUS NONLINEAR PDEs THROUGH ORDER COMPLETION

NORTH-HOLLAND MATHEMATICS STUDIES 181
(Continuation of the Notas de Matemática)

Leopoldo Nachbin passed away in April 1993. As Editor of the Mathematics Studies, he will be succeeded by Saul Lubkin.
The present book was recommended for publication by Leopoldo Nachbin.

NORTH-HOLLAND – AMSTERDAM • LONDON • NEW YORK • TOKYO

SOLUTION OF CONTINUOUS NONLINEAR PDEs THROUGH ORDER COMPLETION

Michael B. OBERGUGGENBERGER

Institut für Mathematik und Geometrie
Universität Innsbruck
Innsbruck, Austria

Elemér E. ROSINGER

Department of Mathematics and Applied Mathematics
University of Pretoria
Pretoria, South Africa

1994

NORTH-HOLLAND – AMSTERDAM • LONDON • NEW YORK • TOKYO

ELSEVIER SCIENCE B.V.
Sara Burgerhartstraat 25
P.O. Box 211, 1000 AE Amsterdam, The Netherlands

ISBN: 0 444 82035 3

This book is printed on acid-free paper.

Printed in the Netherlands.

Dedicated to
Anne-Sophie
and
Myra-Sharon

FOREWORD

This work inaugurates a *new* and *general* solution method for *arbitrary continuous nonlinear* PDEs. The solution method is based on Dedekind *order completion* of usual spaces of smooth functions defined on domains in Euclidean spaces. However, the nonlinear PDEs dealt with need *not* satisfy any kind of monotonicity properties. Moreover, the solution method is completely *type independent*. In other words, it does not assume anything about the nonlinear PDEs, except for the *continuity* of their left hand term, which includes the unknown function. Furthermore, the right hand term of such nonlinear PDEs can in fact be any given *discontinuous* and *measurable* function.

One of the *advantages* and *novelties* of this solution method is that the resulting generalized solutions of arbitrary continuous nonlinear PDEs can be assimilated with usual *measurable functions* defined on Euclidean domains.

By requiring minimal smoothness conditions, the method based on order completion goes quite far beyond all other linear or nonlinear methods developed earlier in the literature in order to provide generalized solutions for PDEs. In addition, owing to the rather ultimate intuitive clarity of the concept of order, this method offers a new and basic insight into the mechanisms involved both in *existence* results and the *structure* of generalized solutions.

Finally, owing to the same clarity of the concept of order, this method leads to a new *numerical solution* method for large classes of nonlinear PDEs. This new numerical method - based on a direct approximation - is presented elsewhere.

The work is divided in three parts, followed by an appendix.

Part I, in Sections 2 - 5, presents the basic idea, as well as the general existence results obtained through the order completion method. Section 2 is consecrated to certain basic and rather elementary - although less than customary - approximation properties related to the local and global solution of arbitrary continuous nonlinear PDEs. In Sections 3 and 4 the relevant spaces of generalized functions are introduced. The basic existence results are presented in Theorems 5.1 - 5.3 and Corollary 5.1 in Section 5.

Section 6 offers a first clarification of certain basic aspects of the order completion method in solving PDEs. For that purpose, a few examples of PDEs and of their generalized solutions are presented, together with comments on the nature and meaning of such solutions.

Part I ends with a rather lengthy Section 7 which is dedicated to the *structure* of the generalized solutions obtained in Section 5. In subsection 7.1, it is shown that such generalized solutions for arbitrary continuous nonlinear PDEs can be assimilated with usual *measurable functions* defined on Euclidean domains.

In Subsection 7.2 a further result on the *structure* of the generalized solutions is presented by showing that the spaces of generalized functions $\hat{\mathcal{M}}^m_T(\Omega)$ have a *flabby sheaf* structure. This is a particularly important point. Indeed, as is known, Kaneko, the space $\mathcal{D}'(\Omega)$ of the L Schwartz distributions is *not* a flabby sheaf, this being one of its significant *disadvantages*. In other words, if Δ is an open subset of Ω and T is a distribution on Δ, that is, $T \in \mathcal{D}'(\Delta)$, then in general, one *cannot* find a distribution $S \in \mathcal{D}'(\Omega)$ such that T is given by S restricted to Δ. Contrary to that situation, the spaces of generalized functions $\hat{\mathcal{M}}^m_T(\Omega)$ are flabby sheaves, this fact being another *advantage* of the order completion method in solving nonlinear PDEs. Finally, Subsection 7.3 offers several technically more involved examples and counterexamples connected with the relationship between measurable functions and generalized solutions.

Part II gives a number of nontrivial applications of the existence results in Section 5.

First, in Section 8, the solution of the Cauchy problem for a large class of first order nonlinear systems is presented. These systems can have any finite number of independent variables. This application shows the *versatility* as well as the *easy* and *straightforward* nature of the order completion method in solving nonlinear PDEs, including systems.

In Section 9 one takes as if a step back, in order to have a deeper insight into the basic abstract mechanism of the order completion method when applied to obtaining *existence of solutions* for rather general equations which need not necessarily be PDEs. A further such insight, albeit a more specific one, is presented in Section 10.

Section 11 returns to nonlinear systems of PDEs and extends recent results to the case of *rough initial data* given by *measures*. These systems include, among others, the well known Carleman system.

In Section 12 a connection and comparison is established between the order completion method and the classical functional analytic method based on the completion of uniform spaces. Again, several important advantages of the order completion method become apparent.

Sections 13 and 14 complete Part II by presenting several extensions of the order completion method. It is particularly important to note that Section 13 introduces a major further *departure* in the use of the order completion method for the solution of arbitrary continuous nonlinear PDEs. Indeed, unlike in the previous Sections, in Section 13 the use of *non 'pull-back'* partial orders is presented, see (13.1), in order to obtain the existence of global generalized solutions in Theorem 13.1. It should be mentioned that the use of such non 'pull-back' partial orders is essential in Section 18, see (18.19). However, an appropriate development of the method of solving arbitrary continuous nonlinear PDEs through order completion in *non 'pull-back'* partial orders will require a more extensive treatment, which will be presented in a subsequent volume.

Part III is dedicated to what amounts to three different and independent approaches to the *group invariance* of *global generalized solutions* of large classes of nonlinear PDEs. The respective approaches achieve for the first time in the literature the following two results which constitute a major aim of Lie's original 1874 project:

- *global generalized solutions* can be obtained for large classes of linear and nonlinear PDEs, and

- the *group invariance* of these global generalized solutions can be studied within *finite dimensional* manifolds.

In this way, one also obtains three different and independent possible solutions to Hilbert's Fifth Problem when this problem is considered in its natural extended sense. A detailed presentation of related issues can be found in Subsection 16.1.

Section 16 presents the group invariance in the case of global generalized solutions of arbitrary *analytic* nonlinear PDEs, obtained through the global Cauchy-Kovalevskaia theorem, Rosinger [5-7]. One major *advantage* of this approach to group invariance is the following. All *projectable* Lie groups of symmetry for classical solutions will automatically extend to symmetry groups of the global generalized solutions. Related applications are given to delta wave solutions of semilinear hyperbolic equations. However, the group invariance goes beyond the case of projectable groups, as the application to Riemann solvers of the nonlinear shock wave equation indicates.

In Section 17 the group invariance is applied in the case of global generalized solutions obtained for large classes of linear or nonlinear PDEs based on the method in Colombeau [1,2,3]. Here however, one encounters limitations owing to the *growth conditions* required by the Colombeau method. Moreover, the mentioned extension of classical symmetry groups to symmetry groups of global generalized solutions cannot be performed in the easy way encountered in Sections 16 and 18.

Finally, Section 18 presents the group invariance of global generalized solutions obtained through the order completion method. This approach benefits from all the advantages of the approach in Section 16, and in addition, it appears to be still more simple and direct.

The work ends with an Appendix in which the basic Dedekind order completion results of 1937, obtained by MacNeille, is presented without proof. The Appendix also contains a number of further general results on order completion developed by the authors, according to the necessities of the theory in Parts I and II of the work.

This work was started in August 1990, during a three month visit of M Oberguggenberger at the Department of Mathematics and Applied Mathematics of the University of Pretoria, when a first version of a good deal of Parts I and II, as well as of the Appendix was completed in collaboration with E E Rosinger. The idea of solving general, not necessarily monotonous nonlinear PDEs through the Dedekind order completion of spaces of smooth functions was suggested by E E Rosinger, who also outlined certain main avenues, as well as details of the order completion method, in particular, the non 'pull-back', approach in Section 13. Most of the specific results in Sections 2-6 of Part I, as well as Sections 8-12 and 14 of Part II were contributed by M Oberguggenberger, who introduced the 'pull-back' partial orders (4.7) which are essential in the general existence results in Section 5, as well as in their further applications in Sections 8-11 and 14. Several revisions and extensions followed. During February - April 1991, H A Biagioni visited the mentioned department and had a thorough discussion of the first version of Part I and the Appendix with E E Rosinger. A number of corrections and improvements resulted. Later in 1991, E E Rosinger visited for five months the Institut für Mathematik und Geometrie, Universität Innsbruck, where in collaboration with M Oberguggenberger, Parts I, II and the Appendix were brought near to their completion. During this period Y-G Wang contributed several useful suggestions to Part I. Further improvements were obtained during a 1992 visit at Innsbruck by E E Rosinger. Certain suggestions related to the flabbiness of several sheaves used in this work, and aimed at improving Subsection 7.2, were contributed by J W de Roewer. Part III was done during 1993 by E E Rosinger, with help from two of his doctoral students, Y E Walus and M Rudolph.

The authors should like to dedicate a most special, kind and grateful thought to the memory of Prof Leopoldo Nachbin, long time editor of the Mathematics Studies of North-Holland. Over the last decade and a half, Prof Nachbin proved to be by far the most ardent supporter and promoter of the newly emerged nonlinear theories of generalized functions. As one indication of his support and promotion of these theories, let us only mention the following. From the nine research monographs published in this field prior to the present one, five were published by Prof Nachbin during 1980 - 1990, in his mentioned North-Holland series of Mathematics Studies.

The authors are particularly grateful to Prof D Laugwitz for his appreciation of the method and results in this work.

The authors owe special thanks and gratitude to Mrs Yvonne Munro at the University of Pretoria for her particularly kind collaboration in word processing a number of versions of the whole manuscript. The role of editors and publishers may be more important in promoting research, however, the first ever - and obviously sine qua non - stage is that of word processing, and thus, of bringing a manuscript to its publishable form. And the kind collaboration of Mrs Munro is of such a nature as to make an author wish to start writing the next book, as soon as the previous one has been completed.

Last, and certainly not least, the authors wish to express their grateful appreciation to Drs Arjen Sevenster, Associate Publisher at Elsevier, whose generous help over the years has contributed to the publication of several of the research monographs on nonlinear theories of generalized functions.

The Authors

Innsbruck, Pretoria
August 1993

TABLE OF CONTENTS

'... Analysis is but Inequalities ...'

J. Dieudonné

'... Equality is also a fraud ...'

attributed to Lenin
by F. Treves

PART I. GENERAL EXISTENCE OF SOLUTIONS THEORY

1. INTRODUCTION

The basic existence and uniqueness results in Theorems 5.1 and 5.2, Corollary 5.1 and Theorem 5.3 in Section 5, obtained through Dedekind order completion, have been announced at the end of Rosinger [6], see p 370, and they come as a first step in a completely new approach which we could call the 'order first' method in the solution of general nonlinear PDEs, which need not be monotonous.

This 'order first' method will roughly mean that for proving the *existence of solutions* of large classes of nonlinear PDEs we shall *not* need functional analysis or algebra, and instead, we shall only use the Dedekind *order completion* of spaces of smooth functions defined on Euclidean spaces. Needless to say, however, that for *other* problems, such as for instance the regularity of solutions of nonlinear PDEs, we shall be ready, according to need and usefulness, to bring in any kind of mathematical methods available.

In Rosinger [1-8], a nonlinear theory of generalized functions, based on an 'algebra first' approach has been introduced and developed for the solution of rather general classes of continuous nonlinear PDEs. A particular, however quite natural and central case of that approach has been extensively pursued in Colombeau [1,2], while in Oberguggenberger [1] the general approach in Rosinger [1-8] was given additional power, by further extending the ranges of its algebraic framework. A recent account of many of the main ideas, methods and applications to linear and nonlinear PDEs of this 'algebra first' approach can be found in Oberguggenberger [2].

In short, the essence of this 'algebra first' method is to embed the L Schwartz distributions into suitable spaces of generalized functions given by quotient algebras, or sometime quotient vector spaces of the respective form

(1.1) $\mathcal{D}'(\Omega) \subset A = \mathcal{A}/\mathcal{I}$

(1.2) $\mathcal{D}'(\Omega) \subset E = \mathcal{S}/\mathcal{V}$

where one may have, for instance, that $\mathcal{A}$ is a subalgebra in $\mathcal{C}^\infty(\Omega))^{\mathbb{N}}$, $\mathcal{I}$ is an ideal in $\mathcal{A}$, while $\mathcal{S}$ and $\mathcal{V}$ are vector subspaces in $(\mathcal{C}^\infty(\Omega))^{\mathbb{N}}$, where $\Omega \subset \mathbb{R}^n$ is any open set.

Given a nonlinear PDE

$$(1.3) \qquad T(x,D)U(x) = f(x), \quad x \in \Omega$$

one can, under rather general conditions, extend the nonlinear partial differential operator $T(x,D)$ so that it will act between the spaces of generalized functions in (1.1) or (1.2), in any of the following ways

$$(1.4) \qquad T(x,D) : A \longrightarrow A'$$

$$(1.5) \qquad T(x,D) : E \longrightarrow A$$

$$(1.6) \qquad T(x,D) : E \longrightarrow E'$$

Here A' and E' are quotient algebras, respectively vector spaces similar to those in (1.1) and (1.2). Once extensions such as in (1.4) - (1.6) have been constructed, the solution of the nonlinear PDE in (1.3) can be obtained in that framework. The important fact to note is that, in view of the embeddings (1.1), (1.2), the solutions of nonlinear PDEs in (1.3) obtained within the frameworks in (1.4) - (1.6) will contain as particular cases the known classical as well as distributional solutions. Furthermore, one obtains solutions for important classes of earlier unsolved or unsolvable linear and nonlinear PDEs. For instance, the method in Rosinger [1-8], which develops the general nonlinear theory of extensions of the type (1.4) and (1.5), yields results such as a global version of the Cauchy-Kovalevskaia theorem, as well as an algebraic characterization for the solvability of arbitrary polynomial nonlinear PDEs with continuous coefficients. The method in Colombeau [1,2] is based on extensions of the type

$$(1.7) \qquad T(x,D) : \mathcal{G}(\Omega) \longrightarrow \mathcal{G}(\Omega)$$

where $\mathcal{G}(\Omega)$ is a specific instance of the quotient algebras A in (1.1) which however enjoys many optimal properties. This method yields the solution of large classes of earlier unsolved or unsolvable linear and nonlinear PDEs, see Rosinger [5, pp 145-192]. In Oberguggenberger [1] instances of the very general extensions (1.6) have been considered in the case of nonlinear PDEs. More precisely, generalized solutions for systems of semilinear wave equations with rough initial data have been obtained in conditions which contain as particular cases some of the important earlier results, such as for instance in Rauch & Reed. Further extensions of these results have recently been presented in Gramchev [1,2] and Oberguggenberger & Wang.

It is important to note that the nonlinear theory of generalized functions in Rosinger [1-8], Colombeau [1,2] and Oberguggenberger [1,2] is nontrivial, since it had to overcome the constraints imposed by the so called Schwartz impossibility result, Schwartz. Furthermore, the interest in the generalized solutions obtained for various classes of nonlinear PDEs is not limited to the sphere of exact solutions. Indeed, the method in Colombeau [1,2] has led to important results in the numerical solution of nonlinear nonconservative systems of PDEs, see Biagioni and the literature cited there.

As mentioned in the Foreword, Chapter 1 and Final Remarks in Rosinger [6], the need for an 'algebra first' approach in the study of generalized solutions of nonlinear PDEs arises from the fact that the so called Schwartz impossibility result has its roots in a rather simple and basic algebraic conflict between discontinuity, multiplication and differentiation. A further detailed study of this basic conflict can be found in Oberguggenberger [2, pp XI, 1-36]. As a consequence of that situation, as well as of the sequential approach to weak and generalized solutions originated by Mikusinski, one is led to the construction of a nonlinear theory of generalized functions along the lines in (1.1) - (1.7). A typical feature of that construction is that functional analytic methods are hardly at all used at the beginning. Instead, one deals with algebraic properties of quotient vector spaces and quotient algebras of sequences of smooth functions on domains in Euclidean spaces.

The mentioned relegation of functional analysis to a secondary role should not come as a surprise. Indeed, the classical Cauchy-Kovalevskaia theorem for instance, has been proved without functional analytic methods, yet it is by far the most general and powerful nonlinear existence, uniqueness and regularity result. In fact, nearly one century of functional analysis has not been able to improve on that theorem on its own terms of nonlinear, type independent generality. And so it comes to pass that the Cauchy-Kowalevskaia theorem remains an unsurpassed maximal point. Still, the only 'hard' mathematics used in the proof of that theorem is the rather obvious manipulation of classical, Abel type majorants for complex power series, and in particular, the ability to sum up a geometric series. By the way, recently, a certain awareness starts to emerge, Evans, about the limitation of functional analytic methods in the study of nonlinear PDEs.

Coming back to the aim of this work, namely, the start of a completely new method in solving general nonlinear PDEs through Dedekind order completion, the following should be noted. Modern mathematics is a multilayered theory, in which successive layers include and enrich by further particularization the more basic and general ones. For instance, we can note the following succession of less and less basic layers:

- set theory
- binary relations
- order
- algebra
- topology
- functional analysis
- etc.

The 'algebra first' approach initiated and developed in Rosinger [1-8], brought with it a certain 'desescalation' from the rather exclusively functional analytic methods in the modern treatment of linear and nonlinear PDEs, to the algebraic methods in rings of continuous or smooth functions.

In this work, as announced in Rosinger [6, pp 369, 370], a further 'desescalation' is initiated, this time to an 'order first' approach.

Concerning this 'order first' approach, two more comments are appropriate.

In the study of linear PDEs, for more than a decade by now, there exists a precedent for the 'order first' approach. Indeed, in Brosowski, the linear Dirichlet problem

$$(1.8) \quad \begin{aligned} &\Delta U(x) = 0, \quad x \in \Omega \\ &U(x) = u(x), \quad x \in \partial\Omega \end{aligned}$$

with $\Omega \subset \mathbb{R}^n$ open and bounded is considered, and a solution method is presented which is based on the Dedekind order completion of the space $\mathcal{C}(\partial\Omega)$ of continuous functions on the compact set $\partial\Omega \subset \mathbb{R}^n$. However, as it stands, this method does not extend to arbitrary linear PDEs, let alone to sufficiently general classes of nonlinear PDEs.

In its essence, the basic existence and uniqueness results in Theorems 5.1 and 5.2, Corollary 5.1 and Theorem 5.3 in Section 5 can be formulated as follows. Given a m-th order nonlinear PDE

$$(1.9) \quad T(x,D)U(x) = f(x), \quad x \in \Omega \subset \mathbb{R}^n$$

if the nonlinear partial differential operator $T(x,D)$ is for instance continuous in $x \in \Omega$, it will define a mapping

$$(1.10) \quad T(x,D) : \mathcal{C}^m(\Omega) \longrightarrow \mathcal{C}^0(\Omega).$$

The problem is that especially in the *nonlinear* case, even for the most simple instances of $T(x,D)$, the mapping (1.10) is unfortunately *not surjective*. This means, of course, that for a given $f \in \mathcal{C}^0(\Omega)$ in (1.9), we *cannot* always find a *classical solution* $U \in \mathcal{C}^m(\Omega)$. The obvious way out of this impasse is to *embed* $\mathcal{C}^m(\Omega)$ into a convenient space X of *generalized functions* and to *extend* $T(x,D)$ to X, so that, although we may happen to have

$$(1.11) \qquad T(x,D)\ \mathcal{C}^m(\Omega) \underset{\neq}{\subset} \mathcal{C}^0(\Omega)$$

we shall nevertheless have the *reversed* inclusion

$$(1.12) \qquad T(x,D)\ X \supseteq \mathcal{C}^0(\Omega).$$

Needless to say, it may be convenient to have reversed inclusions of the stronger type

$$(1.13) \qquad T(x,D)X \supseteq \mathcal{F} \supseteq \mathcal{C}^0(\Omega)$$

where $\mathcal{F}$ is a certain class of functions defined on Ω. In such a case (1.13) means that the nonlinear PDE in (1.9) will have a generalised solution $U \in X$ for every *discontinuous* right hand term $f \in \mathcal{F}$.

As seen in Theorem 5.3 in Section 5, one can indeed obtain a stronger reversed inclusion (1.13) for a class $\mathcal{F}$ of *measurable* functions on Ω.

Therefore, the nonlinear PDEs in (1.9) can be solved for certain *discontinuous measurable* right hand terms f.

The functional analytic methods obtain the spaces X of generalized functions through a *topological completion*. The 'algebra first' method originated and developed in Rosinger [1-8], constructs the spaces X according to (1.1) - (1.5).

The 'order first' method initiated in this paper will construct spaces X of generalized functions through Dedekind type *order completion* of spaces of smooth functions. In other words, basically, we shall have

$$(1.14) \qquad X = (\mathcal{C}^0(\Omega), \leq)^{\wedge}$$

where $\wedge$ is an order completion.

Finally, it is important to recall that so called deep theorems in mathematics, in particular in functional analysis, may in fact involve much more simple and basic mathematical structures than it may appear at first sight. A good example in this regard is the 1936 so called 'spectral theorem' of Freudenthal, see Luxemburg & Zaanen, which is in fact but a theorem about order, and it can be proved in terms of order. Nevertheless, it happens that the celebrated theorems of spectral representation of Hermitian and in fact normal operators in Hilbert space, as well as the Radon-Nikodym theorem in measure theory are its rather immediate corollaries. Furthermore, the Poisson formula for harmonic functions in an open circle is also a particular case of Freudenthal's theorem.

One of the most attractive and promising aspects of the 'order first' approach initiated in this work is that it does not need the introduction of various extraneous and possibly arbitrary entities or structures, beyond those which are inevitably involved, namely $\mathcal{C}^m(\Omega)$, $\mathcal{C}^0(\Omega)$ and of course, the nonlinear partial differential operators $T(x,D)$.

An important effect is that the *structure* of the generalized functions, which are introduced by the order completion of spaces of smooth functions, is surprisingly *simple*. Indeed, these *generalized functions* are in essence just usual *measurable functions* on domains in Euclidean spaces. Moreover, the understanding of the structure of these generalized functions is very much helped by the well known earlier literature on the structure of the Dedekind order completion of spaces of continuous functions, see Zaharov, Zaanen, Mack & Johnson, Nakano & Shimogaki, Dilworth.

This simple and familiar structure of the generalized functions which results from the method of order completion appears to offer significant advantages in the *numerical solution* of nonlinear PDEs, a subject which is to be presented in a subsequent work. These advantages follow from the fact that the respective numerical solutions can be obtained by a straightforward approximation in suitable spaces of functions, without the need for the discretizations usually encountered in finite difference, finite element or spectral methods. This numerical application further shows that the method of solution of nonlinear PDEs through order completion, although it may appear too abstract, does in fact reach to the most applicative aspects involved, and does so in effective and novel ways.

After all, it may indeed be the case that 'analysis is but inequalities'. Inequalities which describe not so much estimates, but rather order ...

2. APPROXIMATION OF SOLUTIONS OF CONTINUOUS NONLINEAR PDEs

The nonlinear PDEs considered in this work are of the form

$$(2.1)\qquad T(x,D)U(x) = f(x), \quad x \in \Omega$$

where $\Omega \subset \mathbb{R}^n$ is open, f is given, U is the unknown function, while

$$(2.2)\qquad T(x,D)U(x) = F(x,U(x),\ldots,D_x^p U(x),\ldots), \quad x \in \Omega, \quad p \in \mathbb{N}^n, \quad |p| \leq m$$

with F being *jointly continuous* in all of its variables.

For convenience, we shall first consider the particular case of (2.2) when Ω is *bounded* and $T(x,D)$ is of the *polynomial* nonlinear form, with continuous coefficients

$$(2.3)\qquad T(x,D) = D_t^m + \sum_{1\leq i\leq h} c_i(x) \prod_{1\leq j\leq k_i} D_t^{p_{ij}} D_y^{q_{ij}}$$

where we have $x = (t,y) \in \Omega$, $t \in \mathbb{R}$, $y \in \mathbb{R}^{n-1}$, $p_{ij} \in \mathbb{N}$, $q_{ij} \in \mathbb{N}^{n-1}$, $p_{ij} < m$ and $p_{ij} + |q_{ij}| \leq m$. The coefficients in (2.3) are supposed to satisfy

$$(2.4)\qquad c_i \in \mathcal{C}^0(\bar{\Omega})$$

where for $\ell \in \bar{\mathbb{N}} = \mathbb{N} \cup \{\infty\}$ we denote $\mathcal{C}^\ell(\bar{\Omega}) = \mathcal{C}^\ell(\mathbb{R}^n)|_{\bar{\Omega}}$.

Although the type of nonlinear PDEs in (2.1), (2,3) covers a particularly large class of evolution type equations of applicative interest, the method of solving nonlinear PDEs through order completion is *not restricted* to bounded domains Ω, see the sequel, and the end of Section 5. It is *not restricted* either to the type of equations with $T(x,D)$ in (2.3), see end of this Section, as well as Section 8 and the sequel. The only reason that, at this first stage, we study equations of type (2.1), (2,3) and consider them on bounded domains comes from the fact that Proposition 2.1, and especially Lemma 2.1 below, are so much easier to prove under these conditions. Therefore, in this way, we can gain a good insight into the essence of the mathematics involved, without being distracted by technicalities.

The basic result on the approximation of solutions is the following, and it is important to note that we do *not* require any kind of *monotonicity* conditions on the nonlinear PDEs in (2.1), (2.3).

Proposition 2.1

Let Ω be bounded and $T(x,D)$ given by (2.3).
If $f \in \mathcal{C}^0(\bar{\Omega})$ then

$$\begin{aligned} &\forall \; \epsilon > 0 : \\ (2.5) \qquad &\exists \; \Gamma_\epsilon \subset \Omega \text{ closed, nowhere dense, } U_\epsilon \in \mathcal{C}^m(\Omega \setminus \Gamma_\epsilon) : \\ &f - \epsilon \leq T(x,D)U_\epsilon \leq f \text{ on } \Omega \setminus \Gamma_\epsilon \end{aligned}$$

Proof

Let us take $a > 0$, such that $\Omega \subset [-a,a]^n$. Let us assume that by continuous extension, we have c_i, $f \in \mathcal{C}^0([-a,a]^n)$. Given now $\epsilon > 0$, from Lemma 2.1 below we obtain $\delta > 0$.

Finally, let us subdivide $[-a,a]$ in intervals $I_1,\ldots,I_\mu$, such that each n-dimensional interval

$$(2.6) \qquad J = I_{\lambda_1} \times \ldots \times I_{\lambda_n}$$

with $1 \leq \lambda_1,\ldots,\lambda_n \leq \mu$, has a diameter not exceeding δ.

If $a_J \in J$ is the centre of the above type of n-dimensional interval J, then according to Lemma 2.1, there exists a polynomial P_{a_J}, such that, for $x \in \text{int } J$, we have

$$f(x) - \epsilon \leq T(x,D)P_{a_J}(x) \leq f(x).$$

Let us now take

$$\Gamma_\epsilon = ([-a,a]^n \setminus \cup \text{ int } J) \cap \Omega$$

where the union ranges over all n-dimensional intervals J in (2.6). If we define $U_\epsilon \in C^m(\Omega \setminus \Gamma_\epsilon)$ by

$$U_\epsilon = P_{a_J} \quad \text{on} \quad \Omega \cap \text{int } J$$

then (2.5) follows.

Remark 2.1

1) The presence of the *closed*, *nowhere dense* singularity set Γ_ϵ in Proposition 2.1 is a quite natural, minimal and unavoidable type of lack of global regularity. Indeed, even in the particular case when both $T(x,D)$ and f are analytic, we cannot expect to have classical solutions on the whole of the domain of definition Ω. In fact, even in that analytic case, the best general global existence result still involves a closed, nowhere dense singularity set, which as in (2.7) below, can in fact have zero Lebesgue measure, see for details Chapter 2 in Rosinger [6].

2) In view of the form of $T(x,D)$ in (2.3), it follows from the proof of Lemma 2.1 that U_ϵ can in fact be chosen on each interval J in (2.6), as polynomial of degree m in t alone, being independent of the variable $y \in \mathbb{R}^{n-1}$, see (2.3).

3) In view of the fact that Γ_ϵ is obtained as a rectangular grid generated by a finite number of hyperplanes, we obviously have

$$(2.7) \qquad \text{mes }(\Gamma_\epsilon) = 0.$$

4) By a similar construction, one can obtain a version of (2.5), in which

$$f \leq T(x,D)U_\epsilon \leq f + \epsilon \quad \text{on} \quad \Omega \setminus \Gamma_\epsilon. \tag{2.8}$$

Lemma 2.1

Given $f \in \mathcal{C}^0(\bar{\Omega})$, then

$$\begin{aligned} &\forall\ \epsilon > 0 : \\ &\exists\ \delta > 0 : \\ &\forall\ x_0 \in \Omega : \\ &\exists\ P_{x_0} \text{ polynomial in } x \in \mathbb{R}^n : \\ &\forall\ x \in \Omega : \\ &\|x - x_0\| \leq \delta \Longrightarrow f(x) - \epsilon \leq T(x,D)P_{x_0}(x) \leq f(x). \end{aligned} \tag{2.9}$$

Remark 2.2

As seen in the proof of Lemma 2.1, we can take

$$\delta = \min \{\tau(\epsilon),\ C,\ 1\}$$

where τ is the modulus of continuity of f and $C > 0$ is suitably chosen, depending on the coefficients c_i in (2.3). Furthermore, we can take

$$P_{x_0}(x) = a_{x_0}(t - t_0)^m$$

where $x = (t,y)$, $x_0 = (t_0,y_0) \in \mathbb{R}^n$, see (2.3), while $a_{x_0} \in \mathbb{R}$ is suitably chosen.

Proof of Lemma 2.1

Since f is uniformly continuous on $\bar{\Omega}$, for given $\epsilon > 0$, we can take $\delta_1 > 0$, such that

$$x, x_0 \in \Omega,\ \|x - x_0\| \le \delta_1 \Longrightarrow |f(x) - f(x_0)| \le \frac{\epsilon}{4} .$$

For given $x_0 = (t_0, y_0) \in \Omega$, let us take

$$P_{x_0}(x) = a(t - t_0)^m, \quad \text{with} \quad x = (t,y) \in \mathbb{R}^n .$$

Then (2.3) yields

$$T(x_0, D) P_{x_0}(x_0) = am!$$

while for $x = (t,y) \in \Omega$, we obtain

$$T(x,D) P_{x_0}(x) = am! + \sum_{1 \le i \le h} c_i(x) \prod_{1 \le j \le k_i} D_t^{p_{ij}} (a(t - t_0)^m) .$$

We note that

$$|D_t^{p_{ij}}(t - t_0)^m| \le m!\ |t - t_0|, \quad \text{for} \quad |t - t_0| \le 1.$$

Let us now take

$$a = (f(x_0) - \epsilon/2)/m!$$

It follows that there exists $C > 0$ depending only on m, $\|f\|_\infty$, $\|c_i\|_\infty$ and the structure of the products in $T(x,D)$, such that, for $x = (t,y) \in \Omega$, we have

$$|T(x,D) P_{x_0}(x) - am!| \le C\ |t - t_0|, \quad \text{for} \quad |t - t_0| \le 1.$$

Let us now take $0 < \delta_2 \le 1$, such that

$$|t - t_0| \le \delta_2 \Longrightarrow C\ |t - t_0| \le \frac{\epsilon}{4} .$$

Then, we can take

$$\delta = \min \{\delta_1, \delta_2\} .$$

Given $x, x_0 \in \Omega$, we have

$$T(x,D)P_{x_0}(x) - f(x) =$$

$$(T(x,D)P_{x_0}(x) - am!) + (am! - f(x_0)) + (f(x_0) - f(x)).$$

But in case $\|x - x_0\| \leq \delta$, it follows that

$$-\frac{\epsilon}{4} \leq T(x,D)P_{x_0}(x) - am! \leq \frac{\epsilon}{4} .$$

Since

$$am! - f(x_0) = -\frac{\epsilon}{2}$$

$$-\frac{\epsilon}{4} \leq f(x_0) - f(x) \leq \frac{\epsilon}{4}$$

this completes the proof of (2.9). ■

Now let us return to the *general* case of the continuous nonlinear PDEs in (2.1), (2.2), and consider them on an arbitrary, possibly *unbounded* domain Ω.

For $x \in \Omega$, let us denote

$$(2.10) \qquad R_x = \left\{ F(x,\xi_0,\ldots,\xi_p,\ldots) \;\middle|\; \begin{array}{l} p \in \mathbb{N}^n, \quad |p| \leq m \\ \xi_p \in \mathbb{R} \end{array} \right\}$$

that is, the range in $\mathbb{R}$ of $F(x,\ldots)$. Since F is jointly continuous in all of its arguments, it follows that R_x must be a bounded or halfbounded interval in $\mathbb{R}$, or possibly, the whole of $\mathbb{R}$ itself. This latter case, which is easier to deal with, see (2.11) below, happens with all of the linear PDEs, as well as with most of the nonlinear PDEs of applicative interest. In particular, it obviously happens with the type of nonlinear partial differential operators in (2.3).

In the case of *systems* of linear or nonlinear PDEs, such as for instance dealt with in Section 8, the range set R_x will obviously be a subset of a corresponding multidimensional Euclidean space.

Given $x \in \Omega$, it is obvious that a *necessary* condition for the existence of a *classical* solution of (2.1), (2.2) in a neighbourhood of x is that

$$(2.11) \qquad f(x) \in R_x.$$

We shall assume that, instead of (2.11), we have satisfied the somewhat stronger condition

$$(2.12) \qquad f(x) \in \text{int } R_x, \quad x \in \Omega .$$

Obviously, when

$$R_x = \mathbb{R}, \quad x \in \Omega$$

then both conditions (2.11) and (2.12) are satisfied. Therefore, the results in the rest of this Section apply to the particular nonlinear PDEs in (2.1), (2.3) as well.

The corresponding general version of Lemma 2.1 above, this time valid for continuous nonlinear PDEs in (2.1), (2.2), on possibly *unbounded* domains Ω, is presented now.

Lemma 2.2

Given $f \in \mathcal{C}^0(\Omega)$, then

$$(2.13) \qquad \begin{aligned} &\forall\ x_0 \in \Omega,\quad \epsilon > 0 :\\ &\exists\ \delta > 0,\quad P \text{ polynomial in } x \in \mathbb{R}^n :\\ &\forall\ x \in \Omega :\\ &\quad \|x - x_0\| \leq \delta \Rightarrow f(x) - \epsilon \leq T(x,D)P(x) \leq f(x) \end{aligned}$$

Proof

Let us take $x_0 \in \Omega$. Then, for $\epsilon > 0$ small enough, (2.12) yields

$$\xi_p \in \mathbb{R}, \quad \text{with} \quad p \in \mathbb{N}^n, \quad |p| \leq m$$

such that

$$F(x_0, \xi_0, \ldots, \xi_p, \ldots) = f(x_0) - \frac{\epsilon}{2} .$$

Let us take a polynomial P in $x \in \mathbb{R}^n$, which satisfies the condition

$$D_x^p P(x_0) = \xi_p, \quad p \in \mathbb{N}^n, \quad |p| \leq m .$$

In this case, obviously, we obtain

$$(2.14) \qquad T(x_0, D) P(x_0) - f(x_0) = -\frac{\epsilon}{2} .$$

Since both $T(x,D)P(x)$ and $f(x)$ are continuous in $x \in \Omega$, property (2.13) follows easily from (2.14). □

The important general approximation result, corresponding to Proposition 2.1 above, and valid for nonlinear partial differential operators of type (2.2), on possibly *unbounded* domains Ω, is given in

Proposition 2.2

If $f \in \mathcal{C}^0(\Omega)$, then

$$(2.15) \qquad \begin{array}{l} \forall \ \epsilon > 0 : \\ \exists \ \Gamma_\epsilon \subset \Omega \ \text{ closed, nowhere dense, } \ U_\epsilon \in \mathcal{C}^m(\Omega \backslash \Gamma_\epsilon) : \\ \qquad f - \epsilon \leq T(x,D) U_\epsilon \leq f \ \text{ on } \ \Omega \backslash \Gamma_\epsilon \end{array}$$

Proof

Let

$$(2.16)\qquad \Omega = \bigcup_{\nu \in \mathbb{N}} K_\nu$$

where, for $\nu \in \mathbb{N}$, the compact sets K_ν are n-dimensional intervals

$$K_\nu = [a_\nu, b_\nu]$$

with $a_\nu = (a_{\nu,1},\ldots,a_{\nu,n})$, $b_\nu = (b_{\nu,1},\ldots,b_{\nu,n}) \in \mathbb{R}^n$, $a_{\nu,1} \leq b_{\nu,1},\ldots,a_{\nu,n} \leq b_{\nu,n}$. We also assume that K_ν, with $\nu \in \mathbb{N}$, are locally finite, that is

$$(2.17)\qquad \begin{array}{l} \forall\ x \in \Omega : \\ \exists\ x \in V_x \subseteq \Omega,\ V_x \text{ neighbourhood of } x : \\ \quad \{v \in \mathbb{N} \mid K_\nu \cap V_x \neq \phi\} \text{ is finite} \end{array}$$

We also assume that the interiors of K_ν, with $\nu \in \mathbb{N}$, are pairwise disjoint. We note that such K_ν, with $\nu \in \mathbb{N}$, exist, see for instance Forster.

Let us take now $\epsilon > 0$ given arbitrarily but fixed. Let us take $\nu \in \mathbb{N}$ and apply Lemma 2.2 to each $x_0 \in K_\nu$. Then we obtain $\delta_{x_0} > 0$ and P_{x_0} polynomial in $x \in \mathbb{R}^n$, such that

$$f(x) - \epsilon \leq T(x,D)P_{x_0}(x) \leq f(x),\quad x \in \Omega \cap \bar{B}(x_0,\delta_{x_0}) .$$

Since K_ν is compact, it follows that

$$(2.18)\qquad \begin{array}{l} \exists\ \delta > 0 : \\ \forall\ x_0 \in K_\nu : \\ \exists\ P_{x_0} \text{ polynomial in } x \in \mathbb{R}^n : \\ \forall\ x \in \Omega : \\ \quad \|x - x_0\| \leq \delta \Rightarrow f(x) - \epsilon \leq T(x,D)P_{x_0}(x) \leq f(x) \end{array}$$

Now, in view of (2.18), within the compact K_ν, we can repeat the argument in the proof of Proposition 2.1 and obtain $\Gamma_{\nu,\epsilon} \subset K_\nu$ a rectangular grid generated by a finite number of hyperplanes. Further, we can define

$$(2.19) \qquad U_{\nu,\epsilon} \in \mathcal{C}^m(K_\nu \backslash \Gamma_{\nu,\epsilon})$$

such that

$$f - \epsilon \leq T(x,D)U_{\nu,\epsilon} \leq f \quad \text{on} \quad K_\nu \backslash \Gamma_{\nu,\epsilon} .$$

But in view of (2.17) it follows that

$$(2.20) \qquad \Gamma_\epsilon = \bigcup_{\nu \in \mathbb{N}} \Gamma_{\nu,\epsilon} \subset \Omega \quad \text{and} \quad \Gamma = \Omega \backslash \bigcup_{\nu \in \mathbb{N}} \text{int } K_\nu \quad \text{are closed and nowhere dense.}$$

Moreover (2.16), (2.19) and (2.20) will yield

$$U_\epsilon \in \mathcal{C}^m(\Omega \backslash (\Gamma_\epsilon \cup \Gamma))$$

which satisfies (2.15). ■

Remark 2.3

1) As in (2.7), in the general case treated in Proposition 2.2, we again have

$$(2.21) \qquad \text{mes } (\Gamma_\epsilon) = 0$$

since according to (2.20), Γ_ϵ is a countable union of rectangular grids, each generated by a finite number of hyperplanes.

2) As in (2.8), by a similar argument, we can obtain a version of (2.15), in which

(2.22) $\quad f \leq T(x,D)U_\epsilon \leq f + \epsilon$ on $\Omega \backslash \Gamma_\epsilon$.

3) It is easy to note, by following the argument in the proof of Lemma 2.2, that (2.11) implies the property

$$\begin{aligned} &\forall\ x \in \Omega : \\ (2.23) \qquad &\exists\ U \in \mathcal{C}^\infty(\Omega) : \\ &\quad T(x,D)U(x) = f(x) \end{aligned}$$

However, a direct application of Lemma 2.2 itself offers the stronger property

$$\begin{aligned} &\forall\ x \in \Omega : \\ &\exists\ \delta > 0 : \\ (2.24) \qquad &\forall\ A \subset \Omega \cap B(x,\delta),\ \ A \text{ finite} : \\ &\exists\ U \in \mathcal{C}^\infty(\Omega) : \\ &\quad T(y,D)U(y) = f(y), \quad y \in A \end{aligned}$$

as well as a proof of it which can have an interest in itself. Indeed, given $x \in \Omega$ and $\epsilon > 0$, we obtain from (2.13)

(2.25) $\quad U_- \in \mathcal{C}^\infty(\Omega)$

such that

(2.26) $\quad T(y,D)U_-(y) \leq f(y), \quad y \in \Omega \cap B(x,\delta)$.

By a similar argument, we can find

(2.27) $\quad U_+ \in \mathcal{C}^m(\Omega)$

for which

(2.28) $\quad f(y) \leq T(y,D)U_+(y), \quad y \in \Omega \cap B(x,\delta)$.

Now, for $\lambda \in [0,1]$ let us define the *convex combination*

$$(2.29) \qquad U_\lambda = (1-\lambda)U_- + \lambda U_+ \in \mathcal{C}^\infty(\Omega)$$

and the *continuous* function

$$\varphi : [0,1] \times \Omega \longrightarrow \mathbb{R}$$

by

$$\varphi(\lambda,y) = T(y,D)U_\lambda(y) - f(y) \ .$$

Then (2.26), (2.28) give

$$\varphi(0,y) \le 0 \le \varphi(1,y), \quad y \in \Omega \cap B(x,\delta) \ .$$

Hence the continuity of φ results in

$$(2.30) \qquad \begin{array}{l} \forall \ y \in \Omega \cap B(x,\delta) : \\ \exists \ \lambda \in [0,1] : \\ \quad T(y,D)U_\lambda(y) = f(y) \end{array}$$

since the above equality just means that $\varphi(\lambda,y) = 0$. Let us now take any finite set $A \subset \Omega \cap B(x,\delta)$ and apply (2.30) to each $a \in A$, obtaining thus $U_{\lambda_a} \in \mathcal{C}^\infty(\Omega)$ which satisfies (2.1) at a. But as A is finite, we can consider on Ω a partition of unity given by $\alpha_a \in \mathcal{C}^\infty(\Omega)$, with $a \in A$, which satisfies

$$(2.31) \qquad \begin{array}{l} \alpha_a = 1 \quad \text{on a neighbourhood of} \quad a \\ 0 \le \alpha_a \le 1 \quad \text{on} \quad \Omega \\ \sum_{a\in A} \alpha_a = 1 \quad \text{on} \quad \Omega \end{array}$$

Let us then define $U \in \mathcal{C}^\infty(\Omega)$ by

$$(2.32) \qquad U = \sum_{a\in A} \alpha_a \cdot U_{\lambda_a} \ .$$

The relations (2.29) - (2.32) will now give (2.24).

The interest in (2.32) is in the fact the function U which it defines, and which satisfies (2.24), is always a *convex combination* of the same two functions U_- and U_+ in (2.25) - (2.28).

Finally we note that in a similar way, one can obtain the following stronger version of (2.24), namely

$$\forall\ A \subset \Omega,\ A \text{ discrete in } \Omega :$$

$$\exists\ U \in \mathcal{C}^\infty(\Omega) \tag{2.33}$$

$$T(x,D)U(x) = f(x),\quad x \in A$$

3. SPACES OF GENERALIZED FUNCTIONS

Let $\Omega \subseteq \mathbb{R}^n$ be an arbitrary nonvoid open set.

In view of Propositions 2.1 and 2.2, it is useful to consider the following spaces of real valued functions. Given $\ell \in \bar{\mathbb{N}}$, we define

$$(3.1) \qquad \mathcal{C}^{\ell}_{nd}(\Omega) = \left\{ u \;\middle|\; \begin{array}{l} \exists\ \Gamma \subset \Omega \ \text{ closed, nowhere dense:} \\ \quad *)\ \ u : \Omega \setminus \Gamma \longrightarrow \mathbb{R} \\ \quad **)\ \ u \in \mathcal{C}^{\ell}(\Omega \setminus \Gamma) \end{array} \right\}$$

Obviously we have the inclusions

$$(3.2) \qquad \mathcal{C}^{\ell}(\Omega) \subset \mathcal{C}^{\ell}_{nd}(\Omega), \quad \ell \in \bar{\mathbb{N}} .$$

An essential role will be played by the Dedekind order completion of a space associated to $\mathcal{C}^0_{nd}(\Omega)$ as follows. Let us define the equivalence relation $\sim$ on $\mathcal{C}^0_{nd}(\Omega)$ by

$$(3.3) \qquad f \sim g \Longleftrightarrow \left[\begin{array}{l} \exists\ \Gamma \subset \Omega, \text{ closed, nowhere dense :} \\ \quad *)\ f,g \in \mathcal{C}^0(\Omega \setminus \Gamma) \\ \quad **)\ f = g \ \text{ on } \ \Omega \setminus \Gamma \end{array} \right]$$

Then, on the quotient space

$$(3.4) \qquad \mathcal{M}^0(\Omega) = \mathcal{C}^0_{nd}(\Omega)/\sim$$

we define the partial order $\leq$ by

$$(3.5) \qquad F \leq G \Longleftrightarrow \left[\begin{array}{l} \exists\ f \in F,\ g \in G,\ \Gamma \subset \Omega \text{ closed nowhere dense :} \\ \quad *)\ f,g \in \mathcal{C}^0(\Omega \setminus \Gamma) \\ \quad **)\ f \leq g \ \text{ on } \ \Omega \setminus \Gamma \end{array} \right]$$

The spaces $\mathcal{C}^0_{nd}(\Omega)$ and $\mathcal{M}^0(\Omega)$ have been considered in Fine et.al., although not in connection with solving PDEs.

It is easy to see that $\mathcal{M}^0(\Omega)$ equipped with this partial order $\le$ is a Riesz space, and also an algebra, see Luxemburg & Zaanen. In particular, it has no maximum or minimum elements. However, in the sequel, we shall not make use of the fact that $(\mathcal{M}^0(\Omega), \le)$ is an algebra or a Riesz space. The reason for this comes from the fact that in the study of rather general *nonlinear* PDEs such as in (2.1), the vector space or algebra structure of $\mathcal{M}^0(\Omega)$ does not at first appear to play an important role. In particular, the *order completion* of $\mathcal{M}^0(\Omega)$ can be made in *most simple* terms, without considering the ways the vector space or algebra structure of $\mathcal{M}^0(\Omega)$ can be extended to that completion. For a treatment of these latter issues see for instance Clifford, Dilworth, Nakano & Shimogaki, or Zaharov.

Therefore, we can apply to $(\mathcal{M}^0(\Omega), \le)$ the most general, abstract Dedekind type order completion due to MacNeille, see Appendix, and obtain the *order complete* space

$$(3.6) \qquad (\hat{\mathcal{M}}^0(\Omega), \le)$$

as well as the embedding

$$(3.7) \qquad \begin{array}{ccc} \mathcal{M}^0(\Omega) & \longrightarrow & \hat{\mathcal{M}}^0(\Omega) \\ F & \longmapsto & <F] \end{array}$$

which preserves suprema and infima, and in addition, it is an *order isomorphical embedding.*

Moreover, in view of MacNeille's construction of the order completion, we can extend (3.7) as follows

$$(3.8) \qquad \begin{array}{ccccc} \mathcal{M}^0(\Omega) & \xrightarrow{\text{inj}} & \hat{\mathcal{M}}^0(\Omega) & \xrightarrow{\text{id}} & \mathcal{P}(\mathcal{M}^0(\Omega)) \\ F & \longmapsto & <F] & \longmapsto & <F] \end{array}$$

For $\ell = 0$, the inclusions (3.2) become

$$(3.9) \qquad \mathcal{C}^0(\Omega) \subset \mathcal{C}^0_{\text{nd}}(\Omega)$$

Let us consider on $\mathcal{C}^0(\Omega)$ the usual order $\leq$, according to which

$$(3.10) \qquad f \leq g \iff \left[\begin{array}{l} \forall x \in \Omega : \\ f(x) \leq g(x) \end{array} \right] .$$

Then we obtain

Lemma 3.1

The embedding

$$(3.11) \qquad \begin{array}{ccc} \mathcal{C}^0(\Omega) & \longrightarrow & \mathcal{M}^0(\Omega) \\ f & \longmapsto & F \end{array}$$

where F is the $\sim$ equivalence class of f, is an order isomorphical embedding of $\mathcal{C}^0(\Omega)$ in $\mathcal{M}^0(\Omega)$.

Proof

First we note that the mapping in (3.11) is indeed injective. For that, take $f,g \in \mathcal{C}^0(\Omega)$ such that $f \sim g$. Then, owing to (3.3), we have

$$(3.12) \qquad f = g \quad \text{on} \quad \Omega \backslash \Gamma$$

for a suitable $\Gamma \subset \Omega$ closed, nowhere dense. It follows that $\Omega\backslash\Gamma$ is open and dense in Ω. Therefore (3.12) and the continuity of both f and g will imply that

$$f = g \quad \text{on} \quad \Omega.$$

Assume now given $f,g \in \mathcal{C}^0(\Omega)$, and let F and G be $\sim$ equivalence classes of f and g respectively. If $f \leq g$ in $\mathcal{C}^0(\Omega)$, then obviously $F \leq G$, since in (3.5) we can take $\Gamma = \phi$. Conversely, let $F \leq G$ in $\mathcal{M}^0(\Omega)$, and assume that $f(x) > g(x)$ for a certain $x \in \Omega$. Then, owing to the continuity of f and g, there exists V, with $x \in V \subset \Omega$, V an open neighbourhood of x, such that $f > g$ on V. On the other hand $F \leq G$ implies that $f \leq g$ on $\Omega\backslash\Gamma$, for a suitable $\Gamma \subset \Omega$ closed, nowhere dense. In this way it must follow that $V \subset \Gamma$, which is absurd.

■

Corollary 3.1

We have the following order isomorphical embeddings

$$(3.13) \qquad \mathcal{C}^0(\Omega) \longrightarrow \mathcal{M}^0(\Omega) \longrightarrow \hat{\mathcal{M}}^0(\Omega) \xrightarrow{\text{id}} \mathcal{P}(\mathcal{M}^0(\Omega))$$

where the embedding $\mathcal{M}^0(\Omega) \longrightarrow \hat{\mathcal{M}}^0(\Omega)$ also preserves suprema and infima.

■

The order complete space $\hat{\mathcal{M}}^0(\Omega)$ will be one of the important spaces of *generalized functions* which are naturally connected with the solution of continuous nonlinear PDEs (2.1), see Theorems 5.1 and 5.2.

A *unique advantage* of the method originated in this work of solving nonlinear PDEs through order completion is precisely in the *very simple* structure of the generalized functions which are elements of $\hat{\mathcal{M}}^0(\Omega)$. In fact, to a good extent these generalized functions are *not* so much generalized since under certain conditions, they can be identified with usual *measurable functions*. Details in this regard are presented in Subsection 7.1.

Remark 3.1

In view of the important role played in the sequel by the spaces of functions

$$\mathcal{C}^\ell_{\text{nd}}(\Omega), \quad \ell \in \bar{\mathbb{N}}$$

it is useful to illustrate the extent to which they are larger than the respective spaces of usual smooth functions

$$\mathcal{C}^\ell(\Omega), \quad \ell \in \bar{\mathbb{N}} \ .$$

A deeper understanding of the role played by the spaces of functions $\mathcal{C}^\ell_{\text{nd}}(\Omega)$, $\ell \in \bar{\mathbb{N}}$, is connected with the *flabbiness* of the sheaves of sections they naturally generate. Details in this regard can be found in Subsection 7.2, and especially in Remark 7.5. Here however, we shall first deal with the simplest, shall we say, real analytic aspects involved.

For that, let be given

(3.14) $\quad \Gamma \subseteq \Omega$ a closed subset

then, Narasimhan, there exists

(3.15) $\quad \gamma \in \mathcal{C}^\infty(\Omega)$

such that

(3.16) $\quad \Gamma = \{x \in \Omega \mid \gamma(x) = 0\}$.

Therefore, given any $g \in \mathcal{C}^\infty(\Omega)$, let us define $f : (\Omega\backslash\Gamma) \longrightarrow \mathbb{R}$ by

(3.17) $\quad f(x) = g(x)/\gamma(x)$ if $x \in \Omega\backslash\Gamma$.

Suppose now that

(3.18) $\quad g(x) \neq 0, \quad x \in \Gamma.$

Then obviously

(3.19) $\quad \Gamma$ nowhere dense $\Longrightarrow f \in \mathcal{C}^\ell_{nd}(\Omega)\backslash\mathcal{C}^\ell(\Omega), \quad \ell \in \bar{\mathbb{N}}$

and f is unbounded in a neighbourhood of every point of Γ.

On the other hand, let us assume that γ in (3.15) and (3.16) is such that

(3.20) $\quad \gamma^{-1}((-\infty,0)), \ \gamma^{-1}((0,\infty)) \neq \phi$.

Then we define the associated Heaviside type function $H_\gamma : (\Omega \backslash \Gamma) \longrightarrow \mathbb{R}$ by

$$(3.21) \qquad H_\gamma(x) = \left| \begin{array}{ll} 0 & \text{if} \quad \gamma(x) < 0 \\ 1 & \text{if} \quad \gamma(x) > 0 \end{array} \right. .$$

Given now any $g, h \in \mathcal{C}^\infty(\Omega)$, we can define $f : (\Omega \backslash \Gamma) \longrightarrow \mathbb{R}$ by

$$(3.22) \qquad f = g + (h - g) \cdot H_\gamma \ .$$

Let us suppose that

$$(3.23) \qquad g(x) \neq h(x), \qquad x \in \Gamma.$$

Then

$$(3.24) \qquad \Gamma \text{ nowhere dense} \Longrightarrow f \in \mathcal{C}^\ell_{nd}(\Omega) \backslash \mathcal{C}^\ell(\Omega), \quad \ell \in \bar{\mathbb{N}}$$

and obviously, f is bounded in a neighbourhood of every point in Ω. Moreover, f is discontinuous at each point of Γ where γ changes sign.

Remark 3.2

We give a simple, *ring theoretic characterization* of the closed, nowhere dense subsets $\Gamma \subseteq \Omega$ which appear in an essential manner in the definition of the spaces $\mathcal{C}^\ell_{nd}(\Omega)$, with $\ell \in \bar{\mathbb{N}}$, as well as of the spaces $\mathcal{M}^0(\Omega)$, $\hat{\mathcal{M}}^0(\Omega)$, and subsequent spaces of generalized functions. Indeed, given any $f \in \mathcal{C}^0(\Omega)$, let us denote its *zero set*, Gillman & Jerison, by

$$(3.25) \qquad Z(f) = \{x \in \Omega \mid f(x) = 0\}.$$

Then, obviously, $Z(f)$ is closed in Ω. The interesting point to note is that

$$(3.26) \qquad \left[\begin{array}{l} Z(f) \text{ is} \\ \text{nowhere dense} \\ \text{in } \Omega \end{array} \right] \Longleftrightarrow \left[\begin{array}{l} f \text{ is } \textit{not} \text{ a} \\ \textit{zero divisor} \\ \text{in } \mathcal{C}^0(\Omega) \end{array} \right] .$$

For the implication '$\Longrightarrow$', let $g \in \mathcal{C}^0(\Omega)$ be such that $f \cdot g = 0$ on Ω. Then clearly

$$Z(g) \supseteq \Omega \setminus Z(f) \ .$$

Hence $Z(g)$ is dense in Ω, which means that $g = 0$ on Ω. Conversely, for the implication '$\Longleftarrow$', assume that $Z(f)$ contains a certain nonvoid, open subset Δ. Then there exists $g \in \mathcal{D}(\Omega)$, $g \neq 0$, such that

$$Z(g) \supseteq \Omega \setminus \Delta \ .$$

Then obviously $f \cdot g = 0$ on Ω, and (3.26) follows. In view of (3.14) - (3.16), we can conclude that, given any

$$\ell \in \bar{\mathbb{N}}$$

then the *closed and nowhere dense* subsets $\Gamma \subsetneq \Omega$ are precisely those sets which are the *zero sets* of functions $f \in \mathcal{C}^\ell(\Omega)$ that are *not zero divisors* in $\mathcal{C}^\ell(\Omega)$.

4. EXTENDING T(x,D) TO THE ORDER COMPLETION OF SPACES OF SMOOTH FUNCTIONS

In view of (2.2) and (3.1) it is clear that we have the mapping

$$(4.1) \qquad T(x,D) : \mathcal{C}^m_{nd}(\Omega) \longrightarrow \mathcal{C}^0_{nd}(\Omega).$$

Moreover, if $u \in \mathcal{C}^m_{nd}(\Omega)$ and $u \in \mathcal{C}^m(\Omega \backslash \Gamma)$ for a suitable closed, nowhere dense $\Gamma \subset \Omega$, then also

$$(4.2) \qquad T(x,D)\, u \in \mathcal{C}^0(\Omega \backslash \Gamma)$$

which means that the singularity set Γ of u will not increase by the application of the nonlinear partial differential operator $T(x,D)$ in (2.2).

The *generalized solutions* of the nonlinear PDEs in (2.1) will be elements of the Dedekind order completion of a space associated to $\mathcal{C}^m_{nd}(\Omega)$ and $T(x,D)$ in the following way. We define the equivalence relation $\sim_T$ on $\mathcal{C}^m_{nd}(\Omega)$ by, see (3.3)

$$(4.3) \qquad u \sim_T v \iff T(x,D)u \sim T(x,D)v.$$

Now, if we define the quotient space

$$(4.4) \qquad \mathcal{M}^m_T(\Omega) = \mathcal{C}^m_{nd}(\Omega)/\sim_T$$

then obviously, the mapping (4.1) can be associated canonically with an *injective* mapping

$$(4.5) \qquad T : \mathcal{M}^m_T(\Omega) \xrightarrow{\text{inj}} \mathcal{M}^0(\Omega)$$

by $T(U) = F$, where F is the unique equivalence class in $\mathcal{M}^0(\Omega)$ of any $T(x,D)u$ where u belongs to the equivalence class U in $\mathcal{M}^m_T(\Omega)$. In other words, we have the *commutative diagram*

$$(4.6) \qquad \begin{array}{ccc} \mathcal{C}^m_{nd}(\Omega) & \xrightarrow{T(x,D)} & \mathcal{C}^0_{nd}(\Omega) \\ \downarrow & & \downarrow \\ \mathcal{M}^m_T(\Omega) & \xrightarrow[T]{inj} & \mathcal{M}^0(\Omega) \end{array}$$

where T is *injective*, while the vertical arrows are the respective canonical mappings onto quotient spaces.

Finally, we can define the partial order $\leq_T$ on $\mathcal{M}^m_T(\Omega)$ as being a 'pull-back' of the partial order $\leq$ on $\mathcal{M}^0(\Omega)$, as follows, see (3.5)

$$(4.7) \qquad U \leq_T V \iff TU \leq TV.$$

As seen in the sequel, this rather straightforward approach in defining the partial order $\leq_T$ has particularly useful consequences. But first let us note that with the above construction, we have:

Lemma 4.1

The mapping

$$(4.8) \qquad \mathcal{M}^m_T(\Omega) \xrightarrow{T} \mathcal{M}^0(\Omega)$$

is an order isomorphical embedding. ∎

The *basic idea* of this work can now be formulated quite simply.

As noted in general, the mappings (1.10) are *not surjective*. Therefore, the same will hold for the mappings (4.1) and (4.8). Let us however extend the mapping (4.8) to the Dedekind *order completion* of both its domain and range. The effect of this extension will be the result on the *existence and uniqueness of solutions* for nonlinear PDEs in (2.1), presented in Theorems 5.1 and 5.2, Corallary 5.1 and Theorem 5.3.

Since the order completion of the range $\mathcal{M}^0(\Omega)$ of the mapping (4.8) was already obtained in Section 3, here we only have to deal with the order completion of its domain $\mathcal{M}_T^m(\Omega)$. For that, we have to comply with the conditions of the general order completion result in the Appendix, namely we need:

Lemma 4.2

The poset $(\mathcal{M}_T^m(\Omega), \leq_T)$ contains no minimum or maximum.

Proof

Given any $U \in \mathcal{M}_T^m(\Omega)$, we shall construct $W \in \mathcal{M}_T^m(\Omega)$, such that

$$(4.9) \qquad W \leq_T U \text{ and } W \neq U.$$

For that, let us take $u \in U$ and $\Gamma \subset \Omega$ closed, nowhere dense, such that $u \in \mathcal{C}^m(\Omega \backslash \Gamma)$. Then, as in (4.2), we have $f = T(x,D)u \in \mathcal{C}^0(\Omega \backslash \Gamma)$. Let us take a nonvoid, bounded open subset $\Delta \subset \Omega \backslash \Gamma$, such that also $\bar{\Delta} \subset \Omega \backslash \Gamma$. Then

$$- \infty < a = \inf \{f(x) \mid x \in \bar{\Delta}\} .$$

We apply now Lemma 2.1 to Δ and $a - 1$ in the role of Ω and f respectively, and obtain for any given $x_0 \in \Delta$ an open neighbourhood $x_0 \in V \subset \Delta$ and a polynomial P_{x_0} in $x \in \mathbb{R}^n$, such that

$$T(x,D)P_{x_0}(x) \leq a - 1, \quad x \in V .$$

Finally, we define $w \in \mathcal{C}^m_{nd}(\Omega)$ by

$$w(x) = \left| \begin{array}{ll} P_{x_0}(x) & \text{if } x \in V \\ u(x) & \text{if } x \in (\Omega \backslash \Gamma) \backslash \bar{V} \end{array} \right.$$

then obviously

$$w \in \mathcal{C}^m(\Omega \backslash (\Gamma \cup \partial V))$$

and

$$T(x,D)w \leq T(x,D)u \quad \text{on} \quad \Omega \backslash (\Gamma \cup \partial V)$$

while

$$T(x,D)w(x) < T(x,D)u(x), \; x \in V.$$

Therefore, the equivalence class W defined by w in $\mathcal{M}_T^m(\Omega)$ will satisfy (4.9). Similarly one can show that $\mathcal{M}_T^m(\Omega)$ does not have a maximum. ■

In view of Lemma 4.2 above, we can again apply the most general, abstract Dedekind type order completion in the Appendix, this time to the poset $(\mathcal{M}_T^m(\Omega), \leq_T)$. In this case we obtain the *order complete* space

$$(4.10) \qquad (\hat{\mathcal{M}}_T^m(\Omega), \leq_T)$$

together with the order isomorphical embedding

$$(4.11) \qquad \begin{array}{ccc} \mathcal{M}_T^m(\Omega) & \longrightarrow & \hat{\mathcal{M}}_T^m(\Omega) \\ U & \longmapsto & <U] \end{array}$$

which also preserves infima and suprema. Furthermore, MacNeille's construction yields

$$(4.12) \qquad \begin{array}{ccccc} \mathcal{M}_T^m(\Omega) & \xrightarrow{\text{inj}} & \hat{\mathcal{M}}_T^m(\Omega) & \xrightarrow{\text{id}} & \mathcal{P}(\mathcal{M}_T^m(\Omega)) \\ U & \longmapsto & <U] & \longmapsto & <U] \end{array}$$

Now, in view of (4.12), we shall define an extension of the mapping T in (4.8), by first defining the mapping, see (A.36)

$$(4.13) \qquad \hat{T} : \mathcal{P}(\mathcal{M}_T^m(\Omega)) \longrightarrow \hat{\mathcal{M}}^0(\Omega)$$

by, see (A.37)

$$(4.14) \qquad \hat{T}(A) = (T(A))^{u\ell} .$$

In view of (4.12), we can restrict the mapping $\hat{T}$ in (4.13) to $\hat{\mathcal{M}}^m_T(\Omega)$, and thus indeed obtain the intended *extension* of the mapping T in (4.8), according to:

Proposition 4.1

We have the commutative diagram

$$(4.15) \qquad \begin{array}{ccc} \mathcal{M}^m_T(\Omega) \ni U & \xrightarrow{\quad T \quad} & T(U) \in \mathcal{M}^0(\Omega) \\ \downarrow & & \downarrow \\ \hat{\mathcal{M}}^m_T(\Omega) \ni <U] & \xrightarrow[\quad \hat{T} \quad]{} & \hat{T}(<U]) = <T(U)] \in \hat{\mathcal{M}}^0(\Omega) \end{array}$$

together with the order isomorphical embedding

$$(4.16) \qquad \hat{\mathcal{M}}^m_T(\Omega) \xrightarrow{\quad \hat{T} \quad} \hat{\mathcal{M}}^0(\Omega)$$

obtained from the restriction of the mapping (4.13) to $\hat{\mathcal{M}}^m_T(\Omega) \subseteq \mathcal{P}(\mathcal{M}^m_T(\Omega))$.

Proof

It follows from Lemma 4.1 above and the Proposition A.1 in Appendix. ∎

Remark 4.1

The following remark is rather fundamental for the understanding of the basic idea of the order completion method in the solution of nonlinear partial differential, as presented in Parts I and II of this work. Namely, suppose given the framework of the mapping, see (4.1)

$$(4.17) \qquad T(x,D) : \mathcal{C}^m_{nd}(\Omega) \longrightarrow \mathcal{C}^0_{nd}(\Omega)$$

still within spaces of classical functions. As seen in (1.10) - (1.13), we shall be interested in extending it to, see (4.6), (4.15)

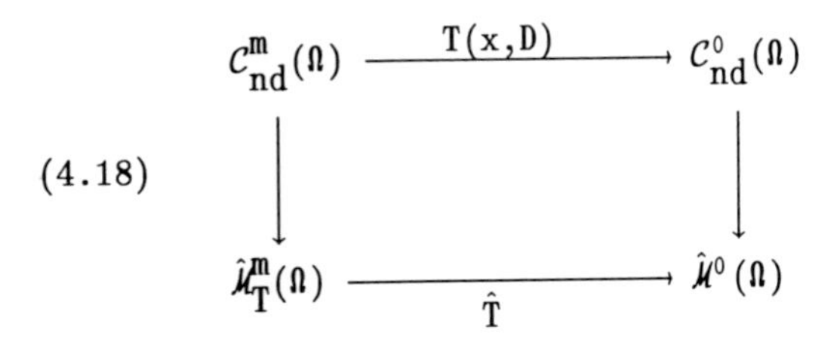

$$(4.18) \qquad \begin{array}{ccc} \mathcal{C}^m_{nd}(\Omega) & \xrightarrow{T(x,D)} & \mathcal{C}^0_{nd}(\Omega) \\ \downarrow & & \downarrow \\ \hat{\mathcal{M}}^m_T(\Omega) & \xrightarrow[\hat{T}]{} & \hat{\mathcal{M}}^0(\Omega) \end{array}$$

with the aim of obtaining at least the inclusion

$$(4.19) \qquad \hat{T}(\hat{\mathcal{M}}^m_T(\Omega)) \supseteq \mathcal{C}^0(\Omega)$$

and thus being able to find a *generalized solution* $U \in \hat{\mathcal{M}}^m_T$ of (2.1), for every $f \in \mathcal{C}^0(\Omega)$. Here however, a crucial role is played precisely by the space

$$(4.20) \qquad \hat{\mathcal{M}}^m_T(\Omega)$$

which will be the space containing the generalized solutions U. And then, the question arises as to what extent is it appropriate to have the definition of that space of generalized functions $\hat{\mathcal{M}}^m_T(\Omega)$ *dependent* on the linear or nonlinear partial differential operator in (2.1) or (4.17). And certainly, in our case, $\hat{\mathcal{M}}^m_T(\Omega)$ does depend on $T(x,D)$, see (4.4).

This dependence, nevertheless, is *not* at all an unusual phenomenon when generalized solutions are constructed for partial differential equations. And even in the case of rather simple and basic *linear* partial differential equations, the respective spaces of generalized functions, which will contain the generalized solutions, are often *dependent* on the partial differential operator under consideration. Indeed, these spaces will be defined by conveniently *pulling back* the topological structure of the range space of the partial differential operators onto their domain of definition. And needless to say, that pulling back is *dependent* on those operators. One such basic example is mentioned in (12.1) - (12.8). Many more can be found in Hörmander, Treves or in Gilbarg & Trudinger.

The problem which arises with such a 'pull-back' procedure is the clarification of the *structure* of the elements of the resulting spaces of generalized functions.

In our case, that is, of the spaces $\hat{\mathcal{M}}^m_T(\Omega)$, a good insight into the structure of their elements will be presented in Section 7. The interest in the respective structural results is, among others, in the fact that they are the *same* for any of the large class of nonlinear partial differential operators in (2.2), as they only depend on the *continuity* properties of these operators, or on the *local* character of their action.

Finally we should note that, as seen in Sections 13 and 18, there exist as well an interest in order structures on spaces of generalized solutions which are *not* of the 'pull-back' type.

5. EXISTENCE OF GENERALIZED SOLUTIONS

First we start with two basic results on the existence and uniqueness of generalized solutions for nonlinear PDEs (2.1) on *bounded* domains Ω which are presented in Theorems 5.1 and 5.2 in this Section. They both hold for arbitrary *continuous nonlinear* partial differential operators (2.2). In the first result the right hand term f in (2.1) can be any continuous function on $\bar{\Omega}$. In the second result the right hand term f in (2.1) can be any *discontinuous measurable* function from the Dedekind order completion $\hat{\mathcal{C}}^0(\bar{\Omega})$ of the space $\mathcal{C}^0(\bar{\Omega})$ of continuous functions on $\bar{\Omega}$.

Fortunately, the existence result in Theorem 5.1 extends in an easy way to *unbounded* domains Ω as well, as seen in Corollary 5.1. The reason for that is the fact that the spaces of *generalized functions* involved, namely

$$\hat{\mathcal{M}}^m_T(\Omega), \quad \hat{\mathcal{M}}^0(\Omega)$$

are *flabby sheaves*, see Remark 5.3 and Subsection 7.2. In this way, we are led to the powerful existence and uniqueness result in Theorem 5.3, a result which contains those in Theorems 5.1 and 5.2, as well as that in Corollary 5.1.

This *flabbiness* of the respective sheaves involved is in fact one of the major advantages of the method of solving linear and nonlinear PDEs through order completion. Indeed, as is well known, Kaneko, *neither* the spaces $\mathcal{C}^\ell(\Omega)$, $\ell \in \mathbb{N}$, of classical solutions, *nor* the space $\mathcal{D}'(\Omega)$ of the Schwartz distributions enjoy the property of flabbiness, this being one of their major disadvantages.

It should be noted, however, that these existence and uniqueness results, as well as other similar ones, see for instance Theorem 8.1, are *in part* consequences of general existence results concerning the solution of abstract equations in order complete structures, presented in Theorems 9.1 and 9.2. Further comments in this regard are presented in Remarks 5.2 and 9.1.

Last but not least, it is important to note the following. The basic existence results in this Section, as well as those in Sections 8 - 11, are obtained through the use of the 'pull-back' type of partial orders defined on the domain of the respective nonlinear partial differential operators, see (4.7). This method, however, can be *extended significantly* by the use of more general, that is, *non 'pull-back'* partial orders defined on the domain of the nonlinear partial differential operators considered. A first presentation of such an extension is given in Section 13, where the corresponding basic existence result in Theorem 13.1 is an obvious extension of Theorem 5.1.

The first basic result on the existence and uniqueness of solutions of continuous nonlinear PDEs of type (2.1), (2.2) in the order completion of spaces of smooth functions can be formulated as follows.

Theorem 5.1

If $\Omega \subset \mathbb{R}^n$ is a bounded open set, then

$$(5.1) \qquad \hat{T}(\hat{\mathcal{M}}_T^m(\Omega)) \supseteq \mathcal{C}^0(\bar{\Omega})$$

that is, for every $f \in \mathcal{C}^0(\bar{\Omega})$, there exists a unique $F \in \hat{\mathcal{M}}_T^m(\Omega)$, such that

$$(5.2) \qquad \hat{T}(F) = \langle f] \quad \text{in} \quad \hat{\mathcal{M}}^0(\Omega)$$

which means that F is the unique generalized solution of the continuous nonlinear PDE in (2.1), (2.2)

$$(5.3) \qquad T(x,D)U(x) = f(x), \quad x \in \Omega.$$

Proof

The uniqueness follows easily from the fact that the mapping $\hat{T}$ in (4.16) is injective.

We take an arbitrary $f \in \mathcal{C}^0(\bar{\Omega})$.

Let $\epsilon > 0$. By the property (2.22), there exist a closed, nowhere dense $\Gamma_\epsilon \subset \Omega$ and $u_\epsilon \in \mathcal{C}^m(\Omega\backslash\Gamma_\epsilon)$, such that

$$(5.4) \qquad f \leq T(x,D)u_\epsilon \leq f + \epsilon \quad \text{on} \quad \Omega\backslash\Gamma_\epsilon .$$

Let U_ϵ be the equivalence class in $\mathcal{M}^m_T(\Omega)$ generated by u_ϵ, see (4.4). Then, in view of (3.13), (3.5), (4.8) and (5.4), we have

$$(5.5) \qquad f \leq T(U_\epsilon) \leq f + \epsilon \quad \text{in} \quad \mathcal{M}^0(\Omega).$$

Let us take now

$$(5.6) \qquad A = \{U_\epsilon \mid \epsilon > 0\} \subseteq \mathcal{M}^m_T(\Omega)$$

and

$$(5.7) \qquad F = A^\ell \subseteq \mathcal{M}^m_T(\Omega).$$

Then, according to (A.16) and (A.12), it follows that

$$(5.8) \qquad F \in \hat{\mathcal{M}}^m_T(\Omega) \ .$$

Furthermore we have

$$(5.9) \qquad \phi \underset{\neq}{\subset} F \underset{\neq}{\subset} \mathcal{M}^m_T(\Omega) \ .$$

Indeed, the first strict inclusion follows from (5.5) and Proposition 2.2, if we note that (2.15) gives $V \in \mathcal{C}^m_{nd}(\Omega)$ such that

$$T(V) \leq f \quad \text{in} \quad \mathcal{M}^0(\Omega)$$

hence (5.5) results in $\mathcal{M}^0(\Omega)$ in the inequalities

$$T(V) \leq T\ (U_\epsilon), \quad \epsilon > 0$$

thus $V \in F$. The second strict inclusion follows from (5.7), (A.8) and the fact that in view of (5.6), we obviously have $A \neq \phi$.

An important property is the following

$$(5.10) \qquad F = \{U \in \mathcal{M}_T^m(\Omega) \mid T(U) \leq f \text{ in } \mathcal{M}^0(\Omega)\}.$$

First we prove the inclusion '$\supseteq$'. Let $U \in \mathcal{M}_T^m(\Omega)$ be such that $T(U) \leq f$ in $\mathcal{M}^0(\Omega)$. Then there exists $u \in U$ and $\Gamma \subseteq \Omega$ closed, nowhere dense, with

$$T(x,D)u \leq f \quad \text{on} \quad \Omega \backslash \Gamma.$$

But then (5.4) gives for all $\epsilon > 0$

$$T(x,D)u \leq T(x,D)u_\epsilon \quad \text{on} \quad \Omega \backslash (\Gamma \cup \Gamma_\epsilon)$$

which means that for $\epsilon > 0$, we have

$$U \leq_T U_\epsilon \quad \text{in} \quad \mathcal{M}_T^m(\Omega)$$

therefore $U \in A^\ell = F$.

For the converse inclusion '$\subseteq$' in (5.10), let $U \in F$ and $u \in U$. In view of (5.7), for every $\epsilon > 0$ there exists a closed, nowhere dense $\Gamma'_\epsilon \subseteq \Omega$ such that

$$(5.11) \qquad T(x,D)u \leq T(x,D)u_\epsilon \leq f + \epsilon \quad \text{on} \quad \Omega \backslash \Gamma'_\epsilon$$

where the second inequality follows from (5.4). Assume that $T(U) \nleq f$ in $\mathcal{M}^0(\Omega)$. Then, for every closed, nowhere dense $\Gamma \subseteq \Omega$, there exists $x_0 \in \Omega \backslash \Gamma$, such that

$$T(x_0,D)u(x_0) > f(x_0).$$

Now, due to the continuity properties of $T(x,D)$, u and f, we can find $x_0 \in \Delta \subseteq \Omega \setminus \Gamma$, Δ open and $\delta > 0$, such that

$$T(x,D)u \geq f + \delta \quad \text{on} \quad \Delta$$

which contradicts (5.11), and completes the proof of (5.10).

Finally, we show that, with F constructed in (5.7) - (5.10), we have indeed

$$(5.12) \qquad \hat{T}(F) = \langle f] \quad \text{in} \quad \hat{\mathcal{M}}^0(\Omega).$$

We note that (5.10) yields

$$T(F) \subseteq \langle f]$$

thus, in view of (4.14), (A.14) and (A.18), we obtain

$$\hat{T}(F) = (T(F))^{u\ell} \subseteq \langle f]^{u\ell} = \langle f].$$

Therefore, in order to prove (5.12), it only remains to show that the inclusion holds

$$(5.13) \qquad \langle f] \subseteq \hat{T}(F).$$

Let us take $g \in \mathcal{C}^0(\bar{\Omega})$, such that

$$(5.14) \qquad g(x) < f(x), \quad x \in \bar{\Omega}.$$

It follows that for suitable $\epsilon > 0$, we obtain

$$g(x) \leq f(x) - \epsilon, \quad x \in \Omega.$$

Then, according to Proposition 2.2, we can find $\Gamma \subseteq \Omega$ closed, nowhere dense and $v \in \mathcal{C}^m(\Omega \setminus \Gamma)$, such that

$$(5.15) \qquad g \leq T(x,D)v \leq f \quad \text{on} \quad \Omega \setminus \Gamma.$$

Let V be the equivalence class of v in $\mathcal{M}^m_T(\Omega)$, see (4.4). Then (5.15) and (4.8) yield

$$(5.16) \qquad g \leq T(V) \leq f \quad \text{in} \quad \mathcal{M}^0(\Omega).$$

In particular, owing to (5.10), we obtain

$$V \in F.$$

Now, according to Lemma 5.1 below, we have in $\mathcal{M}^0(\Omega)$

$$(5.17) \qquad f = \sup \{g \quad \text{in} \quad (5.14)\}.$$

But the embedding

$$\mathcal{M}^0(\Omega) \ni h \longmapsto \langle h] \in \hat{\mathcal{M}}^0(\Omega)$$

preserves infima and suprema, see Corollary 3.1. Therefore, in view of (5.16), we have in $\hat{\mathcal{M}}^0(\Omega)$ the relations

$$\begin{aligned} \langle f] &= \sup \{\langle g] \mid g \;\text{ in } (5.14)\} \leq \\ &\leq \sup \{\langle T(V)] \mid g \;\text{ in } (5.14) \text{ and } V \;\text{ in } (5.16)\} \leq \\ &\leq \sup \{\langle T(U)] \mid U \in F\} \end{aligned}$$

the last inequality resulting from (5.10). We apply now (A.37) in $\hat{\mathcal{M}}^0(\Omega)$ and obtain

$$\langle f] \leq \sup \{\langle T(U)] \mid U \in F\} = \hat{T}(F)$$

which ends the proof of (5.13), and also of (5.12). ■

In the proof of Theorem 5.1 above, we used:

Lemma 5.1

For every $f \in \mathcal{C}^0(\bar{\Omega})$, we have in $\mathcal{M}^0(\Omega)$ the relation

$$(5.18) \qquad f = \sup \left\{ g \;\middle|\; \begin{array}{ll} *) & g \in \mathcal{C}^0(\bar{\Omega}) \\ **) & \forall x \in \bar{\Omega}: \\ & \quad g(x) < f(x) \end{array} \right\}$$

Proof

Obviously, we have the inequality '$\geq$'. We prove now the converse inequality '$\leq$'. For that, assume it is false. Then there exists $H \in \mathcal{M}^0(\Omega)$ such that we have

$$(5.19) \qquad \begin{array}{l} \forall g \in \mathcal{C}^0(\bar{\Omega}) : \\ \left[\begin{array}{l} \forall x \in \bar{\Omega} : \\ \quad g(x) < f(x) \end{array} \right] \Rightarrow g \leq H \end{array}$$

while at the same time

$$(5.20) \qquad f \nleq H.$$

Take $h \in H$ and $\Gamma \subset \Omega$ closed, nowhere dense, such that $h \in \mathcal{C}^0(\Omega\backslash\Gamma)$. Then in view of (5.20), it follows that

$$\exists\, x_0 \in \Omega\backslash\Gamma : h(x_0) < f(x_0).$$

Hence there exists $x_0 \in \Delta \subsetneq \Omega\backslash\Gamma$, Δ open and $\delta > 0$, such that

$$h \leq f - \delta \quad \text{on} \quad \Delta.$$

Let us now take $g = f - \frac{\delta}{2}$, then g satisfies *) and **) in (5.18), while on the other hand

$$(5.21) \qquad h(x) < g(x), \quad x \in \Delta.$$

But in view of (5.19), we obtain

$$(5.22) \qquad g \leq h \quad \text{on} \quad \Omega\backslash\Gamma'$$

for a certain closed, nowhere dense $\Gamma' \subset \Omega$. Now obviously (5.21) and (5.22) contradict each other. ∎

Remark 5.1

1) As seen in the proof of Theorem 5.1, the *essential* property of the *unique* generalized solution F in (5.2), is that it satisfies the relation (5.10), in other words

(5.23) $$F = \{U \in \mathcal{M}_T^m(\Omega) \mid T(U) \leq f \text{ in } \mathcal{M}^0(\Omega)\}.$$

This obviously means that F is the *totality* of classes of *subsolutions* of the continuous nonlinear PDE in (2.1) or (5.3). The way this situation is related to the nonuniqueness of solutions, or to the usual initial and/or boundary value problems associated with the PDEs in (2.1) is discussed in Section 6 next, as well as in Part II of this work.

2) As seen easily, the essential ingredient involving the nonlinear PDE in (2.1), and which is needed in the *existence* result in Theorem 5.1, is the abundance of *local*, *classical subsolutions* granted in Lemma 2.2. The rest is but a quite simple and natural construction related to *partial orders*. In this way, the solution of rather large classes of continuous nonlinear PDEs is reduced to *two separate steps*, both of them quite straightforward: first, the construction of polynomials which provide sufficiently many local, classical subsolutions, followed by an order completion of spaces whose elements are functions patched up from locally smooth functions. This method in which the *global* solution of continuous nonlinear PDEs is reduced to the provision of many enough *local*, *classical subsolutions*, is in essence an extension to continuous nonlinear PDEs of the best possible general result for analytic nonlinear PDEs, which is also *only local*, as granted for instance by the Cauchy-Kovalevskaia theorem. In view of that, it may be more easy to understand now, why such an 'order first' approach is simpler, and at the same time more powerful, than the customary functional analytic methods, which from the very beginning, try to find nothing less than a *global* generalized solution, and do so as if forgetting about what in fact can be expected at best, even in such nice cases as those of analytic nonlinear PDEs.

As mentioned in (1.9) - (1.13), the basic problem facing a theory of generalized solutions for linear or nonlinear PDEs is to *reverse* the strict inclusion (1.11) into an inclusion (1.12) or even (1.13), by a suitable choice of the space X of generalized functions.

As seen in (5.1) in Theorem 5.1 above, one can indeed obtain an inclusion of the type (1.12) for the rather large class of continuous nonlinear PDEs in (2.1) and (2.2), by choosing

$$X = \hat{\mathcal{M}}^m_T(\Omega) \quad . \tag{5.24}$$

In Theorem 5.2 next, we show that with the same choice of X as given in (5.24), the result in (5.1) can be further *strengthened* in a significant manner by replacing $\mathcal{C}^0(\bar{\Omega})$ with the much *larger* space $\hat{\mathcal{C}}^0(\bar{\Omega})$. The extent to which $\hat{\mathcal{C}}^0(\bar{\Omega})$ is indeed larger than $\mathcal{C}^0(\bar{\Omega})$ can be seen in Theorem 7.2 in the sequel. In particular, in view of (7.71), $\hat{\mathcal{C}}^0(\bar{\Omega})$ contains a large class of *discontinuous measurable* functions on Ω. In this way one obtains an inclusion of the type (1.13) with an $\mathcal{F}$ which is significantly larger than $\mathcal{C}^0(\bar{\Omega})$.

Theorem 5.2

If $\Omega \subset \mathbb{R}^n$ is a bounded open set, then

$$\hat{T}(\hat{\mathcal{M}}^m_T(\Omega)) \supseteq \hat{\mathcal{C}}^0(\bar{\Omega}) \tag{5.25}$$

which means that for every $A \in \hat{\mathcal{C}}^0(\bar{\Omega})$, there exists a unique $H \in \hat{\mathcal{M}}^m_T(\Omega)$, such that

$$\hat{T}(H) = A \quad \text{in} \quad \hat{\mathcal{M}}^0(\Omega) \tag{5.26}$$

that is, H is the unique generalized solution of the nonlinear PDE in (2.1), (2.2)

$$T(x,D)U(x) = A, \quad x \in \Omega \tag{5.27}$$

where the right hand term $A \in \hat{\mathcal{C}}^0(\bar{\Omega})$ can be *discontinuous and measurable*.

Proof

In view of (3.13) we have the order isomorphical embedding

$$\begin{array}{ccc} \mathcal{C}^0(\bar{\Omega}) & \xrightarrow{\varphi} & \mathcal{M}^0(\Omega) \\ f & \longmapsto & \varphi(f) \end{array}$$

which according to Proposition A.1 in the Appendix, can be extended to the order isomorphical embedding

$$(5.28) \qquad \begin{array}{ccc} \hat{\mathcal{C}}^0(\bar{\Omega}) & \xrightarrow{\hat{\varphi}} & \hat{\mathcal{M}}^0(\Omega) \\ A & \longmapsto & \hat{\varphi}(A) \end{array}$$

where in view of (A.12) and (A.37), we have

$$A = A^{u\ell} \subseteq \mathcal{C}^0(\bar{\Omega})$$

as well as

$$(5.29) \qquad \hat{\varphi}(A) = \{\varphi(f) \mid f \in A\}^{u\ell} = \sup_{\hat{\mathcal{M}}^0(\Omega)} \{<\varphi(f)] \mid f \in A\} .$$

Therefore, given $f \in A$, from (5.1) we obtain $F_f \in \hat{\mathcal{M}}^m_T(\Omega)$, such that

$$\hat{T}(F_f) = <\varphi(f)] \quad \text{in} \quad \hat{\mathcal{M}}^0(\Omega).$$

Then (5.29) gives

$$\hat{\varphi}(A) = \sup_{\hat{\mathcal{M}}^0(\Omega)} \{\hat{T}(F_f) \mid f \in A\}$$

thus

$$(5.30)\qquad \begin{aligned} \hat{\varphi}(A) &\le \sup_{\hat{\mathcal{M}}^{0}(\Omega)} \{\hat{T}(F) \mid F \in \hat{\mathcal{M}}_T^m(\Omega),\ \hat{T}(F) \subseteq \hat{\varphi}(A)\} \le \\ &\le \hat{T}\,(\sup_{\hat{\mathcal{M}}_T^m(\Omega)} \{F \in \hat{\mathcal{M}}_T^m(\Omega) \mid \hat{T}(F) \subseteq \hat{\varphi}(A)\}) \end{aligned}$$

where the second inequality follows from the fact that $\hat{T}$ is increasing, see (4.16). Moreover, in view of the same (4.16), $\hat{T}$ is in fact an order isomorphical embedding, hence

$$(5.31)\qquad \begin{aligned} &\hat{T}\,(\sup_{\hat{\mathcal{M}}_T^m(\Omega)} \{F \in \hat{\mathcal{M}}_T^m(\Omega) \mid \hat{T}(F) \subseteq \hat{\varphi}(A)\}) \le \\ &\le \hat{T}\,(\inf_{\hat{\mathcal{M}}_T^m(\Omega)} \{G \in \hat{\mathcal{M}}_T^m(\Omega) \mid \hat{\varphi}(A) \subseteq \hat{T}(G)\}) \;. \end{aligned}$$

Now (5.30) and (5.31) give

$$(5.32)\qquad \hat{\varphi}(A) \le \hat{T}\,(\inf_{\hat{\mathcal{M}}_T^m(\Omega)} \{G \in \hat{\mathcal{M}}_T^m(\Omega) \mid \hat{\varphi}(A) \subseteq \hat{T}(G)\}) \;.$$

Let us denote

$$(5.33)\qquad H = \inf_{\hat{\mathcal{M}}_T^m(\Omega)} (G \in \hat{\mathcal{M}}_T^m(\Omega) \mid \hat{\varphi}(A) \subseteq \hat{T}(G)\}$$

then (5.32) implies

$$(5.34)\qquad \hat{\varphi}(A) \le \hat{T}(H) \quad \text{in} \quad \hat{\mathcal{M}}^{0}(\Omega).$$

Our aim is to prove that in (5.34) the opposite inequality $\geq$ holds as well. In view of (5.33) and recalling that $\hat{T}$ is increasing, the inequality (5.32) yields further

$$(5.35)\qquad \hat{T}(H) \le \inf_{\hat{\mathcal{M}}^{0}(\Omega)} \{\hat{T}(G) \mid G \in \hat{\mathcal{M}}_T^m(\Omega),\ \hat{\varphi}(A) \subseteq \hat{T}(G)\} \;.$$

Given now $g \in \mathcal{C}^0(\bar{\Omega})$, from (5.1) we obtain $G_g \in \hat{\mathcal{M}}_T^m(\Omega)$, such that

$$\hat{T}(G_g) = <\varphi(g)] \quad \text{in} \quad \hat{\mathcal{M}}^0(\Omega)$$

therefore (5.35) gives

$$(5.36) \qquad \hat{T}(H) \le \inf_{\hat{\mathcal{M}}^0(\Omega)} \{<\varphi(g)] \,|\, g \in \mathcal{C}^0(\bar{\Omega}),\ \hat{\varphi}(A) \subseteq <\varphi(g)]\} \ .$$

However, given $A \in \hat{\mathcal{C}}^0(\bar{\Omega})$ and $g \in \mathcal{C}^0(\bar{\Omega})$, it follows that

$$\hat{\varphi}(A) \subseteq <\varphi(g)] \quad \text{in} \quad \hat{\mathcal{M}}^0(\Omega) \iff A \subseteq <g] \quad \text{in} \quad \hat{\mathcal{C}}^0(\bar{\Omega})$$

since (5.28) is an order isomorphical embedding. In this way (5.36) yields

$$(5.37) \qquad \hat{T}(H) \le \inf_{\hat{\mathcal{M}}^0(\Omega)} \{<\varphi(g)] \,|\, g \in \mathcal{C}^0(\bar{\Omega}),\ A \subseteq <g]\} \ .$$

Now, it only remains to prove the inequality

$$(5.38) \qquad \inf_{\hat{\mathcal{M}}^0(\Omega)} \{<\varphi(g)] \,|\, g \in \mathcal{C}^0(\bar{\Omega}),\ A \subseteq <g]\} \le \hat{\varphi}(A) \ .$$

This however follows from Lemma 5.2 below. Therefore (5.34), (5.37) and (5.38) finally yield

$$(5.39) \qquad \hat{T}(H) = \hat{\varphi}(A) \quad \text{in} \quad \hat{\mathcal{M}}^0(\Omega)$$

which is precisely (5.26), provided that for the sake of simplicity in notation, we identify the order isomorphical embedding (5.28) with an inclusion

$$(5.40) \qquad \begin{array}{ccc} \hat{\mathcal{C}}^0(\bar{\Omega}) & \longrightarrow & \hat{\mathcal{M}}^0(\Omega) \\ A & \longmapsto & A \end{array}$$

■

As a conclusion to the above proof, it is useful to note that for $A \in \hat{\mathcal{C}}^0(\bar{\Omega})$, we have the implication, see (5.26) and (5.33)

$$(5.41) \qquad \phi \underset{\neq}{\subset} A \underset{\neq}{\subset} \mathcal{C}^0(\bar{\Omega}) \Longrightarrow \phi \underset{\neq}{\subset} H \underset{\neq}{\subset} \mathcal{M}_T^m(\Omega) \ .$$

Indeed, let us take $f \in A$, then (5.2) yields $F \in \hat{\mathcal{M}}_T^m(\Omega)$, such that with the convention in (5.40), we obtain

$$\hat{T}(F) = \langle f] \quad \text{in} \quad \hat{\mathcal{M}}^0(\Omega) \ .$$

But $f \in A \in \hat{\mathcal{C}}^0(\bar{\Omega})$ implies

$$\langle f] \subseteq A$$

hence, for $G \in \hat{\mathcal{M}}_T^m(\Omega)$, with $A \subseteq \hat{T}(G)$, we have

$$(5.42) \qquad \hat{T}(F) = \langle f] \subseteq A \subseteq \hat{T}(G) \ .$$

Now we recall that, according to (4.16), $\hat{T}$ is an order isomorphical embedding, thus (5.42) gives

$$(5.43) \qquad F \leq G \quad \text{in} \quad \hat{\mathcal{M}}_T^m(G) \ .$$

Furthermore, (5.33) and (A.34) imply

$$H = \bigcap_{\substack{G \in \hat{\mathcal{M}}_T^m(\Omega) \\ A \subseteq \hat{T}(G)}} G$$

which together with (5.43) and (5.9) result in

$$\phi \underset{\neq}{\subset} F \subseteq H \ .$$

Let us now assume that

$$(5.44) \qquad H = \mathcal{M}_T^m(\Omega)$$

and let us take $g \in \mathcal{C}^0(\bar{\Omega})\backslash A$. We define

$$B = \{\max\{f,g\} \mid f \in A\}^{u\ell} .$$

Then obviously

$$B \in \hat{\mathcal{C}}^0(\bar{\Omega})$$

and $g \in B\backslash A$ gives

$$A \subsetneq B .$$

Applying (5.26) to B, we obtain $K \in \hat{\mathcal{M}}_T^m(\Omega)$ such that

$$\hat{T}(H) = A \subsetneq B = \hat{T}(K)$$

therefore (4.16) gives

$$H \lneq K \quad \text{in} \quad \hat{\mathcal{M}}_T^m(\Omega)$$

which means that

$$H \subsetneq \mathcal{M}_T^m(\Omega) .$$

In the proof of Theorem 5.2, we needed the following

Lemma 5.2

If $A \in \hat{\mathcal{C}}^0(\bar{\Omega})$ then with the notation in (5.28), we have

$$(5.45)\qquad \begin{aligned}\hat{\varphi}(A) &= \sup_{\hat{\mathcal{M}}^0(\Omega)} \{<\varphi(f)] \mid f \in C^0(\bar{\Omega}),\ <f] \subseteq A\} = \\ &= \inf_{\hat{\mathcal{M}}^0(\Omega)} \{<\varphi(f)] \mid f \in C^0(\bar{\Omega}),\ A \subseteq <f]\} .\end{aligned}$$

Proof

The first equality follows from (A.37) and the fact that A is a cut. Actually, we obtain in addition the relation

$$(5.46)\qquad \hat{\varphi}(A) = (\varphi(A))^{u\ell} .$$

We shall now prove that

$$(5.47)\qquad \inf_{\hat{\mathcal{M}}^0(\Omega)} \{<\varphi(f)] \mid f \in C^0(\bar{\Omega}),\ A \subseteq <f]\} = (\varphi(A^u))^{\ell} .$$

For that we note that according to (A.34), we have

$$(5.48)\qquad \inf_{\hat{\mathcal{M}}^0(\Omega)} \{<\varphi(f)] \mid f \in C^0(\bar{\Omega}),\ A \subseteq <f]\} = \bigcap_{\substack{f \in C^0(\bar{\Omega}) \\ A \subseteq <f]}} <\varphi(f)]$$

while

$$A \subseteq <f] \iff f \in A^u$$

thus (5.48) follows from (A.7).

Now in view of (5.46), it suffices to show that

$$(5.49)\qquad (\varphi(A^u))^{\ell} = (\varphi(A))^{u\ell} .$$

First we note that the inclusion $\supseteq$ in (5.49) follows easily from the monotonicity of φ. Indeed, let us show that

$$(5.50)\qquad \varphi(A^u) \subseteq (\varphi(A))^u .$$

Take $c \in \varphi(A^u)$, then $c = \varphi(b)$ for some $b \in A^u$. It follows that

$$c = \varphi(b) \quad \text{and} \quad a \leq b, \quad \forall\, a \in A$$

thus the monotonicity of φ gives

$$\varphi(a) \leq \varphi(b) = c, \quad \forall\, a \in A$$

and therefore (5.50). Applying (A.14) to (5.50), we obtain

$$(5.51) \qquad (\varphi(A^u))^{\ell} \supseteq (\varphi(A))^{u\ell} .$$

In order to show the converse inclusion

$$(5.52) \qquad (\varphi(A^u))^{\ell} \subseteq (\varphi(A))^{u\ell}$$

we have to make use of the special structure of the spaces $\mathcal{C}^0(\bar{\Omega})$ and $\mathcal{M}^0(\Omega)$.

Let us take any $f \in (\varphi(A^u))^{\ell}$, then

$$(5.53) \qquad f \leq \varphi(b), \quad \forall\, b \in A^u .$$

In order to prove that indeed $f \in (\varphi(A))^{u\ell}$, we have to obtain the implication

$$(5.54) \qquad g \in (\varphi(A))^u \Longrightarrow f \leq g \quad \text{in} \quad \mathcal{M}^0(\Omega) .$$

Assume on the contrary that there exists $g \in (\varphi(A))^u$ for which

$$(5.55) \qquad f \not\leq g \quad \text{in} \quad \mathcal{M}^0(\Omega) .$$

But for this g we will have $g \in \mathcal{M}^0(\Omega)$, as well as

$$(5.56) \qquad \varphi(h) \leq g \quad \text{in} \quad \mathcal{M}^0(\Omega), \quad \forall\, h \in A .$$

Now f and g as elements of $\mathcal{M}^0(\Omega) = \mathcal{C}^0_{nd}(\Omega)\backslash\sim$, see (3.4), are equivalence classes in $\mathcal{C}^0_{nd}(\Omega)$. Let us then take representatives $a \in f$ and $b \in g$. It follows that $a, b \in \mathcal{C}^0_{nd}(\Omega)$, while (5.55) and (3.5) will give a nonvoid, open $\Delta \subsetneq \Omega$, such that $a, b \in \mathcal{C}^0(\bar{\Delta})$ and

$$(5.57) \qquad a(x) > b(x), \quad \forall \; x \in \bar{\Delta} \subsetneq \Omega$$

where $\bar{\Delta}$ is the closure of Δ in $\mathbb{R}^n$.

We note that we can assume

$$(5.58) \qquad A \underset{\neq}{\subset} \mathcal{C}^0(\bar{\Omega}) \; .$$

Indeed, if $A = \mathcal{C}^0(\bar{\Omega})$ then it is easy to see that $\varphi(A)$ is unbounded from above in $\mathcal{M}^0(\Omega)$, hence (A.9) and (A.8) give

$$(\varphi(A))^{u\ell} = \mathcal{M}^0(\Omega)$$

which means that (5.52) will trivially hold. If on the other hand (5.58) is valid, then (A.20) gives a constant $M \in \mathbb{R}$, such that

$$(5.59) \qquad h(x) \leq M, \quad \forall \; h \in A, \; x \in \bar{\Omega} \; .$$

Obviously, we can choose M so that we have in addition

$$(5.60) \qquad b(x) \leq M, \quad \forall \; x \in \bar{\Delta} \; .$$

Now we construct $k \in \mathcal{C}^0(\bar{\Omega})$ such that

$$(5.61) \qquad \begin{array}{lll} k(x) = M & \text{for} & x \in \bar{\Omega}\backslash\Delta \\ k(x) = b(x) & \text{for} & x \in \Delta' \subsetneq \Delta \end{array}$$

while

$$(5.62) \qquad k(x) \geq b(x) \quad \text{for} \quad x \in \bar{\Delta}$$

with a suitably chosen nonvoid, open $\Delta' \subsetneq \Delta$. Then

$$(5.63) \qquad k \in A^u .$$

Indeed, if $h \in A$ then (5.59) gives

$$h(x) \leq M, \quad \forall \; x \in \bar{\Omega}\backslash\Delta$$

while owing to (5.56) and the fact that $b \in \mathcal{C}^0(\bar{\Delta})$, we have

$$h(x) \leq b(x), \quad \forall \; x \in \bar{\Delta}$$

therefore (5.60) - (5.62) imply

$$h \leq k \quad \text{on} \quad \bar{\Omega} .$$

Finally, we note that (5.57) and (5.61) give

$$a(x) > k(x), \quad \forall \; x \in \Delta'$$

which means that

$$f \not\leq \varphi(k) \quad \text{in} \quad \mathcal{M}^0(\Omega)$$

therefore in view of (5.63), we contradicted the assumption that

$$f \in (\varphi(A^u))^\ell .$$

In this way (5.52) in proved, and together with it (5.49) as well. ∎

Remark 5.2

1) In order to appreciate the extent to which Theorem 5.2 is more powerful than Theorem 5.1, we only have to see how much $\hat{\mathcal{C}}^0(\bar{\Omega})$ is larger than $\mathcal{C}^0(\bar{\Omega})$. One answer to this question is given in (7.67). This in particular means that $\hat{\mathcal{C}}^0(\bar{\Omega})$ contains all bounded functions $f : \Omega \longrightarrow \mathbb{R}$ which are *discontinuous* on closed, nowhere dense subsets $\Gamma \subset \Omega$.

2) In view of Theorem 9.2, one may ask whether the result in Theorem 5.2 is not in fact a direct consequence of Theorem 5.1, in which case the above rather lengthy proof of Theorem 5.2 would be superfluous. The fact, however, is that with the notations in Sections 4 and 9, we have, see in particular (4.6) and (9.1)

$$X = \mathcal{C}^m_{nd}(\Omega), \quad Y = \mathcal{M}^0(\Omega)$$

$$\begin{aligned} T : X &\longrightarrow Y \\ u &\longmapsto V \end{aligned}$$

where V is the $\sim$ equivalence class of $T(x,D)u$, see (3.4), (4.6). In this way it follows that, see (4.4), (4.5), (9.4)

$$X_T = \mathcal{M}^m_T(\Omega) \ .$$

Therefore Theorem 5.1 does not prove (9.25), but only the lesser inclusion

$$\hat{T}(\hat{X}_T) \supseteq \mathcal{C}^0(\bar{\Omega})$$

while obviously we have

$$\mathcal{C}^0(\bar{\Omega}) \underset{\neq}{\subset} \mathcal{M}^0(\Omega) = Y \ .$$

Remark 5.3

It is rather surprising how easy it is to extend the existence and uniqueness result in Theorem 5.1 to *unbounded* domains $\Omega \subseteq \mathbb{R}^n$. This is a simple and direct effect of the fact that the spaces of functions or generalized functions

$$\mathcal{C}^0_{nd}(\Omega), \quad \mathcal{M}^0(\Omega), \quad \mathcal{M}^m_T(\Omega), \quad \hat{\mathcal{M}}^0(\Omega), \quad \hat{\mathcal{M}}^m_T(\Omega)$$

are *flabby sheaves*, see Subsection 7.2. Therefore they can be easily and naturally extended to, that is, embedded into the respective spaces corresponding to *any* larger open set which contains Ω. This *advantage* of easy and natural embedding or extension is in sharp contradistinction to the case of the spaces $\mathcal{C}^0(\Omega)$ and $\mathcal{D}'(\Omega)$ which are not flabby sheaves, see Kaneko. In this way the spaces of generalized functions

$$\hat{\mathcal{M}}^0(\Omega), \quad \hat{\mathcal{M}}^m_T(\Omega)$$

used in this work enjoy from the start an *advantage* over the space of distributions $\mathcal{D}'(\Omega)$.

Before presenting Corollary 5.1, we need

Lemma 5.3

Given $\Omega \subseteq \mathbb{R}^n$ open, together with a closed nowhere dense subset $\Gamma \subset \Omega$, the following relations hold for every $\ell \in \bar{\mathbb{N}}$

(5.64) $\quad \mathcal{C}^\ell_{nd}(\Omega \backslash \Gamma) = \mathcal{C}^\ell_{nd}(\Omega)$

(5.65) $\quad \mathcal{M}^0(\Omega \backslash \Gamma) = \mathcal{M}^0(\Omega)$

(5.66) $\quad \mathcal{M}^m_T(\Omega \backslash \Gamma) = \mathcal{M}^m_T(\Omega)$

(5.67) $\quad \hat{\mathcal{M}}^0(\Omega \backslash \Gamma) = \hat{\mathcal{M}}^0(\Omega)$

(5.68) $\quad \hat{\mathcal{M}}^m_T(\Omega \backslash \Gamma) = \hat{\mathcal{M}}^m_T(\Omega)$

Proof

The relation (5.64) is an immediate consequence of (3.1), as well as of the fact that a finite union of closed, nowhere dense sets is again closed, nowhere dense. The relation (5.65) follows from (5.64) and (3.3), (3.4). Relation (5.66) is a direct consequence of (5.64), (5.65) and (4.1) - (4.6).

Now (5.67) will result from (5.65) and (A.12), if we note that in the order relation $\leq$ in (3.5) the size of the closed, nowhere dense set Γ is not important. Finally, (5.68) results from (5.66) and (A.12), based on a similar remark on the order relation $\leq_T$ in (4.7). ■

The extension of Theorem 5.1 to *unbounded* domains $\Omega \subseteq \mathbb{R}^n$ is given in

Corollary 5.1

If $\Omega \subseteq \mathbb{R}^n$ is an arbitrary open set, then

$$(5.69) \qquad \hat{T}(\hat{\mathcal{M}}^m_T(\Omega)) \supseteq \mathcal{C}^0_{nd}(\Omega)$$

that is, for every $f \in \mathcal{C}^0_{nd}(\Omega)$, there exists a unique $F \in \hat{\mathcal{M}}^m_T(\Omega)$, such that

$$(5.70) \qquad \hat{T}(F) = \langle f] \quad \text{in} \quad \hat{\mathcal{M}}^0(\Omega)$$

which means that F is the unique generalized solution of the continuous nonlinear PDE in (2.1), (2.2)

$$(5.71) \qquad T(x,D)U(x) = f(x), \quad x \in \Omega.$$

Proof

Let us take $f \in \mathcal{C}^0_{nd}(\Omega)$ and a closed nowhere dense subset $\Gamma \subset \Omega$, such that

$$(5.72) \qquad f\Big|_{\Omega\setminus\Gamma} \in \mathcal{C}^0(\Omega\setminus\Gamma) .$$

Further, let us take a closed nowhere dense subset $\Sigma \subset \Omega\setminus\Gamma$, such that

$$(5.73) \qquad \Omega\setminus(\Gamma \cup \Sigma) = \bigcup_{\nu\in\mathbb{N}} \Omega_\nu$$

where Ω_ν are pair wise disjoint, bounded open sets, which satisfy

$$(5.74) \qquad \bar{\Omega}_\nu \subset \Omega\setminus\Gamma, \quad \nu \in \mathbb{N}$$

where $\bar{\Omega}_\nu$ is the closure of Ω_ν in $\mathbb{R}^n$.

Obviously, we can also assume that

$$(5.75) \qquad \text{mes}\ (\Sigma) = 0$$

however, this is not necessary in the rest of the proof. Regardless of (5.75), it follows that

$$(5.76) \qquad f\Big|_{\bar{\Omega}_\nu} \in \mathcal{C}^0(\bar{\Omega}_\nu), \quad \nu \in \mathbb{N} .$$

In this way, by applying Theorem 5.1 to each Ω_ν, with $\nu \in \mathbb{N}$, the relation (5.76) will give

$$(5.77) \qquad F_\nu \in \hat{\mathcal{M}}^m_T(\Omega_\nu), \quad v \in \mathbb{N}$$

such that

$$(5.78) \qquad \hat{T}(F_\nu) = \langle f_\nu] \quad \text{in} \quad \hat{\mathcal{M}}^m_T(\Omega_\nu), \quad \nu \in \mathbb{N} .$$

Now (5.72) - (5.74), (5.76), (5.77) together with (5.68) and (7.86), (7.87) will yield

$$(5.79) \qquad F \in \hat{\mathcal{M}}^m_T(\Omega \backslash (\Gamma \cup \Sigma))$$

which in view of (5.67) as well as (4.6) and (4.15), satisfies (5.70).

■

Now based on Corollary 5.1, we are in the position to use the abstract result in Theorem 9.2 and obtain the corresponding ultimate *strengthening* of the existence and uniqueness results presented so far in Theorems 5.1 and 5.2, as well as in Corollary 5.1 itself.

Indeed, for arbitrary, possibly *unbounded* domains $\Omega \subseteq \mathbb{R}^n$, as well as for general continuous nonlinear PDEs in (2.1), (2.2), we obtain

Theorem 5.3

For any open set $\Omega \subseteq \mathbb{R}^n$, we have

(5.80) $$\hat{T}(\hat{\mathcal{M}}_T^m(\Omega)) = \hat{\mathcal{M}}^0(\Omega)$$

In other words, for every $A \in \hat{\mathcal{M}}^0(\Omega)$ there exists a unique $F \in \hat{\mathcal{M}}_T^m(\Omega)$, such that

(5.81) $$\hat{T}(F) = A \text{ in } \mathcal{M}^0(\Omega)$$

In particular

(5.82) $$\hat{\mathcal{M}}_T^m(\Omega) \xrightarrow{\hat{T}} \mathcal{M}^0(\Omega)$$

is an order isomorphism, see (4.16).

Proof

In view of Theorem 9.2 and pct 2) in Remark 5.2, it suffices to prove that

(5.83) $$\hat{T}(\hat{\mathcal{M}}_T^m(\Omega)) \supseteq \mathcal{M}^0(\Omega)$$

However, this is a direct consequence of (5.69), as well as of (3.4) and (4.6). ∎

Remark 5.4

1) In view of (3.13), it is obvious that Theorem 5.3 implies both Theorems 5.1 and 5.2. The critical ingredient, however, which allows the strengthening of Theorem 5.1 to Corollary 5.1, and therefore, to Theorem 5.3, is given in Lemma 5.3. This latter property is obviously related to the *flabbiness* of the five respective sheaves, therefore, it is not connected with the abstract result in Theorem 9.2.

2) In the particular case of continuous nonlinear PDEs in (2.1), (2.3), it is easy to see that Theorem 5.3 can be *further extended* to the case when the coefficients are no longer continuous on $\bar{\Omega}$, and instead they only satisfy the condition

$$(5.84) \qquad c_i \in \mathcal{C}^0_{nd}(\Omega), \quad 1 \leq i \leq h$$

A further similar extension of Theorem 5.3 can be obtained in the case of the general nonlinear PDEs in (2.1) and (2.2), if instead of the continuity of F in all its variables, we shall only require for instance that

$$(5.85) \qquad F \in \mathcal{C}^0((\Omega \backslash \Sigma) \times \mathbb{R}^{\bar{m}})$$

where $\Sigma \subset \Omega$ is closed, nowhere dense and may depend on F, while $\bar{m} \in \mathbb{N}$ is the number of the variables $D^p_x U(x)$ in F, with $p \in \mathbb{N}$, $|p| \leq m$. Obviously the nonlinear PDEs in (2.1), (2.3) which satisfy (5.84) are a particular case of the nonlinear PDEs in (2.1) and (2.2) satisfying (5.85). It is quite obvious, however, that singularities of F which are still more general than in (5.85) could be accommodated in existence and uniqueness results such as that in Theorem 5.3.

3) It is important to note that we have as well *$\mathcal{C}^\infty$-regularity type results* concerning the existence of generalized solutions provided by Theorems 5.1, 5.2, Corollary 5.1 and Theorem 5.3 above. Indeed, in view of (2.13), we can assume in (2.15) and (2.22) that

$$(5.86) \qquad U_\epsilon \in \mathcal{C}^\infty(\Omega \backslash \Gamma_\epsilon) \ .$$

In fact, we can assume that

$$(5.87) \qquad U_\epsilon \text{ analytic on } \Omega \backslash \Gamma_\epsilon$$

since according to (2.13), U_ϵ is a polynomial in a neighbourhood of every point in $\Omega \backslash \Gamma_\epsilon$. In this way (5.1) can be strengthened as follows

$$(5.88) \qquad \hat{T}(\hat{\mathcal{M}}^{\infty}_{T}(\Omega)) \supseteq \mathcal{C}^{0}(\bar{\Omega})$$

in other words, in (5.2) we can assume that the generalized solution F of the *continuous* nonlinear PDEs in (2.1), (2.2) or (5.3) is *$\mathcal{C}^{\infty}$-regular* in the sense of

$$(5.89) \qquad F \in \hat{\mathcal{M}}^{\infty}_{T}(\Omega) \ .$$

Similar $\mathcal{C}^{\infty}$-regularity versions can be obtained for Theorem 5.2, Corollary 5.1 and Theorem 5.3.

These $\mathcal{C}^{\infty}$-regularity results on generalized solutions will be important in the study of the *group invariance* properties of the smooth versions of general nonlinear PDEs in (2.1), (2.2), see Section 18.

4) We should like to give a few first indications about the *meaning* of the *generalized solutions*

$$(5.90) \qquad F \in \hat{\mathcal{M}}^{m}_{T}(\Omega)$$

obtained in Theorems 5.1 - 5.3 and Corollary 5.1. For that purpose, it is useful to give a meaning to the *spaces* of generalized functions $\hat{\mathcal{M}}^{m}_{T}(\Omega)$ themselves. This latter task can be dealt with, provided that here we refer in advance to the two main results in Section 7. Indeed, the nature of the spaces $\hat{\mathcal{M}}^{m}_{T}(\Omega)$ will be clarified along *two directions*. First, it will be shown that $\hat{\mathcal{M}}^{m}_{T}(\Omega)$ can be simply related to the space Mes (Ω) of real valued *measurable* functions on Ω. Second, by using the *sheaf* structures involved, we can equally simply relate $\hat{\mathcal{M}}^{m}_{T}(\Omega)$ to various *classical* spaces of real valued functions on Ω.

In order to relate $\hat{\mathcal{M}}^{m}_{T}(\Omega)$ to the set Mes (Ω) of real valued measurable functions on Ω, let us recall (7.11), according to which we have the order isomorphical embedding

$$\hat{\mathcal{M}}^{m}_{T}(\Omega) \xrightarrow{\ \hat{T}\ } \hat{\mathcal{N}}^{0}(\Omega)$$

where $\mathcal{M}^0(\Omega)$ is defined in (7.6). Therefore, with the notation in (7.26), and in view of (7.25), we obtain the *order isomorphical embedding*

$$(5.91) \qquad \hat{\mathcal{M}}^m_T(\Omega)^* \xrightarrow{\hat{T}} \text{Mes}\ (\Omega)$$

Let us turn now to the relation between $\hat{\mathcal{M}}^m_T(\Omega)^*$ and various classical spaces of real valued functions on Ω. The main result in this regard is obtained in (7.172) according to which we have the following *commutative diagram of sheaf morphisms*

$$(5.92) \qquad \begin{array}{ccccccc} \mathcal{C}^m(\Delta) & \xrightarrow{\text{inj}} & \mathcal{C}^m_{nd}(\Delta) & \xrightarrow{\alpha} & \mathcal{M}^m_T(\Delta) & \xrightarrow{\text{inj}} & \hat{\mathcal{M}}^m_T(\Delta)^* \\ \downarrow T(x,D) & & \downarrow T(x,D) & & T \downarrow \text{inj} & & \hat{T} \downarrow \text{inj} \\ \mathcal{C}^0(\Delta) & \xrightarrow{\text{inj}} & \mathcal{C}^0_{nd}(\Delta) & \xrightarrow{\beta} & \mathcal{M}^0(\Delta) & \xrightarrow{\text{inj}} & \hat{\mathcal{M}}^0(\Delta)^* \end{array}$$

for every nonvoid open subset $\Delta \subseteq \Omega$. Here the mappings α and β are the canonical quotient mappings defined by (4.4) and (3.4), respectively. Further, according to (7.173), all the spaces in (5.92), except for $\mathcal{C}^m(\Delta)$ and $\mathcal{C}^0(\Delta)$, are *flabby sheaves*, see (7.92).

We can note now that, with the exception of $\mathcal{M}^m_T(\Delta)$, $\hat{\mathcal{M}}^m(\Delta)^*$ and $\hat{\mathcal{M}}^0(\Delta)^*$, all the other spaces are *classical* spaces of real valued functions on Ω. Indeed, in view of (7.40), this holds for $\mathcal{M}^0(\Delta)$ as well. On the other hand, in view of (4.4) and (3.4), α and β are surjective, while in view of (A.23) in the Appendix, $\mathcal{M}^m_T(\Delta)$ and $\mathcal{M}^0(\Delta)$ are order dense in $\hat{\mathcal{M}}^m_T(\Delta)^*$ and $\hat{\mathcal{M}}^0(\Delta)^*$ respectively. In this way the sheaf morphisms

$$(5.93) \qquad \mathcal{C}^m(\Delta) \xrightarrow{\text{inj}} \mathcal{C}^m_{nd}(\Delta) \xrightarrow{\alpha} \mathcal{M}^m_T(\Delta) \xrightarrow{\text{inj}} \hat{\mathcal{M}}^m_T(\Delta)^*$$

in (5.92) give a definition of the spaces $\hat{\mathcal{M}}^m_T(\Delta)^*$ in terms of the classical spaces $\mathcal{C}^m_{nd}(\Delta)$, and therefore, in terms of the classical spaces $\mathcal{C}^m(\Delta)$. Indeed, as mentioned in Remark 7.5, the spaces $\mathcal{C}^m_{nd}(\Delta)$ can be seen as the *smallest flabby sheaves* which contain the spaces $\mathcal{C}^m(\Delta)$.

Finally, the *commutativity* of the sheaf morphisms in (5.92) shows that the mappings

$$(5.94)\qquad \mathcal{M}^m_T(\Omega) \xrightarrow{\ T\ } \mathcal{M}^0(\Omega),\quad \hat{\mathcal{M}}^m_T(\Omega) \xrightarrow{\ \hat{T}\ } \hat{\mathcal{M}}^0(\Omega)$$

are *natural extensions* of the *classical* mappings

$$(5.95)\qquad \mathcal{C}^m(\Omega) \xrightarrow{T(x,D)} \mathcal{C}^0(\Omega),\quad \mathcal{C}^m_{nd} \xrightarrow{T(x,D)} \mathcal{C}^0_{nd}(\Omega)$$

5) As mentioned at the beginning of this Section, the existence results in Theorems 5.1, 5.2, 5.3 and Corollary 5.1 - obtained on the basis of the 'pull-back' partial orders (4.7) - can significantly be *extended* by the use of the more general *non 'pull-back'* partial orders (13.11). Here we mention several of the *convenient* features of such extensions.

First, in the case of non 'pull-back' partial orders, it is no longer necessary to make the strong identification in (4.3) which, as we have seen, will automatically lead to the injectivity of the mappings (4.5) and (4.16). Therefore, the non 'pull-back' partial orders (13.11) will - unlike in this Section - lead to existence results which need *not* automatically be uniqueness results as well, see related comments in Section 6.

Further, as seen in Theorem 13.1, and in particular, in Corollary 13.1, nontrivial and significant extensions of the existence result in Theorem 5.1 can be obtained for a *variety* of non 'pull-back' and so called *admissible* partial orders which can be associated with the same given nonlinear partial differential operator $T(x,D)$, see (13.35).

Finally, the nontriviality and significance of the non 'pull-back' partial orders is illustrated through their applications in Section 18.

6. A FEW FIRST EXAMPLES

The wide range of continuous nonlinear PDEs covered by the existence and uniqueness result in Theorems 5.1 and 5.2, Corollary 5.1 and Theorem 5.3 raise the following three questions:

- what is the nature of the generalized solutions obtained?

- how are these generalized solutions connected with the earlier known classical, weak, distributional, etc., solutions?

- what is the meaning of the uniqueness of generalized solutions in the context of initial and/or boundary value problems associated with the PDEs in (2.1), (2.2)?

The first question will be answered in Subsection 7.1, where it is shown that the generalized solutions provided by Theorems 5.1 and 5.2, Corollary 5.1 and Theorem 5.3 can be assimilated with *measurable functions* on $\Omega \subseteq \mathbb{R}^n$.

The second question is given a short answer in this Section, while further details are presented starting with Section 8.

The third question is dealt with in the second part of this Section.

First, let us illustrate the *nontriviality* even of the weakest existence result, namely, that in Theorem 5.1. Let $\Omega \subset \mathbb{R}^2$ be an open and bounded domain and let $f \in \mathcal{C}^0(\mathbb{R})$ be *nondifferentiable* on a dense subset S in $\mathbb{R}$. Let us consider the PDE

$$(6.1) \qquad D_t U(t,y) = f(y), \quad x = (t,y) \in \Omega$$

which is obviously of the form (2.1), (2.3), and in addition, it is also linear.

Proposition 6.1

The continuous PDE

$$(6.2) \qquad D_t U(t,y) = f(y), \quad x = (t,y) \in \Omega$$

cannot have solutions

$$(6.3) \qquad u \in \mathcal{C}^1_{nd}(\Omega)$$

that is, there does *not* exist a closed, nowhere dense $\Gamma \subset \Omega$ and $u \in \mathcal{C}^1(\Omega \backslash \Gamma)$, such that u is a *classical* solution of (6.2) on $\Omega \backslash \Gamma$. In particular, there *cannot* exist solutions $u \in \mathcal{C}^1(\Omega)$ of (6.2) on the whole of Ω.

Proof

Assume a solution (6.3) exists. Let us take

$$I = [a,b] \times [c,d] \subsetneq \Omega \backslash \Gamma$$

then $u \in \mathcal{C}^1(I)$ and u is a classical solution of (6.2) on I. It follows that for $a \leq t \leq b$ and $c \leq y \leq d$, we have

$$\begin{aligned} u(t,y) &= u(a,y) + \int_a^t f(y) d\tau = \\ &= u(a,y) + (t - a) f(y). \end{aligned}$$

In other words

$$(6.4) \qquad u(t,y) = tf(y) + g(y), \quad x = (t,y) \in I$$

for a certain $g \in \mathcal{C}^0([c,d])$.

Let us define $v : J \longrightarrow \mathbb{R}$, where $J = [a,b] \times [a,b] \times [c,d]$, by the relation

$$v(t,t',y) = u(t',y) - u(t,y).$$

But we assumed that $u \in \mathcal{C}^1(I)$, thus

$$(6.5) \qquad v \in \mathcal{C}^1(J).$$

On the other hand (6.4) implies that

$$(6.6) \qquad v(t,t',y) = (t' - t)f(y), \quad (t,t',y) \in J.$$

Now if we recall that S is dense in $\mathbb{R}$, in particular $S \cap [c,d]$ is dense in $[c,d]$, it follows that (6.5) and (6.6) contradict each other.

■

In view of Proposition 6.1 above, it is now obvious that the existence and uniqueness result in Theorem 5.1 gives a unique *generalized solution* $F \in \hat{\mathcal{M}}^1_T(\Omega)$ for the continuous PDE in (6.1), which is *not* and *cannot* be classical. Further comments on the PDE in (6.1) are presented at the end of this Section.

In the context of Theorem 5.1, we now note the following *coherence* property between *classical and generalized solutions*. Whenever *classical solutions*

$$(6.7) \qquad U \in \mathcal{C}^m(\Omega)$$

exist for arbitrary continuous nonlinear PDEs in (2.1), (2,2), these solutions will always be *generalized solutions* in the sense of Theorem 5.1. Indeed, this results easily from the inclusions (3.2) and the commutative diagrams (4.6) and (4.15).

The following more general *coherence* between *classical and generalized* solutions is obtained in a similar way. Suppose given

$$(6.8) \qquad U \in \mathcal{C}^m(\Omega \backslash \Gamma) \text{ with } \Gamma \subset \Omega \text{ closed, nowhere dense.}$$

If U is a *classical solution* on $\Omega \backslash \Gamma$ of any given continuous nonlinear PDE in (2.1), (2.2), then it is also a *generalized solution* in the sense of Theorem 5.1, on the whole of Ω.

It should be mentioned that the relevance of the *extended* type of classical solutions (6.8) has recently come to attention. Indeed, in Rosinger [6] it is shown that an *arbitrary nonlinear analytic* PDE, with associated noncharacteristic analytic Cauchy problem, always has global solutions of type (6.8), where Ω is the whole of the domain of analyticity of the given equation.

Let us now turn to the meaning of the *uniqueness* of generalized solutions constructed in Section 5. For that, it will be instructive to consider the respective construction in some detail in the case of the simplest instance of the PDEs in (2.1), (2.2), namely

(6.9) $\quad D_t U(t) = 0, \quad t \in (-1,1) \subset \mathbb{R}$

whose classical solutions are

(6.10) $\quad U(t) = c, \quad t \in (-1,1)$

with $c \in \mathbb{R}$ given arbitrarily.

We shall now follow in the case of (6.9), the main steps of the construction of the *unique* generalized solution presented in the proof of Theorem 5.1.

Our main interest is to give an explicit expression to F defined in (5.7), since owing to (5.2), F gives the *unique* generalized solution of (6.9). In order to do so we note that in view of (5.10), F contains among others, the set of all the classes $U \in \mathcal{M}^1_T(-1,1)$ generated by functions $u \in \mathcal{C}^1_{nd}(-1,1)$ such that, for $\nu \in \mathbb{N}$, there exist $\epsilon_\nu \leq 0$, $c_\nu \in \mathbb{R}$ and $\Omega_\nu \subseteq (-1,1)$ nonvoid, open, which satisfy

(6.11) $\quad u(t) = \epsilon_\nu t + c_\nu, \quad t \in \Omega_\nu$

while

(6.12) $\quad \bigcup_{\nu \in \mathbb{N}} \Omega_\nu$ is dense in $(-1,1)$

and

(6.13) $(\Omega_\nu \mid \nu \in \mathbb{N})$ are pair wise disjoint.

In particular F contains all the classes $U \in \mathcal{M}_T^1(-1,1)$ generated by functions $u \in C^1(-1,1)$ given by

(6.14) $u(t) = c, \quad t \in (-1,1)$

where $c \in \mathbb{R}$ is fixed arbitrarily. In this way, in spite of the *uniqueness* of F, we recover the *multiplicity* of the classical solutions (6.10), each of which may for instance correspond to an initial value problem, such as

(6.15) $U(0) = c.$

We can conclude that the meaning of the *uniqueness* in Theorem 5.1 is that the generalized solution obtained through Dedekind *order completion* does in fact contain not one or several, but rather the *totality* of all possible solutions of the PDE in (2.1), (2.2), solutions possible within the framework of the given 'order first' approach. Needless to say, a similar remark holds for the respective uniqueness results in Theorems 5.2, 5.3 and Corollary 5.1.

In other words, the order completion method used in Section 5 offers in particular the delivery of the *totality* of *global generalized solutions* of continuous nonlinear PDEs. Therefore, the *uniqueness* is a consequence of the simple fact that there can only be *one single* such a totality, be that within the given, or within any other method. However, in case *initial and/or boundary* value conditions are added to the continuous nonlinear PDEs, that totality of global generalized solutions will be replaced with the appropriate smaller sets of those global generalized solutions which happen to satisfy the given initial and/or boundary requirements. Details in this respect are presented in Sections 8 - 11.

Here, it is important to note that the above type of *uniqueness* which is automatically associated with the basic existence results in Section 5 and with their applications in Sections 8 - 11, comes from the use of the 'pull-back' partial orders (4.7), in particular, from the fact that the mappings T and $\hat{T}$, see (4.8) and (4.16), are *injective*. However, once the more general non 'pull-back' approach in Section 13 is used, the corresponding mappings T_* and $\hat{T}_*$, see (13.14), will *no longer* be necessarily injective. As a consequence, the respective existence results, see Theorem 13.1, for instance, need no longer come with an automatically associated uniqueness property as well.

In view of the above and of Remark 5.1, we come shortly back to the continuous PDE in (6.1). According to (5.10), its *unique generalized solution* $F \in \hat{\mathcal{M}}^1_T(\Omega)$ will be given by

$$(6.16) \qquad F = \{U \in \mathcal{M}^1_T(\Omega) \mid D_t U \leq f \text{ in } \mathcal{M}^0(\Omega)\}$$

where in view of (4.6), we have the commutative diagram

$$(6.17) \qquad \begin{array}{ccc} \mathcal{C}^1_{nd}(\Omega) & \xrightarrow{\;D_t\;} & \mathcal{C}^0_{nd}(\Omega) \\ \downarrow & & \downarrow \\ \mathcal{M}^1_T(\Omega) & \xrightarrow[D_t]{\text{inj}} & \mathcal{M}^0(\Omega) \end{array}$$

where D_t is defined according to (4.5). In other words, F is the set of all the classes $U \in \mathcal{M}^1_T(\Omega)$ generated by functions $u \in \mathcal{C}^1_{nd}(\Omega)$ which have the property that for a suitable $\Gamma \subset \Omega$ closed, nowhere dense, the relations hold

$$(6.18) \qquad u \in \mathcal{C}^1(\Omega \setminus \Gamma), \quad D_t u(t,y) \leq f(y), \quad x = (t,y) \in \Omega \setminus \Gamma.$$

The point in Proposition 6.1 above is that the unique generalized solution F in (6.16) *cannot* be a classical solution even in the extended sense in (6.3). In view of (4.11), this means that

$$(6.19) \qquad F \neq \langle U], \quad \text{for any } U \in \mathcal{M}^1_T(\Omega)$$

where we recall that, according to (4.4), each $U \in \mathcal{M}^1_T(\Omega)$ is the $\sim_T$ equivalence class generated by a certain $u \in \mathcal{C}^1_{nd}(\Omega)$.

The question of the nature of such a nonclassical, generalized solution as for instance $F \in \hat{\mathcal{M}}^1_T(\Omega)$ given in (6.16) above, is part of the first question formulated at the beginning of this Section, and will be dealt with next, in Subsection 7.1.

The result in Proposition 6.1 is but one case of many other similar and more general ones. For instance, we can replace D_t by the directional derivative $D_t + \lambda D_y$, with $\lambda \in \mathbb{R}$ arbitrary but fixed. Then, with the same kind of continuous function f, the PDE

$$(6.20) \qquad (D_t + \lambda D_y)U(t,y) = f(y - \lambda t), \quad x = (t,y) \in \Omega$$

cannot have solutions

$$(6.21) \qquad u \in C^1_{nd}(\Omega) .$$

Indeed, instead of (6.4), we obtain now the representation

$$u(t,y) = u(a,y-\lambda(t-a)) + (t-a)f(y-\lambda t)$$

and then we can follow an argument similar to that in the proof of Proposition 6.1.

In the same way, it follows that the second order hyperbolic PDE

$$(6.22) \qquad (D_t + \lambda D_y)(D_t + \mu D_y)U(t,y) = f(y-\lambda t), \quad x = (t,y) \in \Omega$$

with $\lambda, \mu \in \mathbb{R}$, *cannot* have solutions

$$(6.23) \qquad u \in \mathcal{C}^2_{nd}(\Omega)$$

for arbitrary $f \in \mathcal{C}^0(\mathbb{R})$.

However, the above phenomenon in (6.2) (6.3), (6.20), (6.21) or (6.22), (6.23) is *not* confined to hyperbolic equations. Indeed, it is known, Hellwig [p 173], that there exists $f \in \mathcal{C}^0(\bar{\Omega})$ and a dense subset S of Ω, with Ω the unit ball in $\mathbb{R}^3$, such that whenever u is a weak solution of

$$(6.24) \qquad \Delta U = f \quad \text{in} \quad \Omega$$

then u will fail to be twice differentiable at any of the points of S.

A classical, 1928 result of Perron, see Hellwig [p 83], gives further depth to the above phenomenon. In addition, even in the *linear* case, it dramatically highlights the rather severe limitations of the traditional vision which so often tries to divide and reduce PDEs to the three classes of elliptic, parabolic and hyperbolic.

Proposition 6.2 (O Perron)

Given the system of PDEs

$$(6.25) \qquad \begin{aligned} &D_t U(t,y) - D_y U(t,y) - D_y V(t,y) = 0 \\ &a\, D_y U(t,y) - D_t V(t,y) + D_y V(t,y) + f(t+y) = 0, \quad t \geq 0, \quad y \in \mathbb{R} \end{aligned}$$

and the initial value conditions

$$(6.26) \qquad \left. \begin{aligned} U(0,y) &= 0 \\ V(0,y) &= 0 \end{aligned} \right., \quad y \in \mathbb{R}$$

The system (6.25) is hyperbolic for $a > 0$, parabolic for $a = 0$, and elliptic for $a < 0$. Furthermore, the necessary and sufficient condition for the existence of a solution

$$(6.27) \qquad U,\ V \in \mathcal{C}^1$$

is respectively

(6.28) $\quad f \in \mathcal{C}^0, \quad \text{if} \quad a > 0$

(6.29) $\quad f \in \mathcal{C}^2, \quad \text{if} \quad a = 0$

and

(6.30) $\quad f \quad \text{analytic}, \quad \text{if} \quad a < 0.$ ■

7. GENERALIZED SOLUTIONS AS MEASURABLE FUNCTIONS

7.1 $\hat{\mathcal{M}}^0(\Omega)$ and the Measurable Functions

As mentioned in the Introduction, one of the main novelties of the 'order first' approach in solving continuous nonlinear PDEs is that, in a certain sense, the *generalized solutions* obtained prove to be identifiable with usual *measurable functions* on domains in Euclidean spaces. The details of this identification are the subject of the present Subsection.

In view of the existence results in Section 5, the *generalized functions* we are interested in are elements F in the order complete space $\hat{\mathcal{M}}_T^m(\Omega)$.

An important point to note however, is that in view of the *order isomorphical embedding*, see (4.16)

$$\hat{\mathcal{M}}_T^m(\Omega) \xrightarrow{\hat{T}} \hat{\mathcal{M}}^0(\Omega) \tag{7.1}$$

the problem of the structure of generalized functions in $\hat{\mathcal{M}}_T^m(\Omega)$ can be *reduced* to a good extent to the problem of the structure of generalized functions in $\hat{\mathcal{M}}^0(\Omega)$.

Additional information on the structure of $\hat{\mathcal{M}}_T^m(\Omega)$ is presented in Subsection 7.2, where it is shown that $\hat{\mathcal{M}}_T^m(\Omega)$ is in fact a *flabby sheaf of sections* over Ω.

Fortunately, the problem of the structure of $\hat{\mathcal{M}}^0(\Omega)$ has, at least in part, a well known answer, Zaharov, Nakano & Shimogaki.

For the sake of simplicity and clarity, we shall however start by describing the structure of generalized functions in a space which is more particular than $\hat{\mathcal{M}}^0(\Omega)$, yet it is of comparable interest.

Let us recall, see (2.7) and (4.2), that the closed, nowhere dense subsets $\Gamma \subset \Omega$ which so far have appeared essentially in various stages of the argument can be assumed to satisfy the condition

(7.2) $\quad \operatorname{mes}(\Gamma) = 0.$

More precisely, the existence results in Section 5 remain valid if we replace $\hat{\mathcal{M}}^m_T(\Omega)$ with the order complete space $\hat{\mathcal{N}}^m_T(\Omega)$ defined as follows. Given $\ell \in \bar{\mathbb{N}}$, let us define, see (3.1)

$$(7.3) \qquad C^\ell_z(\Omega) = \left\{ u \;\middle|\; \begin{array}{l} \exists\ \Gamma \subset \Omega \text{ closed, nowhere dense:} \\ \quad *)\ \ u : \Omega\backslash\Gamma \longrightarrow \mathbb{R} \\ \quad **)\ \ u \in C^\ell(\Omega\backslash\Gamma) \\ \quad ***)\ \ \operatorname{mes}(\Gamma) = 0 \end{array} \right\}$$

Then obviously, see (3.2)

$$(7.4) \qquad C^\ell(\Omega) \subset C^\ell_z(\Omega) \subset C^\ell_{nd}(\Omega), \quad \ell \in \bar{\mathbb{N}}.$$

In particular, if we also take into account (4.2), then we obtain the following restriction of the mapping in (4.1)

$$(7.5) \qquad T(x,D) : C^m_z(\Omega) \longrightarrow C^0_z(\Omega).$$

Now, in view of (7.4), the equivalence $\sim$ on $C^0_{nd}(\Omega)$ defined in (3.3) can be restricted to $C^0_z(\Omega)$, in which case we obtain the quotient space, see (3.4)

$$(7.6) \qquad \mathcal{N}^0(\Omega) = C^0_z(\Omega)/\sim .$$

In a similar way, by restricting the equivalence relation $\sim_T$ on $C^m_{nd}(\Omega)$, defined in (4.3), to $C^m_z(\Omega)$, which is possible in view of (7.4), we obtain the quotient space, see (4.4)

$$(7.7) \qquad \mathcal{N}^m_T(\Omega) = C^m_z(\Omega)/\sim_T .$$

Then owing to elementary set theoretic properties of induced equivalence relations, the relations (7.4), (7.6) and (7.7) yield the inclusions

$$(7.8) \qquad \mathcal{N}^m_T(\Omega) \subset \mathcal{M}^m_T(\Omega), \quad \mathcal{N}^0(\Omega) \subset \mathcal{M}^0(\Omega).$$

In this way, the commutative diagram (4.6) is replaced with the commutative diagram

$$(7.9) \qquad \begin{array}{ccc} \mathcal{C}^m_z(\Omega) & \xrightarrow{\quad T(x,D) \quad} & \mathcal{C}^0_z(\Omega) \\ \downarrow & & \downarrow \\ \mathcal{N}^m_T(\Omega) & \xrightarrow[\quad T \quad]{\quad inj \quad} & \mathcal{N}^0(\Omega) \end{array}$$

where the vertical arrows are the respective canonical mappings into quotient spaces.

Finally, owing to (7.8), the order relations $\leq_T$ and $\leq$ defined in (4.7) and (3.5), can be restricted to $\mathcal{N}^m_T(\Omega)$ and $\mathcal{N}^0(\Omega)$ respectively. Then, similar with (4.15) and (4.16), we obtain the commutative diagram

$$(7.10) \qquad \begin{array}{ccc} \mathcal{N}^m_T(\Omega) \ni U & \xrightarrow{\qquad T \qquad} & T(U) \in \mathcal{N}^0(\Omega) \\ \downarrow & & \downarrow \\ \hat{\mathcal{N}}^m_T(\Omega) \ni <U] & \xrightarrow[\qquad \hat{T} \qquad]{} & \hat{T}(<U]) = <T(U)] \in \hat{\mathcal{N}}^0(\Omega) \end{array}$$

and the order isomorphical embedding

$$(7.11) \qquad \hat{\mathcal{N}}^m_T(\Omega) \xrightarrow{\quad \hat{T} \quad} \hat{\mathcal{N}}^0(\Omega)$$

where $\hat{\mathcal{N}}^m_T(\Omega)$ and $\hat{\mathcal{N}}^0(\Omega)$ are the Dedekind order completions of $\mathcal{N}^m_T(\Omega)$ and $\mathcal{N}^0(\Omega)$ respectively.

In this way, instead of directly studying the structure of generalized functions which are elements of $\hat{\mathcal{M}}^0(\Omega)$, we shall first concentrate on the structure of generalized functions which are in $\hat{\mathcal{N}}^0(\Omega)$, and which will prove to be *usual measurable functions* on Ω.

In Corollary 7.1 and Remark 7.3 at the end of this Subsection, we shall see the conditions under which *generalized functions* $U \in \hat{\mathcal{M}}^0(\Omega)$ can be associated with usual *measurable* functions on Ω.

Let us denote by

(7.12) $\quad \mathrm{Mes}(\Omega)$

the set of all Lebesgue measurable functions $f : \Omega \longrightarrow \mathbb{R}$. As usual, we identify in $\mathrm{Mes}(\Omega)$ any two functions which differ on a subset of zero Lebesgue measure.

The important point to note is that $\mathrm{Mes}(\Omega)$ is Dedekind complete with respect to the usual order relation $\leq$ between two measurable functions, namely

(7.13) $\quad f \leq g \Longleftrightarrow \mathrm{mes}\{x \in \Omega \mid f(x) > g(x)\} = 0.$

In other words, every bounded subset of $\mathrm{Mes}\ (\Omega)$ has an infimum and a supremum, Luxemburg & Zaanen.

Moreover, from our point of view, the essential property of the space $\mathcal{M}^0(\Omega)$ is given in:

Lemma 7.1

We have the order isomorphical embedding

(7.14) $\quad \mathcal{M}^0(\Omega) \longrightarrow \mathrm{Mes}\ (\Omega)$

defined as follows: given $f \in \mathcal{C}^0_{\mathrm{z}}(\Omega)$, then the element of $\mathcal{M}^0(\Omega)$ which is the $\sim$ equivalence class of f is mapped into the element of $\mathrm{Mes}\ (\Omega)$, which is the usual equivalence class of measurable functions almost everywhere equal to f on Ω.

Proof

We first prove that the mapping (7.14) is well defined.

Let $f, f' \in F \in \mathcal{N}^0(\Omega)$. We have to show that f and f' are almost everywhere equal on Ω. But $f, f' \in \mathcal{C}^0_z(\Omega)$ and (7.3) imply that

$$(7.15) \qquad f \in \mathcal{C}^0(\Omega \setminus \Gamma), \quad f' \in \mathcal{C}^0(\Omega \setminus \Gamma')$$

where $\Gamma, \Gamma' \subsetneq \Omega$ are closed, nowhere dense and in addition, the satisfy

$$(7.16) \qquad \text{mes } (\Gamma) = \text{mes } (\Gamma') = 0.$$

On the other hand $f, f' \in F$, together with (7.6) and (3.3), give

$$(7.17) \qquad f, f' \in \mathcal{C}^0(\Omega \setminus \Gamma''), \quad f = f' \text{ on } \Omega \setminus \Gamma''$$

for a suitable closed, nowhere dense $\Gamma'' \subsetneq \Omega$. However, we can choose Γ'' so that

$$(7.18) \qquad \text{mes } (\Gamma'') = 0.$$

Indeed, we note that (7.15) and (7.16) imply

$$(7.19) \qquad f, f' \in \mathcal{C}^0(\Omega \setminus \Gamma_0)$$

where we can take $\Gamma_0 = \Gamma \cup \Gamma'$, hence $\Gamma_0 \subsetneq \Omega$ is closed, nowhere dense, and

$$(7.20) \qquad \text{mes } (\Gamma_0) = 0.$$

On the other hand $\Omega \setminus \Gamma''$ is dense in Ω, therefore it is dense in the open set $\Omega \setminus \Gamma_0$. Now, see (7.17)

$$f = f' \text{ on } \Omega \setminus \Gamma''$$

and (7.19) will give

$$f = f' \quad \text{on} \quad \Omega \backslash \Gamma_0 .$$

In this way we can take $\Gamma'' = \Gamma_0$, and then (7.18) follows from (7.20).

It follows therefore that the mapping (7.14) is indeed well defined.

We show now that the mapping (7.14) is also injective. For that, let $f \in F \in \mathcal{N}^0(\Omega)$ be such that $f = 0$ almost everywhere on Ω. Then (7.3) gives

$$f \in \mathcal{C}^0(\Omega \backslash \Gamma)$$

for a suitable closed and nowhere dense $\Gamma \subset \Omega$, which in addition satisfies $\text{mes}\,(\Gamma) = 0$. But by our assumption

$$f = 0 \quad \text{on} \quad \Omega \backslash \Gamma'$$

for a certain subset $\Gamma' \subset \Omega$, with $\text{mes}\,(\Gamma') = 0$. Thus

$$\text{mes}\,\{x \in \Omega \backslash \Gamma \mid f(x) \neq 0\} = 0$$

and by the continuity of f on the open set $\Omega \backslash \Gamma$, it follows that

$$f = 0 \quad \text{on} \quad \Omega \backslash \Gamma.$$

In this way $f \sim 0$, thus $F = 0 \in \mathcal{N}^0(\Omega)$.

An obvious adaptation of the above proof will show that the mapping (7.14) is also an order isomorphical embedding. ■

It follows therefore that, for *generalized functions* which are elements of $\hat{\mathcal{N}}^0(\Omega)$, we have the *representation* property given in:

Theorem 7.1

Given a generalized function $F \in \hat{\mathcal{N}}^{\rho}(\Omega)$. If

(7.21) $\quad F$ is *not* the minimum or the maximum in $\hat{\mathcal{N}}^{\rho}(\Omega)$

then

(7.22) $\quad F \in \text{Mes}\ (\Omega)$.

Proof

We recall that, according to MacNeille's theorem in the Appendix, $\hat{\mathcal{N}}^{\rho}(\Omega)$ is not only Dedekind order complete but it is also order complete, thus, it has a minimum and a maximum. Therefore, the embedding (7.14) does *not* imply that

(7.23) $\quad \hat{\mathcal{N}}^{\rho}(\Omega) \subseteq \text{Mes}\ (\Omega)$

since Mes (Ω) is only Dedekind order complete, and it does not have a minimum or a maximum.

However, (7.21) and (7.22) follow from (7.14), if we apply (A.40) in the Proposition A.1 in Appendix, in which case we obtain the order isomorphical embedding

(7.24) $\quad \hat{\mathcal{N}}^{\rho}(\Omega) \subseteq \widehat{\text{Mes}}\ (\Omega)$. ∎

Remark 7.1

It will be convenient to write (7.21) and (7.22) in the shorter form given by the *order isomorphical embedding*

(7.25) $\quad \hat{\mathcal{N}}^{\rho}(\Omega)^{*} \xrightarrow{\text{inj}} \text{Mes}\ (\Omega)$

where, with the notations in MacNeille's theorem in the Appendix, we define

(7.26) $\quad \hat{X}^{*} = \hat{X} \setminus \{\min \hat{X}, \max \hat{X}\}$.

Indeed, it is important to note that the *order complete* set $\hat{X}$ delivered by the MacNeille Theorem will contain both

$$\min \hat{X} \quad \text{and} \quad \max \hat{X} .$$

However, these two elements of $\hat{X}$ are usually of no interest in the case the initial poset X is a set of real valued functions, since in such a case the above two elements may correspond to functions which are everywhere unbounded, see for instance when $X = \mathcal{C}^0([0,1],\mathbb{R})$ is considered with the usual order. Therefore, in such cases it is useful to consider the above subset $\hat{X}^*$ of $\hat{X}$, which will always be Dedekind order complete.

One important difference between $\hat{X}$ and $\hat{X}^*$ is that $\hat{X}$ is both order complete and Dedekind order complete, while $\hat{X}^*$ is only Dedekind order complete, see Appendix.

We return now to the more general problem of the *representation* of the *generalized functions* in $\hat{\mathcal{M}}^0(\Omega)$, as *measurable functions* on Ω.

The *special interest* in the more general space $\hat{\mathcal{M}}^0(\Omega)$ comes from the fact that we have the mappings, see (3.4) and (3.7)

$$(7.27) \qquad \mathcal{C}^0_{nd}(\Omega) \longrightarrow \mathcal{M}^0(\Omega) \longrightarrow \hat{\mathcal{M}}^0(\Omega)$$

where the first arrow is the canonical mapping into a quotient space, while the second one is an order isomorphical embedding. In view of (7.27), the space $\hat{\mathcal{M}}^0(\Omega)$ contains all the functions

$$(7.28) \qquad u \in \mathcal{C}^0(\Omega \setminus \Gamma), \quad \Gamma \subset \Omega \text{ closed, nowhere dense}$$

and as is well known, Oxtoby, when $\text{mes}(\Omega) < \infty$, then the sets of *singularities* Γ such as in (7.28) can have *arbitrary positive measures*, subject only to the condition

$$(7.29) \qquad \text{mes}(\Gamma) < \text{mes}(\Omega).$$

In this way, such singularity sets Γ can model not only *shocks* but possibly Cantor set type *strange attractors* as well, which have recently been connected with *turbulence* or *chaos*, Temam.

To start with, we define a mapping

$$(7.30) \qquad \mathcal{C}^0_{nd}(\Omega) \ni u \longmapsto \bar{u} \in \mathcal{C}^0_{nd}(\Omega)$$

where $\bar{u}$ is called the *maximal regularization* of u, as follows. Let $U \in \mathcal{U}^0(\Omega)$ be the $\sim$ equivalence class of u. Let

$$(7.31) \qquad \bar{\Gamma} = \bigcap_{v \in U} \Sigma$$

where for each $v \in U$, we have that $\Sigma \subset \Omega$ is closed, nowhere dense and $v \in \mathcal{C}^0(\Omega \setminus \Sigma)$. Then obviously $\bar{\Gamma} \subset \Omega$ is also closed and nowhere dense. Now we define

$$(7.32) \qquad \bar{u} : \Omega \setminus \bar{\Gamma} \longrightarrow \mathbb{R}, \quad \bar{u} \in \mathcal{C}^0(\Omega \setminus \bar{\Gamma})$$

by

$$(7.33) \qquad \bar{u}(x) = v(x) \quad \text{if} \quad x \in \Omega \setminus \Sigma$$

where $v \in U$ and $\Sigma \subset \Omega$ corresponds to it as above. We note that (7.33) is a correct definition. Indeed, clearly we have

$$\Omega \setminus \bar{\Gamma} = \bigcup_{v \in U} (\Omega \setminus \Sigma).$$

Further, if $x \in (\Omega \setminus \Sigma) \cap (\Omega \setminus \Sigma')$, with Σ and Σ' corresponding to $v, v' \in U$ respectively, then $v \in \mathcal{C}^0(\Omega \setminus \Sigma)$ and $v' \in \mathcal{C}^0(\Omega \setminus \Sigma')$, hence $v, v' \in \mathcal{C}^0(\Omega \setminus (\Sigma \cup \Sigma'))$. But $x \in (\Omega \setminus \Sigma) \cap (\Omega \setminus \Sigma') = \Omega \setminus (\Sigma \cup \Sigma')$, thus both v and v' are continuous at x. It follows that $v(x) = v'(x)$, since $v, v' \in U$, that is $v \sim v'$.

In this way, the mapping (7.30) is well defined, and in addition, we obviously have

$$(7.34) \qquad \mathcal{C}^0_{nd}(\Omega) \ni u \longmapsto \bar{u} \in U \in \mathcal{M}^0(\Omega)$$

where as above, U is the $\sim$ equivalence class of u. It is easy to note from (7.31) - (7.33), that for $u,v \in \mathcal{C}^0_{nd}(\Omega)$, we have

$$(7.35) \qquad u \sim v \Longleftrightarrow \bar{u} = \bar{v}.$$

Now we define the mapping

$$(7.36) \qquad \mathcal{C}^0_{nd}(\Omega) \ni u \longmapsto \mu u \in \mathrm{Mes}(\Omega), \quad \mu u \in \mathcal{C}^0(\Omega \setminus \bar{\Gamma})$$

by

$$(7.37) \qquad (\mu u)(x) = \left| \begin{array}{ll} \bar{u}(x) & \text{if} \quad x \in \Omega \setminus \bar{\Gamma} \\ 0 & \text{if} \quad x \in \bar{\Gamma} \end{array} \right.$$

where $\bar{\Gamma} \subset \Omega$ is associated with u according to (7.31).

In view of (7.34) and (7.37), it is clear that

$$(7.38) \qquad \mathcal{C}^0_{nd}(\Omega) \ni u \longmapsto \mu u \in U \in \mathcal{M}^0(\Omega)$$

where U is the $\sim$ equivalence class of u. Further, (7.35) and (7.37) yield for $u,v \in \mathcal{C}^0_{nd}(\Omega)$ the equivalence

$$(7.39) \qquad u \sim v \Longleftrightarrow \mu u = \mu v \ .$$

Proposition 7.1

The mapping

$$(7.40) \qquad \mathcal{M}^0(\Omega) \ni U \longmapsto \mu u \in \mathrm{Mes}\ (\Omega)$$

where $u \in U$, is an embedding.

Proof

In view of (7.39), the mapping (7.40) is well defined. In addition, it is injective. Indeed, assume given $U, V \in \mathcal{M}^0(\Omega)$ and $u \in U$, $v \in V$, such that

$$(7.41) \qquad \mu u = \mu v \text{ almost everywhere on } \Omega.$$

But (7.32), (7.37) imply

$$\bar{u} \in \mathcal{C}^0(\Omega \backslash \bar{\Gamma}), \quad \bar{v} \in \mathcal{C}^0(\Omega \backslash \bar{\Sigma})$$

where $\bar{\Gamma}, \bar{\Sigma} \subsetneq \Omega$ are closed and nowhere dense, and are associated with $\bar{u}$ and $\bar{v}$ respectively, according to (7.31).

If now $U \neq V$, then in view of (7.34), we obtain that

$$\bar{u} \sim \bar{v} \text{ does not hold.}$$

Thus, for a certain $x \in \Omega \backslash (\bar{\Gamma} \cup \bar{\Sigma})$, we shall have

$$\bar{u}(x) \neq \bar{v}(x).$$

But $\bar{u}, \bar{v} \in \mathcal{C}^0(\Omega \backslash (\bar{\Gamma} \cup \bar{\Sigma}))$, hence there exists $x \in W \subset \Omega \backslash (\bar{\Gamma} \cup \bar{\Sigma})$, W open, such that

$$\bar{u}(x) \neq \bar{v}(x), \quad x \in V$$

which in view of (7.37), will give

$$\mu u(x) \neq \mu v(x), \quad x \in V$$

thus contradicting (7.41). ∎

Remark 7.2

In view of the Proposition A.1 in Appendix, the embedding, see (7.40)

$$(7.42) \qquad \mathcal{M}^0(\Omega) \xrightarrow{\text{inj}} \text{Mes}\ (\Omega)$$

can be extended to an *increasing* mapping

$$(7.43) \qquad \hat{\mathcal{M}}^0(\Omega)^* \longrightarrow \text{Mes}\ (\Omega).$$

However, since the embedding (7.42) is *not* an increasing mapping, we *cannot* conclude that the increasing mapping (7.43) is an embedding, or in particular, an order isomorphical embedding. In order to see that (7.42) is not increasing, we can use the following simple construction. Let $\Gamma \subsetneq \Omega$ be a closed, nowhere dense set, which has a positive Lebesgue measure. First we note that in (3.14) - (3.16) we can always assume

$$0 \leq \gamma < 1$$

by replacing γ with

$$\gamma' = \frac{\gamma^2}{1 + \gamma^2} .$$

Let us define $u, v \in \mathcal{C}^0_{nd}(\Omega)$ by

$$(7.44) \qquad \begin{aligned} u(x) &= \left| \begin{array}{lll} 2/\gamma(x) & \text{if} & x \in \Omega\backslash\Gamma \\ 0 & \text{if} & x \in \Gamma \end{array} \right. \\ v &= 1 \quad \text{on} \quad \Omega \end{aligned}$$

Then it is clear that, with the notation in (7.31) and (7.32), we obtain

$$(7.45) \qquad \bar{u} = u, \quad \bar{\Gamma} = \Gamma, \quad \bar{v} = v, \quad \bar{\Sigma} = \phi.$$

Now (7.44) and (7.37) will give

$$(7.46) \qquad \mu u = 0 < \mu v \quad \text{on} \quad \bar{\Gamma}.$$

However (7.44) and (7.45) imply that

$$(7.47) \qquad \bar{v} < \bar{u} \quad \text{on} \quad \Omega\backslash\Gamma.$$

The relations (7.46) and (7.47) imply that we *cannot* have

(7.48) $\quad \mu v \leq \mu u$ almost everywhere on Ω

although (7.47) yields

(7.49) $\quad V \leq U$

where U and V are the $\sim$ equivalence classes of u and v respectively.

We shall use now known results in Zaharov, Nakano & Shimogaki, see (7.55) and (7.56) below, in order to come *nearer* to an order isomorphical embedding, see for details (7.77) below and Subsection 7.3

(7.50) $$\hat{\mathcal{M}}^{0}(\Omega)^{*} \xrightarrow{\text{inj}} \text{Mes}\,(\Omega).$$

Let us now assume that the open subset $\Omega \subset \mathbb{R}^n$ is *bounded*.

Let us then define on the compact subset $\bar{\Omega} \subset \mathbb{R}^n$, the space of functions

(7.51) $$\mathcal{C}^{0}_{Q}(\bar{\Omega}) = \left\{ f : \bar{\Omega} \longrightarrow \mathbb{R} \;\middle|\; \begin{array}{rl} *) & f \text{ bounded} \\ **) & \exists\, \Gamma \subset \bar{\Omega} \text{ of first category :} \\ & f \in \mathcal{C}^{0}(\bar{\Omega}\backslash\Gamma) \end{array} \right\}$$

and consider the equivalence relation $\sim_{Q}$ on $\mathcal{C}^{0}_{Q}(\bar{\Omega})$, given by

(7.52) $$f \sim_{Q} g \Longleftrightarrow \left[\begin{array}{l} \exists\, \Gamma \subset \bar{\Omega} \text{ of first category :} \\ f = g \text{ on } \bar{\Omega}\backslash\Gamma \end{array} \right]$$

Let us consider on $\mathcal{C}^{0}(\bar{\Omega})$ the usual order $\leq$, while on $\mathcal{C}^{0}_{Q}(\bar{\Omega})/\sim_{Q}$, the order $\leq_{Q}$ defined as follows, see (3.5) for a similar definition

(7.53) $$F \leq_{Q} G \Longleftrightarrow \left[\begin{array}{l} \exists\, f \in F,\ g \in G,\ \Gamma \subset \bar{\Omega} \text{ of first category :} \\ f \leq g \text{ on } \bar{\Omega}\backslash\Gamma \end{array} \right]$$

Then we obtain the order isomorphical embedding

$$(7.54) \qquad \mathcal{C}^0(\bar{\Omega}) \xrightarrow{\text{inj}} \mathcal{C}^0_Q(\bar{\Omega})/\sim_Q$$

defined by $f \longmapsto F$, where F is the $\sim_Q$ class generated by $f \in \mathcal{C}^0(\bar{\Omega}) \subset \mathcal{C}^0_Q(\bar{\Omega})$, since the complementary of a first category subset of $\bar{\Omega}$ is dense in $\bar{\Omega}$.

The *important* property is that (7.54) can be extended to the *order isomorphism*

$$(7.55) \qquad \hat{\mathcal{C}}^0(\bar{\Omega})^* \xrightarrow{\text{bij}} \mathcal{C}^0_Q(\bar{\Omega})\backslash\sim_Q$$

in particular

$$(7.56) \qquad \mathcal{C}^0_Q(\bar{\Omega})/\sim_Q \text{ is a Dedekind complete Riesz space}$$

see for details Nakano & Shimogaki.

In order to relate (7.55) to a possible version of (7.50), let us define, see (3.1)

$$(7.57) \qquad \mathcal{C}^0_A(\Omega) = \left\{ f : \Omega \longrightarrow \mathbb{R} \;\middle|\; \begin{array}{l} *) \;\; f \text{ bounded} \\ **) \;\; \exists\, \Gamma \subset \Omega \text{ closed, nowhere dense :} \\ \qquad f \in \mathcal{C}^0(\Omega\backslash\Gamma) \end{array} \right\}$$

together with the equivalence relation $\sim_A$ on $\mathcal{C}^0_A(\Omega)$, see (3.3), given by

$$(7.58) \qquad f \sim_A g \iff \left[\begin{array}{l} \exists\, \Gamma \subset \Omega \text{ closed, nowhere dense :} \\ f = g \text{ on } \Omega\backslash\Gamma \end{array} \right]$$

as well as the order $\leq_A$ on $\mathcal{C}^0_A(\Omega)/\sim_A$ defined by, see (3.5)

$$(7.59) \qquad F \leq_A G \iff \left[\begin{array}{l} \exists\, f \in F,\; g \in G,\; \Gamma \subset \Omega \text{ closed, nowhere dense :} \\ f \leq g \text{ on } \Omega\backslash\Gamma \end{array} \right]$$

It is easy to see that

$$(7.60) \qquad \mathcal{C}^0_A(\Omega) \subsetneq \mathcal{C}^0_{nd}(\Omega)$$

moreover $\sim_A$ and $\leq_A$ are the restrictions to $\mathcal{C}^0_A(\Omega)$ of $\sim$ and $\leq$ respectively, as defined in (3.3) and (3.5). Therefore, similar to (3.13), we obtain the order isomorphical embeddings

$$(7.61) \qquad \mathcal{C}^0(\bar{\Omega}) \xrightarrow{\text{inj}} \mathcal{C}^0_A(\Omega)/\sim_A \xrightarrow{\text{id}} \mathcal{M}^0(\Omega) \ .$$

A further useful connection, this time involving the spaces $\mathcal{C}^0_A(\Omega)$ and $\mathcal{C}^0_Q(\bar{\Omega})$, is given in:

Lemma 7.2

We have

$$(7.62) \qquad \mathcal{C}^0_A(\Omega) \subsetneq \mathcal{C}^0_Q(\bar{\Omega}).$$

Further, given $f, g \in \mathcal{C}^0_A(\Omega)$, then

$$(7.63) \qquad f \sim_A g \Longleftrightarrow f \sim_Q g \ .$$

Proof

Take $f \in \mathcal{C}^0_A(\Omega)$, then

$$f \in \mathcal{C}^0(\Omega \backslash \Gamma)$$

for a suitable closed, nowhere dense set $\Gamma \subset \Omega$. It follows that

$$f \in \mathcal{C}^0(\bar{\Omega} \backslash (\Gamma \cup \partial \Omega))$$

and obviously $\Gamma \cup \partial \Omega$ is again closed and nowhere dense.

The implication '$\Longrightarrow$' in (7.63) follows easily from (7.62) and the fact that a closed, nowhere dense set is of first category.

For the implication '$\Longleftarrow$' in (7.63), let us assume that $f \in \mathcal{C}^0(\Omega \backslash \Gamma_f)$ and $g \in \mathcal{C}^0(\Omega \backslash \Gamma_g)$, for suitable closed, nowhere dense sets $\Gamma_f, \Gamma_g \subset \Omega$. Now $f \sim_Q g$ means that for a certain first category set $\Gamma \subset \bar{\Omega}$, we have $f = g$ on $\bar{\Omega} \backslash \Gamma$. But obviously

$$f, g \in \mathcal{C}^0(\Omega \backslash (\Gamma_f \cup \Gamma_g))$$

while

$$\Omega \backslash \Gamma = \bar{\Omega} \backslash (\Gamma \cup \partial \Omega) \text{ is dense in } \Omega$$

since Γ is of first category and $\partial \Omega$ is closed, nowhere dense. In this way

$$f = g \text{ on } \Omega \backslash (\Gamma_f \cup \Gamma_g)$$

which yields $f \sim_A g$, since $\Gamma_f \cup \Gamma_g$ is again closed and nowhere dense in Ω. ■

It follows that we can define the embedding

$$(7.64) \qquad \mathcal{C}^0_A(\Omega)/\sim_A \xrightarrow[\text{inj}]{\psi} \mathcal{C}^0_Q(\bar{\Omega})/\sim_Q$$

by sending the $\sim_A$ equivalence class of each $f \in \mathcal{C}^0_A(\Omega)$ into the $\sim_Q$ equivalence class of f. In fact, we have the stronger result:

Proposition 7.2

The mapping in (7.64), namely

$$(7.65) \qquad \mathcal{C}^0_A(\Omega)/\sim_A \xrightarrow[\text{inj}]{\psi} \mathcal{C}^0_Q(\bar{\Omega})/\sim_Q$$

is an order isomorphical embedding.

Proof

Let $F,G \in \mathcal{C}^0_A(\Omega)/\sim_A$, $f \in F$, $g \in G$ and $f \leq g$ on $\Omega\setminus\Gamma$, with Γ closed, nowhere dense. Then clearly $\psi(F) \leq_Q \psi(G)$. Conversely, assume given $F,G \in \mathcal{C}^0_A(\Omega)/\sim_A$ such that $\psi(F) \leq_Q \psi(G)$. Let $f \in F$, $g \in G$ be such that

$$f \leq g \quad \text{on} \quad \bar{\Omega}\setminus\Gamma$$

for a suitable first category set $\Gamma \subset \bar{\Omega}$. But $f \in \mathcal{C}^0(\Omega\setminus\Gamma_f)$ and $g \in \mathcal{C}^0(\Omega\setminus\Gamma_g)$ for certain closed, nowhere dense sets $\Gamma_f, \Gamma_g \subset \Omega$, hence

$$f,g \in \mathcal{C}^0(\Omega\setminus(\Gamma_f \cup \Gamma_g))$$

while

$$\Omega\setminus\Gamma \text{ is dense in } \Omega, \quad \Omega\setminus(\Gamma_f \cup \Gamma_g) \text{ is open and dense in } \Omega.$$

Therefore

$$f \leq g \quad \text{on} \quad \Omega\setminus(\Gamma_f \cup \Gamma_g)$$

and since $\Gamma_f \cup \Gamma_g$ is closed and nowhere dense in Ω, the inequality $F \leq_A G$ follows. ■

Now we can formulate a first result on the size and the nature of two of the sets of generalized functions encountered so far.

Theorem 7.2

Given a bounded open set $\Omega \subset \mathbb{R}^n$, we have the commutative diagram of order isomorphisms

(7.66)

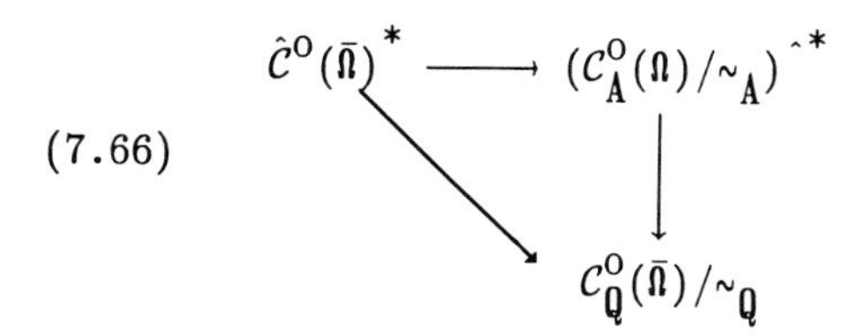

obtained through the extension of the order isomorphical embeddings (7.61) and (7.64) to the respective order completions, see (A.40).

Being order isomorphisms, the mappings in (7.66) preserve infima and suprema.

Proof

We note that the diagram defined by the order isomorphical embeddings (7.61), (7.65) and (7.54), namely

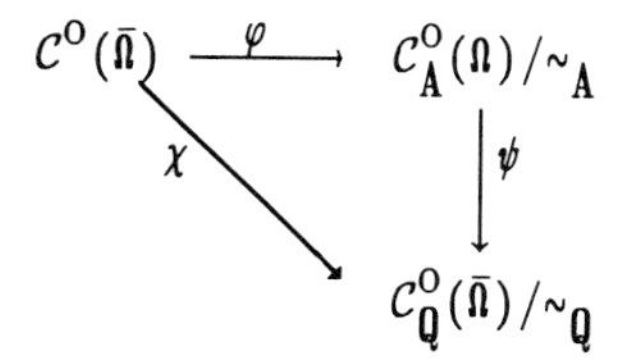

is commutative. Therefore, in view of (7.55), (7.56), as well as of (A.59) and (A.60) in Corollary A.2 in the Appendix, we can extend it to the commutative diagram

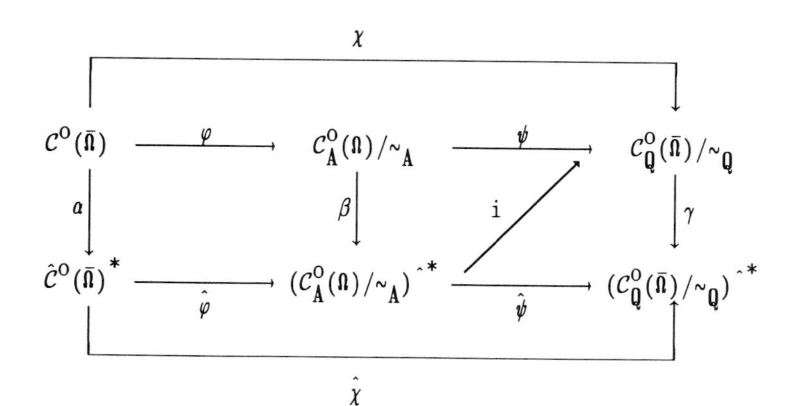

where $\hat{\varphi}$, $\hat{\psi}$, $\hat{\chi}$, i are order isomorphisms. ■

The interest in (7.66) is due to the order isomorphism

$$(7.67) \qquad \hat{\mathcal{C}}^0(\bar{\Omega})^* \longrightarrow (\mathcal{C}^0_A(\Omega)/\sim_A)^{\hat{}*} .$$

Indeed, as seen in (7.61), we have the order isomorphical embedding, in fact inclusion

$$(7.68) \qquad \mathcal{C}^0_A(\Omega)/\sim_A \longrightarrow \mathcal{M}^0(\Omega) .$$

Therefore (A.40) in Proposition A.1 in Appendix implies that (7.68) extends to the order isomorphical embedding

$$(7.69) \qquad (\mathcal{C}^0_A(\Omega)/\sim_A)^{\hat{}*} \longrightarrow \hat{\mathcal{M}}^0(\Omega)^* .$$

Now, by the same argument, we can extend the order isomorphical embedding

$$(7.70) \qquad \mathcal{C}^0(\bar{\Omega}) \longrightarrow \text{Mes }(\Omega)$$

to the order isomorphical embedding

$$(7.71) \qquad \hat{\mathcal{C}}^0(\bar{\Omega})^* \longrightarrow \text{Mes }(\Omega) .$$

Finally, by comparing (7.67), (7.69) and (7.71), we obtain:

Corollary 7.1

We have the order isomorphical embeddings

$$(7.72) \qquad \hat{\mathcal{M}}^0(\Omega)^* \longleftarrow (\mathcal{C}^0_A(\Omega)/\sim_A)^{\hat{}*} \longrightarrow \text{Mes }(\Omega)$$ ■

Remark 7.3

Obviously, it is quite natural to ask whether the order isomorphical embedding, see (7.69) and (7.72)

$$(7.73) \qquad (\mathcal{C}^0_A(\Omega)/\sim_A)^{\wedge *} \longrightarrow \hat{\mathcal{M}}^0(\Omega)^*$$

is in fact an order isomorphism? Or, one may ask about the weaker property, namely, whether, see (7.50), there exists an order isomorphical embedding

$$(7.74) \qquad \hat{\mathcal{M}}^0(\Omega)^* \longrightarrow \text{Mes}\ (\Omega)\ .$$

In view of the existence of the embedding

$$\mathcal{M}^0(\Omega) \longrightarrow \text{Mes}\ (\Omega)$$

given in (7.40), the possibility of a positive answer to (7.74) may appear to be more likely.

The answer to (7.73) is *negative*. Namely, in the sense of the order isomorphical embedding in (7.72), we have

$$(7.75) \qquad (\mathcal{C}^0_A(\Omega)/\sim_A)^{\wedge} \subsetneq \hat{\mathcal{M}}^0(\Omega).$$

Furthermore, the following *partially negative* answer to (7.74) holds as well. Given *any* mapping

$$(7.76) \qquad \mathcal{M}^0(\Omega) \xrightarrow{\ \varphi\ } \text{Mes}\ (\Omega)$$

which coincides with the mapping (7.40), when restricted to elements of $\mathcal{M}^0(\Omega)$ generated by piecewise constant functions or continuous functions on Ω, then the extension of φ

$$(7.77) \qquad \hat{\mathcal{M}}^0(\Omega) \xrightarrow{\ \hat{\varphi}\ } \hat{\text{Mes}}\ (\Omega)$$

constructed according to (A.37), *cannot* be injective.

The proofs of (7.75) and (7.77) are presented in Subsection 7.3.

7.2 The Flabby Sheaf Structure of $\hat{\mathcal{M}}^m_T(\Omega)^*$

7.2.1 The results in Theorem 7.1, Corollary 7.1 and Remark 7.3 show in a clear way that, under the respective conditions, the generalized functions in $\hat{\mathcal{N}}^0(\Omega)$ or $\hat{\mathcal{M}}^0(\Omega)$ are indeed usual measurable functions on Ω, or are related to such functions. However, the mere existence of the order isomorphical embedding, see (7.1)

$$(7.78) \qquad \hat{\mathcal{M}}^m_T(\Omega) \xrightarrow{\hat{T}} \hat{\mathcal{M}}^0(\Omega)$$

does not by itself give enough information about the way the elements $U \in \hat{\mathcal{M}}^m_T(\Omega)$ can themselves be seen as related to functions - in particular measurable ones - on Ω. And in Section 5, it is precisely such elements $U \in \hat{\mathcal{M}}^m_T(\Omega)$ which are the generalized solutions of the continuous nonlinear PDEs in (2.1). Therefore, next we present additional properties of the order isomorphical embedding (7.78), which show the way in which $\hat{\mathcal{M}}^m_T(\Omega)$ is connected with spaces of usual continuous or more smooth functions on Ω.

In our case, a particularly relevant way for relating $\hat{\mathcal{M}}^m_T(\Omega)$ to spaces of usual functions is by relating their respective *sheaf structures*. Indeed, an arbitrary continuous nonlinear partial differential operator (1.10), corresponding to (2.2), is in fact a *local* operator. In other words, given any nonvoid open subset $\Delta \subseteq \Omega$ and $U, V \in \mathcal{C}^m(\Omega)$, we obviously have the localization property

$$U\big|_\Delta = V\big|_\Delta \Longrightarrow T(x,D)U\big|_\Delta = T(x,D)V\big|_\Delta$$

It follows that the structures of *sheaves of sections* over Ω of various spaces of usual or generalized functions are essentially related to the nonlinear partial differential operators under consideration.

In view of that, in this Subsection, we shall prove two basic *sheaf* properties related to the spaces of generalized functions $\hat{\mathcal{M}}^m_T(\Omega)^*$.

First, we shall show that $\hat{\mathcal{M}}^m_T(\Omega)^*$ is a *sheaf of sections* over Ω, constituted from Dedekind complete partially ordered sets and as such, it relates through *sheaf morphisms* to various sheaves of usual functions.

Furthermore, we show that $\hat{\mathcal{M}}^m_T(\Omega)^*$ is a *flabby sheaf of sections* over Ω. In particular, for $m = 0$ and T the identity mapping in (4.5), it follows that $\hat{\mathcal{M}}^0(\Omega)^*$ is also a sheaf of sections over Ω.

Both of these sheaf related results are presented in Theorem 7.3.

Since this is a rather lengthy Subsection and its results are not used in the rest of the work, it can be omitted at a first reading.

7.2.2 Coming back to (7.78), first we note that, in view of (3.2), (4.4), (4.11), (4.15) and (7.78) we have the mappings

$$(7.79) \qquad \mathcal{C}^m(\Omega) \xrightarrow{\text{inj}} \mathcal{C}^m_{nd}(\Omega) \longrightarrow \mathcal{M}^m_T(\Omega) \xrightarrow{\text{inj}} \hat{\mathcal{M}}^m_T(\Omega)^* \xrightarrow{\hat{T}} \hat{\mathcal{M}}^0(\Omega)^*$$

where the second arrow from the left is the canonical mapping into a quotient space, while the last arrow on the right is obtained from (7.78) as follows. In view of (7.26), we only have to prove that for every $\mathcal{F} \in \hat{\mathcal{M}}^m_T(\Omega)$, the implication holds

$$\phi \neq \mathcal{F} \subsetneq \mathcal{M}^m_T(\Omega) \Longrightarrow \phi \neq \hat{T}\mathcal{F} \subsetneq \mathcal{M}^0(\Omega) \ .$$

But (A.20) gives

$$<A] \subseteq \mathcal{F} \subseteq <B]$$

for certain $A, B \in \mathcal{M}^m_T(\Omega)$. Thus in view of (4.15) we have

$$<TA] \subseteq \hat{T}\mathcal{F} \subseteq <TB]$$

which proves the above implication.

Furthermore, the mappings in (7.79) are *consistent* with the nonlinear partial differential operator (2.2), in the sense that we have the *commutative diagram*, see (4.6)

$$(7.80)\qquad \begin{array}{ccccccc} \mathcal{C}^m(\Omega) & \xrightarrow{\text{inj}} & \mathcal{C}^m_{nd}(\Omega) & \longrightarrow & \mathcal{M}^m_T(\Omega) & \xrightarrow{\text{inj}} & \hat{\mathcal{M}}^m_T(\Omega)^* \\ \Big\downarrow T(x,D) & & \Big\downarrow T(x,D) & & \Big\downarrow T \text{ inj} & & \Big\downarrow \hat{T} \text{ inj} \\ \mathcal{C}^0(\Omega) & \xrightarrow{\text{inj}} & \mathcal{C}^0_{nd}(\Omega) & \longrightarrow & \mathcal{M}^0(\Omega) & \xrightarrow{\text{inj}} & \hat{\mathcal{M}}^0(\Omega)^* \end{array}$$

As the main result of this Subsection, we show that $\hat{\mathcal{M}}^m_T(\Omega)^*$ has the structure of a *flabby sheaf of sections* over Ω. In addition, the mappings (7.80) generate a commutative diagram of *sheaf morphisms*, see Theorem 7.3.

Here we should like to mention the following two points. The flabby sheaf structure of $\hat{\mathcal{M}}^m_T(\Omega)^*$ and the existence of the sheaf morphism (7.79) have their own interest related to the fact that, in view of the existence results in Section 5, they offer a better understanding of the structure of generalized solutions of arbitrary continuous nonlinear PDEs (2.1), (2.2). Moreover, the detailed proofs of these sheaf properties, presented in this Subsection, have their own additional interest. Indeed, they offer a better understanding of the Dedekind cuts which define the order completion of quotient spaces of smooth functions $\hat{\mathcal{M}}^m_T(\Omega)^*$, that is, the spaces of generalized functions. And to the extent that order completions such as $\mathcal{M}^m_T(\Omega)^*$ are a first in the literature, the proofs in this Subsection offer a useful opportunity for becoming familiar with their structure.

7.2.3 For convenience, let us recall here these sheaf related notions, see for details Kaneko or Seebach, et. al. Suppose given a set $\mathcal{S}$ of spaces S, as well as a mapping

$$(7.81)\qquad \Omega \supset \Delta \text{ open} \stackrel{\sigma}{\longmapsto} \sigma(\Delta) = S \in \mathcal{S}$$

where $S = \sigma(\Delta)$ is called the *set of sections* over Δ, while each $F \in S = \sigma(\Delta)$ is called a *section* over Δ.

Further, suppose that for each pair of open sets $\Delta_1 \subset \Delta_2 \subset \Omega$, we have a *restriction* mapping

(7.82) $$\rho_{\Delta_1,\Delta_2} : \sigma(\Delta_2) \longrightarrow \sigma(\Delta_1).$$

Then (σ,ρ) is called a *sheaf of sections* over Ω, if and only if the following *four* conditions (7.83) - (7.87) are satisfied:

For every open set $\Delta \subset \Omega$, we have

(7.83) $$\rho_{\Delta,\Delta} = \mathrm{id}_{\sigma(\Delta)} : \sigma(\Delta) \longrightarrow \sigma(\Delta).$$

For every open sets $\Delta_1 \subset \Delta_2 \subset \Delta_3 \subset \Omega$, we have

(7.84) $$\rho_{\Delta_1,\Delta_2} \circ \rho_{\Delta_2,\Delta_3} = \rho_{\Delta_1,\Delta_3}.$$

Given any family $(\Delta_i | i \in I)$ of open sets $\Delta_i \subset \Omega$, let us denote $\Delta = \bigcup_{i \in I} \Delta_i$.

If $F,\ G \in \sigma(\Delta)$, then

(7.85) $$\left[\begin{array}{l} \forall\ i \in I : \\ \rho_{\Delta_i,\Delta} F = \rho_{\Delta_i,\Delta} G \end{array}\right] \Rightarrow F = G.$$

Finally, given $F_i \in \sigma(\Delta_i)$, with $i \in I$, if we have

(7.86) $$\begin{array}{l} \forall\ i,j \in I : \\ \Delta_i \cap \Delta_j \neq \phi \Rightarrow \rho_{\Delta_i \cap \Delta_j,\Delta_i} F_i = \rho_{\Delta_i \cap \Delta_j,\Delta_j} F_j \end{array}$$

then it follows that

(7.87) $$\begin{array}{l} \exists\ F \in \sigma(\Delta) : \\ \forall\ i \in I : \\ \rho_{\Delta_i,\Delta} F = F_i \end{array}$$

Given (σ,ρ) and (σ',ρ') two sheaves of sections over Ω, a *sheaf morphism*

$$(7.88) \qquad (\sigma,\rho) \xrightarrow{\alpha} (\sigma',\rho')$$

is a family of mappings

$$(7.89) \qquad \sigma(\Delta) \xrightarrow{\alpha_\Delta} \sigma'(\Delta)$$

with $\Delta \subset \Omega$, Δ open, which *commute* with the *restriction* mappings, that is, for each pair of open sets $\Delta_1 \subset \Delta_2 \subset \Omega$, we have the commutative diagram

$$(7.90) \qquad \begin{array}{ccc} \sigma(\Delta_2) & \xrightarrow{\alpha_{\Delta_2}} & \sigma'(\Delta_2) \\ \rho_{\Delta_1,\Delta_2} \downarrow & & \downarrow \rho'_{\Delta_1,\Delta_2} \\ \sigma(\Delta_1) & \xrightarrow[\alpha_{\Delta_1}]{} & \sigma'(\Delta_1) \end{array}$$

We shall use the following simplification in notation: given open sets $\Delta_1 \subset \Delta_2 \subset \Omega$ and $F \in \sigma(\Delta_2)$, we shall denote

$$(7.91) \qquad F|_{\Delta_1} = \rho_{\Delta_1,\Delta_2} F.$$

Further, the mappings α_Δ in (7.89), defining a sheaf morphism, will simply be denoted by α.

Finally, we say that the sheaf of sections (σ,ρ) over Ω is *flabby*, if and only if for every open set $\Delta \subset \Omega$, the restriction mapping is *surjective*, that is

$$(7.92) \qquad \rho_{\Delta,\Omega} : \sigma(\Omega) \xrightarrow{\text{sur}} \sigma(\Delta).$$

Obviously (7.92) means the following *extension property*

$$(7.93) \qquad \begin{array}{l} \forall \;\; F \in \sigma(\Delta) \; : \\ \exists \;\; G \in \sigma(\Omega) \; : \\ \quad F = G|_{\Delta} \end{array}$$

Moreover (7.92) implies that for every two open sets $\Delta_1 \subset \Delta_2 \subset \Omega$, the restriction mapping is *surjective*, that is

$$(7.94) \qquad \rho_{\Delta_1,\Delta_2} : \sigma(\Delta_2) \xrightarrow{\text{sur}} \sigma(\Delta_1) \; .$$

This follows in an *extension property* similar to that in (7.93), namely

$$(7.95) \qquad \begin{array}{l} \forall \;\; F \in \sigma(\Delta_1) \; : \\ \exists \;\; G \in \sigma(\Delta_2) \; : \\ \quad F = G|_{\Delta_1} \end{array}$$

7.2.4 Now, it is easy to see that each of the spaces $\mathcal{C}^m(\Omega)$, $\mathcal{C}^m_{nd}(\Omega)$ and $\mathcal{M}^m_T(\Omega)$ in (7.79) has a natural structure of sheaf of sections over Ω. Indeed, we can define the respective *sections* σ, σ' and σ'' in the natural way

$$(7.96) \qquad \begin{array}{l} \Omega \supset \Delta \text{ open} \longmapsto \sigma(\Delta) = \mathcal{C}^m(\Delta) \\ \Omega \supset \Delta \text{ open} \longmapsto \sigma'(\Delta) = \mathcal{C}^m_{nd}(\Delta) \\ \Omega \supset \Delta \text{ open} \longmapsto \sigma''(\Delta) = \mathcal{M}^m_T(\Delta) \end{array}$$

Similarly, for open sets $\Delta_1 \subset \Delta_2 \subset \Omega$, we can define the *restrictions* ρ and ρ' in the natural way

$$(7.97) \qquad \begin{array}{l} \mathcal{C}^m(\Delta_2) \ni f \longmapsto \rho_{\Delta_1,\Delta_2} f = f|_{\Delta_1} \\ \mathcal{C}^m_{nd}(\Delta_2) \ni f \longmapsto \rho'_{\Delta_1,\Delta_2} f = f|_{\Delta_1} \end{array}$$

while the restriction ρ'' is defined as follows

$$(7.98) \qquad \mathcal{M}^m_T(\Delta_2) \ni F \longmapsto \rho''_{\Delta_1,\Delta_2} F = G \in \mathcal{M}^m_T(\Delta_1)$$

where G is the $\sim_T$ equivalence class in $\mathcal{M}_T^m(\Delta_1)$, generated by $f|_{\Delta_1}$, with $f \in F$, that is, f being an element of the $\sim_T$ equivalence class F in $\mathcal{M}_T^m(\Delta_2)$. It is easy to check that G will not depend on the choice of $f \in F$.

With the above definitions of sections and restrictions, we shall establish that $\mathcal{C}^m(\Omega)$, $\mathcal{C}_{nd}^m(\Omega)$ and $\mathcal{M}_T^m(\Omega)$ are sheaves of sections over Ω, with the latter two being in fact *flabby* sheaves, see Theorem 7.3. Moreover $\mathcal{C}_{nd}^m(\Omega)$ is the *smallest* flabby sheaf containing $\mathcal{C}^m(\Omega)$, see Remark 7.5. Taking again $m = 0$ and T the identity mapping in (4.5), it follows in particular that $\mathcal{M}^0(\Omega)$ is also a sheaf of sections over Ω. Moreover, we shall see in Theorem 7.3 that the mappings generated by (7.79)

$$(7.99) \qquad \mathcal{C}^m(\Delta) \xrightarrow{\text{inj}} \mathcal{C}_{nd}^m(\Delta) \longrightarrow \mathcal{M}_T^m(\Delta) \xrightarrow{\text{inj}} \hat{\mathcal{M}}_T^m(\Delta)^* \xrightarrow{\hat{T}} \hat{\mathcal{M}}^0(\Omega)^*$$

with $\Delta \subset \Omega$, Δ open, are sheaf morphisms. The fact that (σ', ρ') and (σ'', ρ'') satisfy (7.86), (7.87) will be explicitely shown below, being useful in the proof of the sheaf structure of $\hat{\mathcal{M}}_T^m(\Omega)^*$ as well.

7.2.5 Now we come to defining on $\hat{\mathcal{M}}_T^m(\Omega)^*$ a structure of sheaf of sections over Ω. For that, first we have to define the mappings (7.81). Here we can again proceed in the natural way, as for instance in (7.96), and take

$$(7.100) \qquad \Omega \supset \Delta \text{ open} \longmapsto \sigma'''(\Delta) = \hat{\mathcal{M}}_T^m(\Delta)^* .$$

A bit more care is needed in order to define the restriction mapping ρ''' corresponding to $\hat{\mathcal{M}}_T^m(\Omega)^*$, as requested by (7.82). For that, let us take two open sets $\Delta_1 \subset \Delta_2 \subset \Omega$ and $\mathcal{F} \in \sigma'''(\Delta_2) = \hat{\mathcal{M}}_T^m(\Delta_2)^*$. Then in view of (7.26), Lemma 4.2 and (A.12), we have

$$(7.101) \qquad \phi \underset{\neq}{\subset} \mathcal{F} \underset{\neq}{\subset} \mathcal{M}_T^m(\Delta_2), \quad \mathcal{F}^{u\ell} = \mathcal{F}.$$

Therefore, in view of (A.20), it follows in particular that

$$(7.102) \qquad \mathcal{F} \text{ is bounded from above in } \mathcal{M}^m_T(\Delta_2).$$

Now, the moment we have secured for every $\mathcal{F} \in \sigma'''(\Delta_2)$ the inclusion, see (7.101)

$$\mathcal{F} \subseteq \mathcal{M}^m_T(\Delta_2)$$

we can go back and use the restriction mapping ρ'' in (7.98), in order to define the restriction mapping ρ''', as follows

$$(7.103) \qquad \sigma'''(\Delta_2) \ni \mathcal{F} \longmapsto \rho'''_{\Delta_1,\Delta_2} \mathcal{F} = \mathcal{G} \in \sigma'''(\Delta_1)$$

where $\mathcal{G}$ is the cut in $\mathcal{M}^m_T(\Delta_1) = \sigma'''(\Delta_1)$ generated by

$$(7.104) \qquad \rho''_{\Delta_1,\Delta_2} \mathcal{F} = \{\rho''_{\Delta_1,\Delta_2} F \mid F \in \mathcal{F}\} \subseteq \mathcal{M}^m_T(\Delta_1).$$

In other words, in view of (A.12), (A.15) - (A.17), we have

$$(7.105) \qquad \rho'''_{\Delta_1,\Delta_2} \mathcal{F} = \{\rho''_{\Delta_1,\Delta_2} F \mid F \in \mathcal{F}\}^{u\ell} \subseteq \mathcal{M}^m_T(\Delta_1).$$

Now we note that owing to (7.101), we clearly have $\rho'''_{\Delta_1,\Delta_2} \mathcal{F} \neq \phi$. Therefore, in view of (7.100) and (7.26), in order to show that ρ''' in (7.103) is well defined, it only remains to prove that in (7.105), we have in fact

$$(7.106) \qquad \rho'''_{\Delta_1,\Delta_2} \mathcal{F} \subsetneqq \mathcal{M}^m_T(\Delta_1).$$

This however follows easily from (7.102), since according to (A.8), (A.9), we have $A^{u\ell} = X \iff A^u = \phi \iff A$ unbounded from above.

7.2.6 Once (σ''',ρ''') was defined, it remains to prove that it satisfies the conditions (7.83) - (7.87).

Condition (7.83) follows directly from (7.104). For condition (7.84), we note that (7.105) can be replaced by the stronger property

$$(7.107) \qquad \rho'''_{\Delta_1,\Delta_2} \mathcal{F} = \{\rho''_{\Delta_1,\Delta_2} F \mid F \in \mathcal{F}\}$$

since the right hand term in (7.107) already happens to be a cut in $\mathcal{M}_T^m(\Delta_1)$.

Indeed, in view of (7.105) and (A.15), the inclusion '$\supseteq$' is obvious. For the converse inclusion '$\subseteq$' we note that the following diagrams are *commutative*:

$$(7.108) \qquad \begin{array}{ccc} \mathcal{M}_T^m(\Delta_2) \ni F & \longmapsto & \langle F] \subset \mathcal{M}_T^m(\Delta_2) \\ \downarrow & & \downarrow \\ \mathcal{M}_T^m(\Delta_1) \ni F|_{\Delta_1} & \longmapsto & \langle F|_{\Delta_1}] = \langle F]|_{\Delta_1} \subset \mathcal{M}_T^m(\Delta_1) \end{array}$$

and dually

$$(7.109) \qquad \begin{array}{ccc} \mathcal{M}_T^m(\Delta_2) \ni F & \longmapsto & [F\rangle \subset \mathcal{M}_T^m(\Delta_2) \\ \downarrow & & \downarrow \\ \mathcal{M}_T^m(\Delta_1) \ni F|_{\Delta_1} & \longmapsto & [F|_{\Delta_1}\rangle = [F\rangle|_{\Delta_1} \subset \mathcal{M}_T^m(\Delta_1) \end{array}$$

which can be written in the equivalent form: if $F \in \mathcal{M}_T^m(\Delta_2)$ then

$$(7.110) \qquad \langle \rho''_{\Delta_1,\Delta_2} F] = \rho''_{\Delta_1,\Delta_2} \langle F]$$

$$(7.111) \qquad [\rho''_{\Delta_1,\Delta_2} F\rangle = \rho''_{\Delta_1,\Delta_2} [F\rangle \, .$$

Indeed, the inclusion '$\supseteq$' in (7.110) follows quite easily. For the inclusion '$\subseteq$' in (7.110) let $G \in \langle F|_{\Delta_1}]$. Let us apply Corollary 7.3

below. Then it follows that one can find $H \in \langle F] \subset \mathcal{M}_T^m(\Delta_2)$, such that $G = H|_{\Delta_1}$, therefore indeed

$$G \in \langle F]\,|_{\Delta_1}$$

and the proof of (7.110) is completed. The relation (7.111) follows in a similar way, thus (7.108) and (7.109) hold.

Based on (7.108) - (7.111), we can prove now the following two properties. Given $\mathcal{F} \subseteq \mathcal{M}_T^m(\Delta_2)$, if $\mathcal{F}$ is bounded from above in $\mathcal{M}_T^m(\Delta_2)$, then

$$(7.112) \qquad \mathcal{F}^u|_{\Delta_1} = (\mathcal{F}|_{\Delta_1})^u$$

similarly, if $\mathcal{F}$ is bounded from below in $\mathcal{M}_T^m(\Delta_2)$, then

$$(7.113) \qquad \mathcal{F}^\ell|_{\Delta_1} = (\mathcal{F}|_{\Delta_1})^\ell.$$

Indeed, in view of (A.6) and (7.108), we have

$$(7.114) \qquad \begin{aligned} \mathcal{F}^u|_{\Delta_1} &= (\bigcap_{F\in\mathcal{F}} [F\rangle)|_{\Delta_1} \subseteq \bigcap_{F\in\mathcal{F}} ([F\rangle|_{\Delta_1}) = \\ &= \bigcap_{F\in\mathcal{F}} [F|_{\Delta_1}\rangle \subseteq \bigcap_{F_1\in\mathcal{F}|_{\Delta_1}} [F_1\rangle = (\mathcal{F}|_{\Delta_1})^u \end{aligned}$$

Thus in order to prove (7.112), we only have to show that the following two inclusions hold

$$(7.115) \qquad \bigcap_{F\in\mathcal{F}} ([F\rangle|_{\Delta_1}) \subseteq (\bigcap_{F\in\mathcal{F}} [F\rangle)|_{\Delta_1}$$

and

$$(7.116) \qquad \bigcap_{F_1\in\mathcal{F}|_{\Delta_1}} [F_1\rangle \subseteq \bigcap_{F\in\mathcal{F}} [F|_{\Delta_1}\rangle.$$

Let G be an element in the left hand term of (7.115), then

$$F|_{\Delta_1} \le_T G, \quad \forall\ F \in \mathcal{F}.$$

But $\mathcal{F}$ is bounded from above in $\mathcal{M}_T^m(\Delta_2)$, hence we have

$$F \le_T W, \quad \forall\ F \in \mathcal{F}$$

for a certain $W \in \mathcal{M}_T^m(\Delta_2)$. Then, in view of Corollary 7.3 below, it follows that

$$\exists\ H \in \mathcal{M}_T^m(\Delta_2) :$$
$$*)\ G = H|_{\Delta_1}$$
$$**)\ F \le_T H, \quad \forall\ F \in \mathcal{F}.$$

In this way G belongs to the right hand term in (7.115), thus the inclusion in (7.115) holds. Let now G be an element in the left hand term of (7.116), then

$$F_1 \le_T G, \quad \forall\ F_1 \in \mathcal{F}|_{\Delta_1}.$$

Take an arbitrary $F \in \mathcal{F}$. Then $F_1 = F|_{\Delta_1} \in \mathcal{F}|_{\Delta_1}$, hence $F|_{\Delta_1} \le G$, and the inclusion (7.116) is indeed valid.

In view of (7.114) - (7.116), one obtains the relation (7.112). The proof of (7.113) is similar.

Now we can return and end the proof of the inclusion '$\subseteq$' in (7.107). Since $\mathcal{F}$ is a cut in $\mathcal{M}_T^m(\Delta_2)$, we have

$$\mathcal{F}|_{\Delta_1} = \mathcal{F}^{u\ell}|_{\Delta_1}.$$

But (7.101) implies that $\mathcal{F}^{u}$ is bounded from below in $\mathcal{M}_T^m(\Delta_2)$, thus (7.113) gives

$$\mathcal{F}^{u\ell}\big|_{\Delta_1} = (\mathcal{F}^{u}\big|_{\Delta_1})^{\ell}.$$

On the other hand, (7.102) and (7.112) yield

$$\mathcal{F}^{u}\big|_{\Delta_1} = (\mathcal{F}\big|_{\Delta_1})^{u} .$$

In this way

$$\mathcal{F}\big|_{\Delta_1} = \mathcal{F}^{u\ell}\big|_{\Delta_1} = (\mathcal{F}^{u}\big|_{\Delta_1})^{\ell} = (\mathcal{F}\big|_{\Delta_1})^{u\ell}$$

and the proof of (7.107) is completed.

In view of (7.107), it is now obvious that (σ''',ρ''') satisfies condition (7.84) in the definition of a sheaf of sections on Ω.

We note that (7.85) holds for (σ',ρ') and (σ'',ρ''). This can be easily checked by using the maximal regularization in (7.30).

Now, we show that (σ',ρ'), (σ'',ρ'') and (σ''',ρ''') satisfy (7.86), (7.87).

First we start with (σ',ρ'). Let $(\Delta_i \mid i \in I)$ be a family of open sets $\Delta_i \subset \Omega$, and let

$$(7.117) \qquad \Delta = \bigcup_{i\in I} \Delta_i .$$

Assume given $f_i \in \sigma'(\Delta_i) = \mathcal{C}_{nd}^m(\Delta_i)$, with $i \in I$, which satisfies (7.86), that is, for $i,j \in I$, we have

$$(7.118) \qquad \Delta_i \cap \Delta_j \neq \phi \Longrightarrow f_i\big|_{\Delta_i\cap\Delta_j} = f_j\big|_{\Delta_i\cap\Delta_j}$$

Obviously, we have

(7.119) $f_i \in \mathcal{C}^m(\Delta_i \backslash \Gamma_i)$, $i \in I$

for suitable closed, nowhere dense sets $\Gamma_i \subset \Delta_i$, with $i \in I$. We show now that

(7.120) $\Gamma = \bigcup_{i \in I} \Gamma_i$ is closed and nowhere dense in Δ.

Indeed (7.118) and (7.119) imply

(7.121) $\Gamma_i \cap \Delta_i \cap \Delta_j = \Gamma_j \cap \Delta_i \cap \Delta_j$, $i, j \in I$

therefore, given $j \in I$, we obtain

$$\Gamma \cap \Delta_j = \bigcup_{i \in I} (\Gamma_i \cap \Delta_j) = \bigcup_{i \in I} (\Gamma_i \cap \Delta_i \cap \Delta_j)$$

and in view of (7.121) it follows that

$$\Gamma \cap \Delta_j = \bigcup_{i \in I} (\Gamma_j \cap \Delta_i \cap \Delta_j) = \Gamma_j \cap \bigcup_{i \in I} \Delta_i = \Gamma_j$$

in other words

(7.122) $\Gamma \cap \Delta_j = \Gamma_j$, $j \in I$

which in view of (7.117) will clearly give (7.120).

Now we can define $f : \Delta \backslash \Gamma \longrightarrow \mathbb{R}$ by

(7.123) $f(x) = f_i(x)$ if $x \in \Delta_i \backslash \Gamma_i$

which in view of (7.117), (7.118) and (7.122) proves to be a correct definition. Then owing to (7.117) and (7.119), we obtain that

$$f \in \mathcal{C}^m(\Delta \backslash \Gamma)$$

while (7.120) gives

$$f \in \mathcal{C}^m_{nd}(\Delta).$$

The fact that (7.87) is satisfied by f follows from (7.123). In this way (σ', ρ') does indeed satisfy (7.86), (7.87).

We turn now to showing that (σ'', ρ'') also satisfies (7.86), (7.87). For that, assume given $F_i \in \sigma''(\Delta_i) = \mathcal{M}^m_T(\Delta_i)$, with $i \in I$, such that (7.86) holds, that is, for $i, j \in I$, we have

$$(7.124) \quad \Delta_i \cap \Delta_j \neq \phi \Longrightarrow F_i \Big|_{\Delta_i \cap \Delta_j} = F_j \Big|_{\Delta_i \cap \Delta_j}.$$

For each $i \in I$, let us take $f_i \in F_i$. Then, in view of (4.4), we again have (7.119).

The difficulty this time is that (7.118), and therefore (7.121), need no longer hold. However, it is known, de Rham, that there exists a finite or countable family of bounded open sets $\Delta'_\nu \subset \Delta$, with $\nu \in \mathbb{M} \subseteq \mathbb{N}$, such that

$$(7.125) \quad \bigcup_{\nu \in \mathbb{M}} \Delta'_\nu = \Delta$$

and

$$(7.126) \quad \forall \ \nu \in \mathbb{M} : \exists \ i \in I : \overline{\Delta'_\nu} \subset \Delta_i$$

while also

$$(7.127) \quad \begin{array}{l} \forall \ K \subset \Delta, \ \ K \ \text{ compact :} \\ \{\nu \in \mathbb{M} \mid K \cap \Delta'_\nu \neq \phi\} \ \text{ is finite.} \end{array}$$

In view of (7.125), we can define the mapping

$$\Delta \ni x \longmapsto \nu_x \in \mathbb{M} \subseteq \mathbb{N}$$

where

$$\nu_x = \min \{\nu \in \mathbb{M} \mid x \in \Delta'_\nu\}.$$

Also (7.126) gives a mapping

$$\mathbb{M} \ni \nu \longmapsto i_\nu \in I$$

such that

$$(7.128) \quad \overline{\Delta'_\nu} \subset \Delta_{i_\nu}.$$

Then obviously

$$(7.129) \quad x \in \Delta'_{\nu_x} \subset \overline{\Delta'_{\nu_x}} \subset \Delta_{i_{\nu_x}}, \quad x \in \Delta .$$

Now, instead of (7.120), let us show that

$$(7.130) \quad \Gamma = \bigcup_{\nu \in \mathbb{M}} ((\Gamma_{i_\nu} \cap \overline{\Delta'_\nu}) \cup \partial \Delta'_\nu) \quad \text{is closed and nowhere dense in } \Delta.$$

This follows easily from (7.127), (7.128) and the fact that the boundary ∂A of any set $A \subset \Delta$ is always closed and nowhere dense in Δ.

We can now define $f : \Delta \backslash \Gamma \longrightarrow \mathbb{R}$ by

$$(7.131) \quad f(x) = \left| \; f_{i_{\nu_x}}(x) \quad \text{if} \quad x \in \Delta'_{\nu_x} \backslash \Gamma \right.$$

which owing to (7.129), is clearly a correct definition.

Let us further show that

(7.132) $\quad f \in \mathcal{C}^m(\Delta\backslash\Gamma)$.

For that, assume given $x \in \Delta\backslash\Gamma$. Then (7.129) and (7.130) give

(7.133) $\quad x \in \Delta'_{\nu_x}\ , \quad x \in \Delta_{i_{\nu_x}}\backslash\Gamma_{i_{\nu_x}}\ , \quad x \notin \bigcup_{\nu\in M} \partial\Delta'_\nu$.

The second relation in (7.133), together with (7.127), imply that

(7.134) $\quad x \in V \subsetneqq (\Delta'_{\nu_x}\backslash\Gamma) \cap (\Delta_{i_{\nu_x}}\backslash\Gamma_{i_{\nu_x}})$

for a certain neighbourhood V of x, which in addition can be assumed to satisfy, in view of the third relation in (7.133)

(7.135) $\quad \nu_y = \nu_x\ , \quad y \in V$.

Now (7.119), (7.131), (7.134) and (7.135) will give indeed (7.132).

Let us define $F \in \mathcal{M}^m_T(\Delta)$ as being the $\sim_T$ equivalence class in $\mathcal{C}^m_{nd}(\Delta)$ generated by f. It only remains to show that F satisfies (7.87). For that, assume given $i \in I$. Further, let

(7.136) $\quad x \in \Delta_i\backslash(\Gamma \cup \Gamma_i)$

then in view of (7.134), we obtain

(7.137) $\quad x \in V \subsetneqq (\Delta'_{\nu_x}\backslash\Gamma) \cap (\Delta_i\backslash(\Gamma \cup \Gamma_i)) \cap (\Delta_{i_{\nu_x}}\backslash\Gamma_{i_{\nu_x}})$

for a suitable neighbourhood V of x.

From (7.131) and (7.135) - (7.137) together with (7.86) it follows that

(7.138) $\quad f|_V = f_{i_{\nu_x}}|_V \sim_T f_i|_V$.

Recapitulating (7.136) - (7.138), we obtain

$$(7.139) \quad \begin{aligned} &\forall\ i \in I,\ \ x \in \Delta_i \backslash (\Gamma \cup \Gamma_i) : \\ &\exists\ x \in V \subset \Delta_i \backslash (\Gamma \cup \Gamma_i),\ \ V \text{ neighbourhood of } x : \\ &f|_V \sim_T f_i|_V \end{aligned}$$

But (7.132) and (7.119) give

$$f,\ f_i \in \mathcal{C}^m(\Delta_i \backslash (\Gamma \cup \Gamma_i))$$

therefore (7.139) clearly implies

$$f|_{\Delta_i \backslash (\Gamma \cup \Gamma_i)} = f_i|_{\Delta_i \backslash (\Gamma \cup \Gamma_i)}\ ,\ i \in I.$$

And, since $\Gamma \cup \Gamma_i$ is closed, nowhere dense in Δ_i, we obtain

$$f|_{\Delta_i} \sim_T f_i,\ \ i \in I.$$

In this way (σ'',ρ'') satisfies (7.86), (7.87).

Now we have to show that (σ''', ρ''') satisfies (7.85). For that purpose, let us take $\mathcal{F},\mathcal{G} \in \sigma'''(\Delta) = \hat{\mathcal{M}}^m_T(\Delta)^*$ such that

$$(7.140) \quad \mathcal{F}|_{\Delta_i} = \mathcal{G}|_{\Delta_i}\ ,\ \ i \in I.$$

In order to prove (7.85), we then have to show that

$$(7.141) \quad \mathcal{F} = \mathcal{G}.$$

The easy way to obtain (7.141) is in the following round about manner. We note that, given $\mathcal{F} \in \sigma'''(\Delta) = \hat{\mathcal{M}}^m_T(\Delta)^*$, then in view of (7.107), (7.103), we can associate with it the family

$$(7.142) \qquad \mathcal{F}_i = \mathcal{F}|_{\Delta_i} \in \sigma'''(\Delta_i) = \hat{\mathcal{M}}_T^m(\Delta_i)^*, \quad i \in I.$$

Now we can define

$$(7.143) \qquad \mathcal{F}^o = \left\{ F \in \mathcal{M}_T^m(\Delta) \;\middle|\; \begin{array}{l} \forall\ i \in I : \\ F|_{\Delta_i} \in \mathcal{F}_i \end{array} \right\}$$

The important fact for us is the relation

$$(7.144) \qquad \mathcal{F}^o = \mathcal{F}.$$

Indeed, if we apply the procedure in (7.142) and (7.143) to $\mathcal{G}$ as well, then we obtain $\mathcal{G}^o = \mathcal{G}$. But in view of (7.140), it is clear that $\mathcal{F}^o = \mathcal{G}^o$. Therefore (7.141) follows.

We return now to the proof of (7.144). The inclusion '$\supseteq$' is an easy consequence of (7.142).

For the converse inclusion '$\subseteq$' in (7.144), let us assume

$$(7.145) \qquad F^o \in \mathcal{F}^o \backslash \mathcal{F}.$$

But $\mathcal{F}^{u\ell} = \mathcal{F}$, since $\mathcal{F} \in \sigma'''(\Delta) = \hat{\mathcal{M}}_T^m(\Delta)^*$.

Hence $F^o \notin \mathcal{F}$ implies that

$$(7.146) \qquad \begin{array}{l} \exists\ G \in \mathcal{M}_T^m(\Delta) : \\ \quad *) \ \ \forall\ H \in \mathcal{F} : H \leq_T G \\ \quad **) \ \ F^o \leq_T G \ \text{ does not hold.} \end{array}$$

Let us take $f^o \in F^o$, $g \in G$, then $**)$ in (7.146) yields a nonvoid open set $V \subset \Delta$, such that

$$f^o|_V, \quad g|_V \in \mathcal{C}^m(V)$$

and

$$T(x,D)f^o > T(x,D)g \quad \text{on} \quad V.$$

Obviously, we can assume that for a suitable $i \in I$, we have $V \subsetneq \Delta_i$. It follows then that

(7.147) $\quad F^o|_{\Delta_i} \leq_T G|_{\Delta_i}$ does not hold.

On the other hand we have

(7.148) $\quad \forall\, H_i \in \mathcal{F}_i : H_i \leq_T G|_{\Delta_i}.$

Indeed, take

$$H_i \in \mathcal{F}_i = \mathcal{F}|_{\Delta_i}$$

then

(7.149) $\quad H_i = H|_{\Delta_i}$

for a certain $H \in \mathcal{F}$. But *) in (7.146) gives then $H \leq_T G$, which together with (7.149) will yield (7.148).

Now, the assumption $F^o \in \mathcal{F}^o$ will in view of (7.143) give

(7.150) $\quad F^o|_{\Delta_i} \in \mathcal{F}_i$.

In the same time (7.142) implies that $\mathcal{F}_i^{u\ell} = \mathcal{F}_i$. In this way (7.147), (7.148) and (7.150) contradict each other, and thus the assumption (7.145) is false. This completes the proof of (7.144), and finally, that of (7.141) as well.

At last, we come to prove that (σ''',ρ''') also satisfies (7.86), (7.87). Assume then given $\mathcal{F}_i \in \sigma'''(\Delta_i) = \hat{\mathcal{M}}_T^m(\Delta_i)^*$, with $i \in I$, such that (7.86) holds, which means that for $i,j \in I$, we have

$$(7.151) \qquad \Delta_i \cap \Delta_j \neq \phi \Longrightarrow \mathcal{F}_i\Big|_{\Delta_i \cap \Delta_j} = \mathcal{F}_j\Big|_{\Delta_i \cap \Delta_j} .$$

In view of (7.125) - (7.127), and in particular, since each open set Δ'_ν, with $\nu \in \mathbf{M}$, is bounded, we can construct a finite or countable family of nonvoid, bounded open sets Δ''_μ, $\mu \in P \subseteq \mathbb{N}$, as follows. Given $x \in \Delta$, then

$$\{\nu \in \mathbf{M} \mid x \in \Delta'_\nu\} \text{ is finite}$$

hence

$$(7.152) \qquad \Delta(x) = \Big(\bigcap_{\substack{\nu \in \mathbf{M} \\ x \in \Delta'_\nu}} \Delta'_\nu\Big) \setminus \overline{\Big(\bigcup_{\substack{\lambda \in \mathbf{M} \\ x \notin \Delta'_\lambda}} \Delta'_\lambda\Big)}$$

is possibly void, or it is bounded and open in Δ. And now, Δ''_μ, with $\mu \in P \subseteq \mathbb{N}$, will be given by the family of all nonvoid $\Delta(x)$ in (7.152), with $x \in \Delta$. Then

$$(7.153) \qquad \Delta \setminus \bigcup_{\mu \in P} \Delta''_\mu = \Gamma \text{ is closed, nowhere dense in } \Delta$$

owing to (7.127). Moreover, we have the particularly convenient property

$$(7.154) \qquad \lambda, \mu \in P, \quad \lambda \neq \mu \Rightarrow \Delta''_\lambda \cap \Delta''_\mu = \phi .$$

In view of (7.152), we can define

$$P \ni \mu \longmapsto \nu_\mu \in \mathbf{M}$$

by

$$\nu_\mu = \min \{\nu \in \mathbf{M} \mid \Delta''_\mu \subseteq \Delta'_\nu\}.$$

And then, taking (7.128) and (7.107) into account, we can define for $\mu \in P$, the set

$$(7.155) \qquad \mathcal{F}''_\mu = \mathcal{F}_{i_{\nu_\mu}}\Big|_{\Delta''_\mu} \in \hat{\mathcal{M}}^m_T(\Delta''_\mu)^*.$$

We note that (7.153) - (7.155) give

$$(7.156) \qquad \begin{array}{l} \forall \ (F''_\mu \mid \mu \in P) \in \prod_{\mu \in P} \mathcal{F}''_\mu : \\ \exists \ F \in \mathcal{M}^m_T(\Delta) : \\ \forall \ \mu \in P : \\ \quad F\big|_{\Delta''_\mu} = F''_\mu \end{array}$$

Indeed, if we take $f''_\mu \in F''_\mu$, with $\mu \in P$, then

$$f''_\mu \in \mathcal{C}^m(\Delta''_\mu \setminus \Gamma''_\mu)$$

where $\Gamma''_\mu \subsetneq \Delta''_\mu$ is closed, nowhere dense in Δ''_μ. In view of (7.154), it follows that

$$\Gamma'' = \bigcup_{\mu \in P} \Gamma''_\mu \ \text{ is closed, nowhere dense in } \bigcup_{\mu \in P} \Delta''_\mu .$$

Therefore (7.154) allows us to define

$$f \in \mathcal{C}^m((\bigcup_{\mu \in P} \Delta''_\mu) \setminus \Gamma'')$$

by

$$f(x) = | \ f''_\mu(x) \quad \text{if} \quad x \in \Delta''_\mu \setminus \Gamma''.$$

In this case, owing to (7.153) we can take $F \in \mathcal{M}^m_T(\Delta)$ as the $\sim_T$ equivalence class generated by f.

The important property for us is the following. Given any $F \in \mathcal{M}_T^m(\Delta)$, such that

(7.157) $\quad F|_{\Delta''_\mu} \in \mathcal{F}''_\mu \,, \quad \mu \in P$

then we shall also have

(7.158) $\quad F|_{\Delta_i} \in \mathcal{F}_i, \quad i \in I.$

Indeed, take any given and fixed $i \in I$, then (7.153) implies that

(7.159) $$\Delta_i \setminus \Gamma = \bigcup_{\mu \in P_i} (\Delta''_\mu \cap (\Delta_i \setminus \Gamma))$$

with

$$P_i = \{\mu \in P \mid \Delta''_\mu \cap (\Delta_i \setminus \Gamma) \neq \phi\}.$$

Let us now start with

(7.160) $$\mathcal{H} = \mathcal{F}_i|_{\Delta_i \setminus \Gamma} \in \sigma'''(\Delta_i \setminus \Gamma) = \hat{\mathcal{M}}_T^m(\Delta_i \setminus \Gamma)^*$$

and corresponding to (7.159), with the family

(7.161) $$\mathcal{H}_\mu = \mathcal{H}|_{\Delta''_\mu \cap (\Delta_i \setminus \Gamma)} \,, \quad \mu \in P_i.$$

Then we can apply (7.142) - (7.144) to (7.160) and (7.161), by replacing $\mathcal{F}$ with $\mathcal{H}$, $\mathcal{F}_i$ with $\mathcal{H}_\mu$, and $i \in I$ with $\mu \in P_i$. In that case we obtain the relation

$$\mathcal{H} = \left\{ H \in \mathcal{M}_T^m(\Delta_i \setminus \Gamma) \;\middle|\; \begin{array}{l} \forall \; \mu \in P_i \;: \\ H|_{\Delta''_\mu \cap (\Delta_i \setminus \Gamma)} \in \mathcal{H}_\mu \end{array} \right\}$$

which rewritten in terms of (7.160), (7.161) will become

$$(7.162)\qquad \mathcal{F}_i\big|_{\Delta_i\setminus\Gamma} = \left\{F_i \in \mathcal{M}_T^m(\Delta_i\setminus\Gamma) \;\middle|\; \begin{array}{l} \forall\ \mu \in P_i : \\ F_i\big|_{\Delta_\mu'' \cap (\Delta_i\setminus\Gamma)} \in \mathcal{F}_i\big|_{\Delta_\mu'' \cap (\Delta_i\setminus\Gamma)} \end{array}\right\}.$$

Let us take now any $F \in \mathcal{M}_T^m(\Delta)$ which satisfies (7.157). If $\mu \in P_i$ then obviously

$$(7.163)\qquad \Delta_\mu'' \cap (\Delta_i\setminus\Gamma) \subsetneq \Delta_{i_{\nu_\mu}} \cap \Delta_i$$

thus (7.155), (7.157) and (7.151) give

$$F\big|_{\Delta_\mu'' \cap (\Delta_i\setminus\Gamma)} \in \mathcal{F}_{i_{\nu_\mu}}\big|_{\Delta_\mu'' \cap (\Delta_i\setminus\Gamma)} = \mathcal{F}_i\big|_{\Delta_\mu'' \cap (\Delta_i\setminus\Gamma)}\ .$$

In this way, owing to (7.162), we obtain

$$F\big|_{\Delta_i\setminus\Gamma} \in \mathcal{F}_i\big|_{\Delta_i\setminus\Gamma}$$

and the proof of (7.158) is completed if we note that for $F \in \mathcal{M}_T^m(\Delta)$ we have

$$F\big|_{\Delta_i\setminus\Gamma} \in \mathcal{F}_i\big|_{\Delta_i\setminus\Gamma} \Longleftrightarrow F\big|_{\Delta_i\setminus\Gamma} \in \mathcal{F}_i\big|_{\Delta_i} = \mathcal{F}_i$$

since Γ is closed and nowhere dense.

Based on (7.156) - (7.158), we can now define

$$(7.164)\qquad \mathcal{F} = \left\{F \in \mathcal{M}_T^m(\Delta) \;\middle|\; \begin{array}{l} \forall\ \mu \in P : \\ F\big|_{\Delta_\mu''} \in \mathcal{F}_\mu'' \end{array}\right\}$$

and obtain

$$(7.165)\qquad \mathcal{F}\big|_{\Delta_i} = \mathcal{F}_i\ ,\quad i \in I.$$

Indeed, the inclusion '$\subseteq$' in (7.165) follows from (7.158). For the converse inclusion '$\supseteq$' let us take an arbitrary but fixed $i \in I$. If $F_i \in \mathcal{F}_i$ and $\mu \in P_i$ then (7.162), (7.163), (7.155) and (7.151) give

$$F_i\big|_{\Delta_\mu''\cap(\Delta_i\setminus\Gamma)} \in \mathcal{F}_i\big|_{\Delta_\mu''\cap(\Delta_i\setminus\Gamma)} = \mathcal{F}_{i_{\nu_\mu}}\big|_{\Delta_\mu''\cap(\Delta_i\setminus\Gamma)} = \mathcal{F}_\mu''\big|_{\Delta_\mu''\cap(\Delta_i\setminus\Gamma)} .$$

It follows that

$$F_i\big|_{\Delta_\mu''\ \cap\ \Delta_i} \in \mathcal{F}_\mu''\big|_{\Delta_\mu''\ \cap\ \Delta_i}\ ,\quad \mu \in P_i$$

therefore

$$(7.166)\qquad \begin{array}{l} \forall\ \mu \in P_i\ : \\ \exists\ F_\mu'' \in \mathcal{F}_\mu''\ : \\ \quad F_i\big|_{\Delta_\mu''\ \cap\ \Delta_i} = F_\mu''\big|_{\Delta_\mu''\ \cap\ \Delta_i} \end{array}$$

For $\mu \in P\setminus P_i$ we can choose any $F_\mu'' \in \mathcal{F}_\mu''$ and thus, together with (7.166), obtain

$$(F_\mu''\ \mid\ \mu \in P) \in \prod_{\mu\in P} \mathcal{F}_\mu''\ .$$

Then (7.156) implies that

$$(7.167)\qquad \begin{array}{l} \exists\ F \in \mathcal{M}_T^m(\Delta)\ : \\ \forall\ \mu \in P\ : \\ \quad F\big|_{\Delta_\mu''} = F_\mu'' \end{array}$$

hence according to (7.164), we obtain

$$F \in \mathcal{F}.$$

But (7.166) and (7.167) give

$$F\big|_{\Delta''_\mu \cap \Delta_i} = F''_\mu\big|_{\Delta''_\mu \cap \Delta_i} = F_i\big|_{\Delta''_\mu \cap \Delta_i}, \quad \mu \in P_i$$

which in view of (7.159) results in

$$F\big|_{\Delta_i} = F_i$$

ending thus the proof of (7.165).

The last thing to prove is that

$$(7.168) \qquad \mathcal{F} \in \hat{\mathcal{M}}^m_T(\Delta)^*$$

in other words that $\mathcal{F}$ is a cut in $\mathcal{M}^m_T(\Delta)$ and moreover

$$(7.169) \qquad \phi \underset{\neq}{\subset} \mathcal{F} \underset{\neq}{\subset} \mathcal{M}^m_T(\Delta).$$

Now, the fact that $\mathcal{F}$ is a cut follows directly from (7.164), (7.154) and (7.155). Then (7.155) and (7.156) clearly give that $\mathcal{F} \neq \phi$. Finally (7.155) implies that each $\mathcal{F}''_\mu$, with $\mu \in P$, is bounded from above in $\mathcal{M}^m_T(\Delta''_\mu)$. Therefore (7.154) and (7.164) show that $\mathcal{F}$ is bounded from above in $\mathcal{M}^m_T(\Delta)$. We conclude that (7.169) holds, and thus (7.168) as well.

7.2.7 In this way we obtained that (σ''', ρ''') is also a sheaf of sections over Ω.

Remark 7.4

We can particularize the above constructions for

$$(7.170) \qquad m = 0, \quad T(x,D) = \mathrm{id}_{\mathcal{C}^0(\Omega)}$$

and obtain the sheaf morphisms

$$(7.171) \quad \mathcal{C}^0(\Delta) \xrightarrow{\text{inj}} \mathcal{C}^0_{\text{nd}}(\Delta) \longrightarrow \mathcal{M}^0(\Delta) \xrightarrow{\text{inj}} \hat{\mathcal{M}}^0(\Delta)^*$$

for nonvoid open $\Delta \subseteq \Omega$, where the second arrow from the left is the canonical mapping into the quotient space (3.4).

For the sake of precision, in case (7.170) holds, the sheaves of sections (σ,ρ), (σ',ρ'), (σ'',ρ'') and (σ''',ρ''') defined in (7.96) - (7.98), (7.100), (7.103) - (7.107) will respectively be denoted by (σ_0,ρ_0), (σ_0',ρ_0'), (σ_0'',ρ_0'') and (σ_0''',ρ_0'''), thus they will specify $\mathcal{C}^0(\Delta)$, $\mathcal{C}^0_{\text{nd}}(\Delta)$, $\mathcal{M}^0(\Delta)$ and $\hat{\mathcal{M}}^0(\Delta)^*$ respectively, where $\Delta \subset \Omega$ is any open set.

Theorem 7.3

The mappings, see (7.80)

$$(7.172) \quad \begin{array}{ccccccc} \mathcal{C}^m(\Delta) & \xrightarrow{\text{inj}} & \mathcal{C}^m_{\text{nd}}(\Delta) & \longrightarrow & \mathcal{M}^m_T(\Delta) & \xrightarrow{\text{inj}} & \hat{\mathcal{M}}^m_T(\Delta)^* \\ T(x,D)\downarrow & & T(x,D)\downarrow & & T\downarrow\text{inj} & & \hat{T}\downarrow\text{inj} \\ \mathcal{C}^0(\Delta) & \xrightarrow{\text{inj}} & \mathcal{C}^0_{\text{nd}}(\Delta) & \longrightarrow & \mathcal{M}^0(\Delta) & \xrightarrow{\text{inj}} & \hat{\mathcal{M}}^0(\Delta)^* \end{array}$$

for nonvoid open sets $\Delta \subseteq \Omega$, are sheaf morphisms.

Moreover, the sheaves of sections

$$(7.173) \quad \mathcal{C}^m_{\text{nd}}(\Delta),\ \mathcal{M}^m_T(\Delta),\ \hat{\mathcal{M}}^m_T(\Delta)^*,\ \mathcal{M}^0(\Delta),\ \hat{\mathcal{M}}^0(\Delta)^*$$

are flabby.

Proof

In view of the part of Subsection 7.2 which precedes Theorem 7.3, the proof of (7.172) reduces to showing that, see (7.79)

$$(7.174) \quad \hat{\mathcal{M}}^m_T(\Delta)^* \xrightarrow{\hat{T}} \hat{\mathcal{M}}^0(\Delta)^*$$

is indeed a sheaf morphism, see (7.90). Therefore, let us take a pair of open sets $\Delta_1 \subset \Delta_2 \subset \Omega$ and show that we have the commutative diagram

$$(7.175)\qquad \begin{array}{ccc} \sigma'''(\Delta_2) & \xrightarrow{\hat{T}} & \sigma_0'''(\Delta_2) \\ \rho'''_{\Delta_1,\Delta_2}\downarrow & & \downarrow \rho'''_{0\,\Delta_1,\Delta_2} \\ \sigma'''(\Delta_1) & \xrightarrow[\hat{T}]{} & \sigma_0'''(\Delta_1) \end{array}$$

Given $\mathcal{F} \in \sigma'''(\Delta_2) = \hat{\mathcal{M}}^m_T(\Delta_2)^*$, see (7.100), then (4.15), (7.79), (A.37) and Remark 7.4 yield

$$\hat{T}\mathcal{F} = \sup \{<T(F)] \mid F \in \mathcal{F}\} \in \hat{\mathcal{M}}^0(\Delta_2)^* = \sigma_0'''(\Delta_2)$$

hence

$$\rho'''_{0\,\Delta_1,\Delta_2} \hat{T}\mathcal{F} = \rho'''_{0\,\Delta_1,\Delta_2} \sup \{<T(F)] \mid F \in \mathcal{F}\} \in \sigma_0'''(\Delta_1) = \hat{\mathcal{M}}^0(\Delta_1)^*$$

where both sup are taken in $\hat{\mathcal{M}}^0(\Delta_2)$. On the other hand, according to (7.107), we obtain

$$\rho'''_{\Delta_1,\Delta_2} \mathcal{F} = \{\rho''_{\Delta_1,\Delta_2} F \mid F \in \mathcal{F}\} \in \sigma'''(\Delta_1) = \hat{\mathcal{M}}^m_T(\Delta_1)^*$$

thus

$$\hat{T}\rho'''_{\Delta_1,\Delta_2} \mathcal{F} = \sup \{<T(\rho''_{\Delta_1,\Delta_2} F)] \mid F \in \mathcal{F}\} \in \sigma_0'''(\Delta_1) = \hat{\mathcal{M}}^0(\Delta_1)^*$$

with sup taken in $\hat{\mathcal{M}}^0(\Delta_1)$.

It follows that in order to prove (7.175), we have to establish in $\hat{\mathcal{M}}^0(\Delta_1)^*$ the relation

$$(7.176)\qquad \rho'''_{0\,\Delta_1,\Delta_2}\ \sup\,\{<T(F)]\,|\,F\in\mathcal{F}\} = \sup\,\{<T(\rho''_{\Delta_1,\Delta_2}F)]\,|\,F\in\mathcal{F}\}$$

For that first we note the commutativity of the diagram

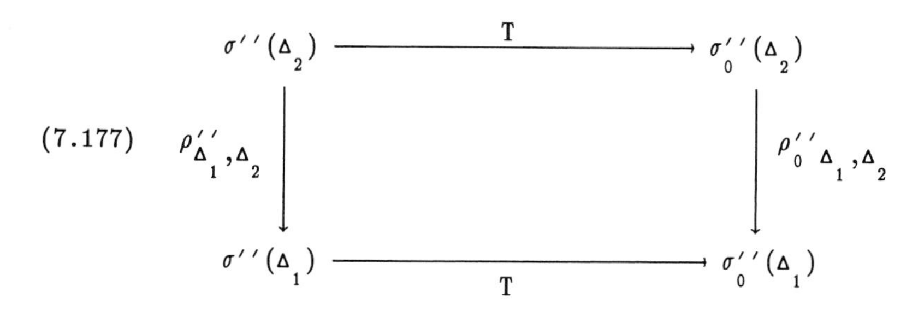

which follows from (4.6) through direct verification. Therefore (7.176) can take the equivalent form

$$(7.178)\qquad \rho'''_{0\,\Delta_1,\Delta_2}\ \sup\,\{<T(F)]\,|\,F\in\mathcal{F}\} = \sup\,\{<\rho''_{0\,\Delta_1,\Delta_2}T(F)]\,|\,F\in\mathcal{F}\}$$

since (7.177) yields

$$T(\rho''_{\Delta_1,\Delta_2}F) = \rho''_{0\,\Delta_1,\Delta_2}T(F).$$

But we note that in view of (7.79), the subset $\{T(F)\,|\,F\in\mathcal{F}\}\subset\mathcal{M}^0(\Delta_2)$ satisfies the condition of Corollary 7.4 in the case of (7.170), since by our assumption $\mathcal{F}\in\hat{\mathcal{M}}^m_T(\Delta_2)^*$. Thus (7.178) follows and the proof of (7.175) is completed.

Let us now prove (7.173). In this regard it suffices to show that each of the spaces in (7.173) satisfy (7.93). For that purpose, let us take an open set $\Delta\subset\Omega$. If now $f\in\mathcal{C}^m_{nd}(\Delta)$, then (3.1) gives

$$(7.179)\qquad f|_{\Delta\setminus\Gamma}\in\mathcal{C}^m(\Delta\setminus\Gamma)$$

for a certain closed, nowhere dense subset $\Gamma\subset\Delta$. Let us define

$$(7.180)\qquad \Delta' = \text{int}\,(\Omega\setminus\Delta),\quad \Sigma = \Omega\setminus(\Delta\cup\Delta')$$

then obviously

(7.181) Δ' is open, $\Sigma \subset \Omega$ is closed, nowhere dense.

Therefore, if we define $g : \Omega \longrightarrow \mathbb{R}$ by

$$(7.182) \quad g = \left| \begin{array}{ll} f & \text{on } \Delta \\ 0 & \text{on } \Omega \setminus \Delta \end{array} \right.$$

the relations (7.179) - (7.182) will give $g \in \mathcal{C}^m_{nd}(\Omega)$. In this way we have secured the condition (7.93) for $\mathcal{C}^m_{nd}$.

Let us now take $F \in \mathcal{M}^m_T(\Delta)$. In view of (4.4), for every $f \in F$, we have $f \in \mathcal{C}^m_{nd}(\Delta)$. Let us fix $f \in F$ and apply to it (7.182), thus obtaining $g \in \mathcal{C}^m_{nd}(\Omega)$. Now we can take $G \in \mathcal{M}^m_T(\Omega)$ as being the $\sim_T$ equivalence class of g, see (4.3). It is easy to see that G does not depend on the choice of f. Furthermore, see (7.98), we have

$$(7.183) \quad F = G|_\Delta \ .$$

Thus the condition (7.93) is satisfied by $\mathcal{M}^m_T$.

Particularizing this result in the case of (7.170), it follows that $\mathcal{M}^0$ will also satisfy (7.93).

Let us turn to $\hat{\mathcal{M}}^{m*}_T$. Given $\mathcal{F} \in \hat{\mathcal{M}}^m_T(\Delta)^*$, then in view of (7.107), in order to prove (7.93), we have to find $\mathcal{G} \in \hat{\mathcal{M}}^m_T(\Omega)^*$ such that

$$(7.184) \quad \mathcal{F} = \{G|_\Delta \mid G \in \mathcal{G}\} \ .$$

For that purpose, let us take

$$(7.185) \quad \mathcal{G} = \left\{ G \in \mathcal{M}^m_T(\Omega) \;\middle|\; \begin{array}{ll} *) & G|_\Delta \in \mathcal{F} \\ **) & G|_{\Omega \setminus \bar{\Delta}} \leq 0 \end{array} \right\}$$

where $\bar{\Delta}$ is the closure of Δ in Ω. Then in view of (7.183), we obviously have the inclusion $\subseteq$ in (7.184). In particular $\mathcal{G} \neq \phi$. Furthermore, assuming that we are in the nontrivial case of $\Omega \setminus \bar{\Delta} \neq \phi$, then owing to **) in (7.185), it is clear that $\mathcal{G} \underset{\neq}{\subset} \mathcal{M}_T^m(\Omega)$. In fact, $\mathcal{G}$ is bounded from above in $\mathcal{M}_T^m(\Omega)$. Indeed, $\mathcal{F}$ is bounded from above, since $\mathcal{F} \in \hat{\mathcal{M}}_T^m(\Delta)^*$. Hence in view of *) and **) in (7.185) both $\mathcal{G}|_\Delta$ and $\mathcal{G}|_{\Omega \setminus \bar{\Delta}}$ are bounded from above. And then, owing to Corollary 7.3, $\mathcal{G}$ is bounded from above. Now we can note that in view of *) in (7.185), we also have the inclusion $\supseteq$ in (7.184). Therefore, in order to be able to use the equality in (7.184) for proving (7.93) in the case of $\hat{\mathcal{M}}_T^{m^*}$, it only remains to show that

$$(7.186) \qquad \mathcal{G}^{u\ell} \subseteq \mathcal{G}$$

in other words, that $\mathcal{G}$ is a cut in $\mathcal{M}_T^m(\Omega)$. For that purpose we apply (7.112) to $\mathcal{G}$ and obtain in view of (7.184)

$$(7.187) \qquad \mathcal{G}^u|_\Delta = (\mathcal{G}|_\Delta)^u = \mathcal{F}^u .$$

But in view of (A.10), $\mathcal{F}^u$ is bounded from below, since

$$\phi \neq \mathcal{F} = (\mathcal{F}^u)^\ell .$$

Thus by applying (7.113) to (7.187), we obtain

$$\mathcal{G}|_\Delta = \mathcal{F} = \mathcal{F}^{u\ell} = (\mathcal{G}^u|_\Delta)^\ell = \mathcal{G}^{u\ell}|_\Delta$$

with the first equality being (7.184). In this way

$$(7.188) \qquad \mathcal{G}^{u\ell}|_\Delta = \mathcal{G}|_\Delta .$$

On the other hand, in view of **) in (7.185), it is clear that

$$(7.189) \qquad \mathcal{G}^{u\ell}|_{\Omega \setminus \bar{\Delta}} = \mathcal{G}|_{\Omega \setminus \bar{\Delta}} .$$

And then (7.188), (7.189) and Corollary 7.3 will give (7.186).

Finally, particularizing the above results on $\mathcal{M}_T^m$ and $\hat{\mathcal{M}}_T^{m*}$ to the case of (7.170), the proof of (7.173) is completed. ■

7.2.8 In the proof of Theorem 7.3 we have made use of:

Lemma 7.3

Given an open set $\omega \subseteq \Omega$ and a family of open sets $\omega_\ell \subseteq \omega$, with $\ell \in L$, such that

(7.190) ω_ℓ, with $\ell \in L$, are pairwise disjoint

and

(7.191) $\bigcup_{\ell \in L} \omega_\ell$ is dense in ω.

Further, given a cut $\mathcal{X} \subseteq \mathcal{M}_T^m(\omega)$ and a family

(7.192) $H_\ell \in \mathcal{X}|_{\omega_\ell}$, $\ell \in L$.

Then there exists $H \in \mathcal{X}$, such that

(7.193) $H|_{\omega_\ell} = H_\ell$, $\ell \in L$.

Proof

We start by noting the following transitivity property of denseness in an arbitrary topological space X. Given $A \subseteq B \subseteq C \subseteq X$, then

(7.194) $\begin{bmatrix} A \text{ dense in } B \\ B \text{ dense in } C \end{bmatrix} \Longrightarrow A$ dense in C.

Now in view of (7.192), we can take $h_\ell \in H_\ell$, with $\ell \in L$, such that

(7.195) $h_\ell \in \mathcal{C}^m(\omega_\ell \backslash \gamma_\ell)$

where $\gamma_\ell \subsetneq \omega_\ell$ is closed and nowhere dense in ω_ℓ. Then (7.194) implies that

$$\bigcup_{\ell \in L} (\omega_\ell \backslash \gamma_\ell) \text{ is dense in } \omega$$

therefore, we can define

$$h \in \mathcal{C}^m_{nd}(\omega)$$

by

(7.196) $h(x) = h_\ell(x)$ if $x \in \omega_\ell \backslash \gamma_\ell$.

Let H be the $\sim_T$ equivalence class in $\mathcal{M}^m_T(\omega)$ generated by h. Then owing to (7.190), we obviously have (7.193).

Now, it only remains to show that

(7.197) $H \in \mathcal{X}$.

Let us assume that (7.197) does not hold. Then, as $\mathcal{X}$ is a cut in $\mathcal{M}^m_T(\omega)$, it follows that

(7.198)
$$\begin{aligned} &\exists \ G \in \mathcal{M}^m_T(\omega) : \\ &\quad *) \ \forall \ K \in \mathcal{X} : K \leq_T G \\ &\quad **) \ H \leq_T G \text{ does not hold} \end{aligned}$$

But in view of (7.191), it is obvious that **) in (7.198) means that, for a certain $\ell \in L$

(7.199) $\quad H|_{\omega_\ell} \leq_T G|_{\omega_\ell}$ does not hold.

However, we have

(7.200) $\quad \forall\, K_\ell \in \mathcal{X}|_{\omega_\ell} : K_\ell \leq_T G|_{\omega_\ell}$.

Indeed, if

$$K_\ell \in \mathcal{X}|_{\omega_\ell}$$

then obviously

$$\exists\, K \in \mathcal{X} : K_\ell = K|_{\omega_\ell}.$$

Now (7.199) and (7.200) imply that

$$H|_{\omega_\ell} \notin \left[\mathcal{X}|_{\omega_\ell}\right]^{u\ell} .$$

In particular (A.15) gives

(7.201) $\quad H|_{\omega_\ell} \notin \mathcal{X}|_{\omega_\ell}$

Since (7.201), (7.193) and (7.192) contradict each other the proof of (7.197) is completed.

■

Corollary 7.2

Given open sets $\omega \in \Omega$ and $\omega_\ell \subsetneq \Omega$, with $\ell \in L$, satisfying (7.190) and (7.191). Given further $H_\ell \in \mathcal{M}_T^m(\omega_\ell)$, with $\ell \in L$. Then there exists $H \in \mathcal{M}_T^m(\omega)$, such that

$$H|_{\omega_\ell} = H_\ell \ , \quad \ell \in L.$$

Proof

Obviously $\mathcal{K} = \mathcal{M}_T^m(\omega)$ is a cut in $\mathcal{M}_T^m(\omega)$, and we have

$$\mathcal{K}|_{\omega_\ell} \supset \mathcal{M}_T^m(\omega_\ell), \quad \ell \in L.$$

Now we can apply Lemma 7.3. ■

Corollary 7.3

Given open sets $\Delta_1 \subsetneq \Delta_2 \subsetneq \Omega$ and $F_1 \in \mathcal{M}_T^m(\Delta_1)$, $F_2 \in \mathcal{M}_T^m(\Delta_2)$. Then there exists $F \in \mathcal{M}_T^m(\Delta_2)$, such that

$$F|_{\Delta_1} = F_1 \ , \quad F|_{\Delta_2 \setminus \bar{\Delta}_1} = F_2|_{\Delta_2 \setminus \bar{\Delta}_1}$$

where $\bar{\Delta}_1$ denotes the closure of Δ_1 in Δ_2.

Proof

We can take

$$\omega = \Delta_2, \quad \omega_1 = \Delta_1, \quad \omega_2 = \Delta_2 \setminus \bar{\Delta}_1, \quad L = \{1,2\}$$

and apply Corollary 7.2. ■

Corollary 7.4

Given open sets $\Delta_1 \subset \Delta_2 \subset \Omega$. Let $\mathcal{F}$ be a subset of $\mathcal{M}_T^m(\Delta_2)$ which is bounded from above, that is, $\mathcal{F} \subsetneq \langle W]$, for suitable $W \in \mathcal{M}_T^m(\Delta_2)$. Then the diagram is commutative

$$(7.202)\qquad \begin{array}{ccc} \mathcal{M}_T^m(\Delta_2) \supset \mathcal{F} & \longrightarrow & \sup \mathcal{F} \in \hat{\mathcal{M}}_T^m(\Delta_2)^* \\ \rho''_{\Delta_1,\Delta_2} \Big\downarrow & & \Big\downarrow \rho'''_{\Delta_1,\Delta_2} \\ \mathcal{M}_T^m(\Delta_1) \supset \rho''_{\Delta_1,\Delta_2} \mathcal{F} & \longmapsto & \sup \rho''_{\Delta_1,\Delta_2} \mathcal{F} = \rho'''_{\Delta_1,\Delta_2} \sup \mathcal{F} \in \mathcal{M}_T^m(\Delta_1)^* \end{array}$$

Proof

Let $F_0 \in \mathcal{F}$ then applying (7.108), we obtain

$$<\rho''_{\Delta_1,\Delta_2} F_0] = \rho'''_{\Delta_1,\Delta_2} <F_0] \ .$$

But obviously

$$\rho'''_{\Delta_1,\Delta_2} <F_0] \le \rho'''_{\Delta_1,\Delta_2} \sup \{<F] \,|\, F \in \mathcal{F}\}$$

while

$$\sup \{<F] \,|\, F \in \mathcal{F}\} = \sup \mathcal{F} \ .$$

Therefore

$$(7.203)\qquad \sup \rho''_{\Delta_1,\Delta_2} \mathcal{F} \le \rho'''_{\Delta_1,\Delta_2} \sup \mathcal{F} \ .$$

Let us now prove the converse inequality $\ge$ in (7.203). Let us take any $G \in \mathcal{M}_T^m(\Delta_1)$ for which

$$(7.204)\qquad \rho''_{\Delta_1,\Delta_2} F \le G, \quad F \in \mathcal{F} \ .$$

Then according to Corollary 7.3, we can take $H \in \mathcal{M}_T^m(\Delta_2)$ such that

$$(7.205)\qquad H\big|_{\Delta_1} = G, \quad H\big|_{\Delta_2 \setminus \bar{\Delta}_1} = W\big|_{\Delta_2 \setminus \bar{\Delta}_1} \ .$$

Therefore we obtain in $\mathcal{M}^m_T(\Delta_2)$ the inequalities

$$F \leq H, \quad F \in \mathcal{F} .$$

In this way in $\hat{\mathcal{M}}^m_T(\Delta_2)^*$ the inequality holds

$$\sup \mathcal{F} \leq \langle H]$$

and then obviously

$$\rho'''_{\Delta_1,\Delta_2} \sup \mathcal{F} \leq \rho'''_{\Delta_1,\Delta_2} \langle H] .$$

But in view of (7.108), we have

$$\rho'''_{\Delta_1,\Delta_2} \langle H] = \langle \rho''_{\Delta_1,\Delta_2} H] = \langle G]$$

the second equality following from (7.205). We can conclude that in $\hat{\mathcal{M}}^m_T(\Delta_1)^*$ the inequality holds

$$\rho'''_{\Delta_1,\Delta_2} \sup \mathcal{F} \leq \langle G]$$

for arbitrary $G \in \mathcal{M}^m_T(\Delta_1)$ which satisfies (7.204). Therefore, according to (A.23), we proved the inequality $\geq$ in (7.203). ∎

7.2.9 In view of the limited literature on the structure of Dedekind cuts in spaces of differentiable functions, it may be useful for a further familiarization with such structures to present a second, alternative proof of the fact that (σ''',ρ''') satisfies the conditions (7.86) and (7.87) in the definition of a sheaf of sections. It should be noted however that, unlike the first proof, this second one uses the Axiom of Choice.

Let us again assume given a family of open sets $\Delta_i \subseteq \Omega$, with $i \in I$, and let us denote

$$\Delta = \bigcup_{i \in I} \Delta_i \ . \tag{7.206}$$

Assume also given $\mathcal{F}_i \in \sigma'''(\Delta_i) = \hat{\mathcal{M}}_T^m(\Delta_i)^*$, with $i \in I$, satisfying (7.86). Then, for $i,j \in I$, we have

$$\Delta_i \cap \Delta_j \neq \phi \Longrightarrow \mathcal{F}_i\big|_{\Delta_i \cap \Delta_j} = \mathcal{F}_j\big|_{\Delta_i \cap \Delta_j} \ . \tag{7.207}$$

In this way, in order to prove (7.87), we have to find $\mathcal{F} \in \sigma'''(\Delta) = \hat{\mathcal{M}}_T^m(\Delta)^*$ which satisfies the relations

$$\mathcal{F}\big|_{\Delta_i} = \mathcal{F}_i \ , \quad i \in I. \tag{7.208}$$

For that purpose let us denote

$$J = \left\{ (i,j) \in I \times I \ \middle| \ \begin{array}{ll} 1) & i \neq j \\ 2) & \Delta_i \cap \Delta_j \neq \phi \end{array} \right\} . \tag{7.209}$$

Further, let us define the mapping

$$\prod_{i \in I} \mathcal{F}_i \ni F = (F_i \mid i \in I) \longmapsto J_F \subseteq J$$

by

$$J_F = \left\{ (i,j) \in J \ \middle| \ F_i\big|_{\Delta_i \cap \Delta_j} = F_j\big|_{\Delta_i \cap \Delta_j} \right\} . \tag{7.210}$$

We intend to show that

$$\begin{array}{l} \exists \ F \in \prod_{i \in I} \mathcal{F}_i \ : \\ \quad J_F = J \end{array} \tag{7.211}$$

using an argument based on the Zorn lemma. For that purpose, let us define

$$(7.212) \qquad \mathcal{A} = \left\{ A = (F,K) \;\middle|\; \begin{array}{ll} 1) & F \in \prod_{i \in I} \mathcal{F}_i \\ 2) & K \subseteq I \\ 3) & K_d \subseteq J_F \end{array} \right\}$$

where we denoted

$$(7.213) \qquad K_d = \left\{ (i,j) \;\middle|\; \begin{array}{ll} 1) & i,j \in K \\ 2) & i \neq j \end{array} \right\} .$$

Now, on $\mathcal{A}$ we consider the reflexive and transitive binary relation $\leq$, defined as follows. Given $A = (F,K)$, $A' = (F',K') \in \mathcal{A}$, then

$$(7.214) \qquad A \leq A' \iff \left[\begin{array}{ll} 1) & K \subseteq K' \\ 2) & \forall \; i \in K : \\ & \quad F'_i = F_i \end{array} \right]$$

where we have assumed that

$$F = (F_i \mid i \in I), \quad F' = (F'_i \mid i \in I) .$$

Since $\leq$ in (7.214) may fail to be antisymmetric, therefore, it may happen not to be a partial order on $\mathcal{A}$, we define the equivalence relation $\sim$ on $\mathcal{A}$, by

$$(7.215) \qquad A \sim A' \iff A \leq A' \leq A$$

and then, we consider the quotient set

$$(7.216) \qquad \mathcal{B} = \mathcal{A}/\sim$$

and define on it the partial order $\leq\leq$ generated by (7.214) and (7.215) as follows. Given $B, B' \in \mathcal{B}$, then

$$(7.217) \qquad B \leq\leq B' \iff \left[\begin{array}{l} \exists \;\; A \in B, \;\; A' \in B' : \\ \quad A \leq A' \end{array} \right] .$$

In this case it is obvious that we have the property

$$(7.218) \quad B \underset{=}{\ll} B' \Longleftrightarrow \left[\begin{array}{c} \forall \quad A \in B, \quad A' \in B' : \\ A \leq A' \end{array}\right].$$

Now we prove that every chain in $(\mathcal{B}, \underset{=}{\ll})$ has a majorant. Indeed, let $B^\lambda \in \mathcal{B}$, with $\lambda \in \Lambda$, be such a chain, and let us take

$$A^\lambda \in B^\lambda, \quad \lambda \in \Lambda$$

where, for $\lambda \in \Lambda$, we shall assume that

$$A^\lambda = (F^\lambda, K^\lambda) \quad \text{and} \quad F^\lambda = (F_i^\lambda \mid i \in I) .$$

Then, a majorant $\bar{B} \in \mathcal{B}$ for the chain $(B^\lambda \mid \lambda \in \Lambda)$ will be defined as the $\sim$ equivalence class generated by

$$(7.219) \quad \bar{A} = (\bar{F}, \bar{K}) \in \mathcal{A}$$

where we take

$$(7.220) \quad \bar{K} = \bigcup_{\lambda \in \Lambda} K^\lambda$$

while

$$\bar{F} = (\bar{F}_i \mid i \in I) \in \prod_{i \in I} \mathcal{F}_i$$

is such that

$$(7.221) \quad \bar{F}_i = F_i^\lambda \quad \text{if} \quad i \in K^\lambda$$

whereas, for $i \in I \backslash \bar{K}$, we can choose $F_i \in \mathcal{F}_i$ arbitrarily.

Indeed, first we note that, in view of (7.218) and pct. 2) in (7.214), the definition in (7.221) is correct, since K^λ, with $\lambda \in \Lambda$, is a chain of subsets of I when considered with respect to the inclusion $\subseteq$. In this way, $\bar{A}$ in (7.219) satisfies pcts. 1) and 2) in (7.212). Let us show that

$\bar{A}$ satisfies pct. 3) in (7.212) as well. For that, let $(i,j) \in \bar{K}_d$. Then (7.213) and (7.220) yield

(7.222) $\quad i \neq j$ and $i \in K^\lambda, \quad j \in K^\mu$

for suitable $\lambda, \mu \in \Lambda$. But as noticed above, we can assume that, for instance, $K^\lambda \subseteq K^\mu$, hence in view (7.222) and pct. 3) in (7.212) applied to $A^\mu = (F^\mu, K^\mu)$, we obtain

(7.223) $\quad (i,j) \in K^\mu_d \subseteq J_{F^\mu}$.

But (7.221) and $i,j \in K^\mu$ yield

(7.224) $\quad \bar{F}_i = F^\mu_i, \quad \bar{F}_j = F^\mu_j$.

Now (7.223), (7.210) and (7.224) give

$$\bar{F}_i\big|_{\Delta_i \cap \Delta_j} = F^\mu_i\big|_{\Delta_i \cap \Delta_j} = F^\mu_j\big|_{\Delta_i \cap \Delta_j} = \bar{F}_j\big|_{\Delta_i \cap \Delta_j}$$

which means that

$$(i,j) \in J_{\bar{F}} \ .$$

In this way, we have obtained indeed

$$\bar{K}_d \subseteq J_{\bar{F}}$$

thus (7.219) holds.

It is easy to see now that, in view of (7.214), the relations (7.220) and (7.221) give

$$A^\lambda \leq \bar{A}, \quad \lambda \in \Lambda$$

and then, according to (7.218), we obtain

$$(7.225) \qquad B^\lambda \le \bar{B}, \quad \lambda \in \Lambda .$$

Now, applying Zorn's lemma, it follows that

$$(7.226) \qquad \begin{array}{l} \forall \ B \in \mathcal{B} : \\ \exists \ \bar{B} \in \mathcal{B} : \\ \quad *) \ B \le\le \bar{B} \\ \quad **) \ \bar{B} \text{ maximal in } (\mathcal{B}, \le\le) \end{array}$$

Our interest is in fact in the following consequence of (7.226), namely

$$(7.227) \qquad \begin{array}{l} \forall \ A = (F,K) \in B \in \mathcal{B} : \\ \quad B \text{ maximal in } (\mathcal{B}, \le\le) \Longrightarrow J_F = J \end{array}$$

since that will prove (7.211).

Therefore, assume that

$$(7.228) \qquad J_F \underset{\neq}{\subset} J$$

for a certain $A = (F,K) \in B \in \mathcal{B}$, with B maximal in $(\mathcal{B}, \le\le)$. Then, let us take any

$$(7.229) \qquad (p,q) \in J \backslash J_F$$

and let us assume that $F = (F_i \mid i \in I) \in \prod_{i \in I} \mathcal{F}_i$. Our aim is to construct

$$(7.230) \qquad A' = (F',K') \in \mathcal{A}$$

with $F' = (F'_i \mid i \in I) \in \prod_{i \in I} \mathcal{F}_i$ and $K' \subseteq I$, such that

$$(7.231) \qquad A \le A', \text{ while } A \sim A' \text{ does not hold.}$$

In that case, denoting by $B' \in \mathcal{B}$ the $\sim$ equivalence class of A', it would follow that

$$(7.232) \qquad B \le\le B' \text{ and } B \neq B'$$

thus contradicting the assumed maximality of B in $(\mathcal{B}, \leq\leq)$.

Now, in view of (7.229) and pct. 3) in (7.212), we have

$$p,q \in K \Longrightarrow (p,q) \in K_d \subseteq J_F$$

which would contradict (7.229). Therefore, we can assume that $p \notin K$, in which case we can take

$$(7.233) \qquad K' = K \cup \{p\}$$

and obtain

$$(7.234) \qquad K \underset{\neq}{\subset} K' .$$

It only remains to define F' in (7.230). For that purpose, we proceed as follows. Let us take

$$(7.235) \qquad F'_i = \left| \begin{array}{ll} F_i & \text{if} \quad i \in K \\ G_i & \text{if} \quad i \in I \setminus K' \end{array} \right.$$

where $G_i \in \mathcal{F}_i$ can be chosen arbitrarily. In this way, we are left with defining F'_p alone. Let us then denote

$$L = \{i \in K \mid \Delta_p \cap \Delta_i \neq \phi\}$$

and consider the open subset of Δ given by

$$D = \bigcup_{i \in L} \Delta_i .$$

We shall distinguish two possibilities. First, if $\Delta_p \cap D$ is dense in Δ_p. Then we take

$$\omega = \Delta_p, \quad \omega_1 = \Delta_p \cap D .$$

We recall now that $L \subsetneq K$ and $K_d \subsetneq J_F$, see pct. 3) in (7.212), therefore, the family

$$F_i\Big|_{\Delta_p \cap \Delta_i} \in \mathcal{M}_T^m(\Delta_p \cap \Delta_i), \quad i \in L$$

satisfies (7.86). And since (σ'',ρ'') is a sheaf of sections, we obtain $H_1 \in \mathcal{M}_T^m(\Delta_p \cap D)$ such that

$$(7.236) \quad H_1\Big|_{\Delta_p \cap \Delta_i} = F_i\Big|_{\Delta_p \cap \Delta_i}, \quad i \in L .$$

Applying Lemma 7.3 above to ω, ω_1 and $H_1 \in \mathcal{M}_T^m(\omega_1)$, we obtain

$$(7.237) \quad F_p \in \mathcal{M}_T^m(\Delta_p)$$

such that

$$(7.238) \quad F_p\Big|_{\Delta_p \cap \Delta_i} = F_i\Big|_{\Delta_p \cap \Delta_i}, \quad i \in L .$$

The second case to consider is when

$$\Delta_p \backslash \overline{(\Delta_p \cap D)} \neq \phi$$

where the closure is taken in Δ. Then we take

$$\omega = \Delta_p, \quad \omega_1 = \Delta_p \cap D, \quad \omega_2 = \Delta_p \backslash \overline{(\Delta_p \cap D)} .$$

In this case, we can use again $H_1 \in \mathcal{M}_T^m(\omega_1)$ in (7.236). Further, we can take any $H_2 \in \mathcal{M}_T^m(\omega_2)$. Then, applying Lemma 7.3 above to ω, ω_1, ω_2, H_1 and H_2, we shall obtain again (7.237) and (7.238). In this way, together with (7.235), the definition of F' in (7.230) is completed, since in view of (7.238), we obviously have

$$K'_d \subsetneq J_{F'} \ .$$

We can therefore conclude that (7.230), (7.235) and (7.234) will indeed lead to (7.231), and hence to (7.232). Thus the proof of (7.227) is completed.

In this way, as mentioned above, we obtain in particular (7.211).

Now we can note that

$$\forall\ F = (F_i \mid i \in I) \in \prod_{i \in I} \mathcal{F}_i \ :$$

$$(7.239) \qquad J_F = J \Longrightarrow \left[\begin{array}{l} \exists\ \bar{F} \in \mathcal{M}^m_T(\Delta) \ : \\ \forall\ i \in I \ : \\ \quad F_i = \bar{F}\big|_{\Delta_i} \end{array}\right]$$

Indeed $J_F = J$ means that (7.124) holds for F. In this way (7.239) follows from the fact that (σ'', ρ'') satisfies (7.86), (7.87).

Based on (7.239), we can now give the definition

$$(7.240) \qquad \mathcal{F} = \left\{ F \in \mathcal{M}^m_T(\Delta) \ \middle| \ \begin{array}{l} \exists\ (F_i \mid i \in I) \in \prod_{i \in I} \mathcal{F}_i \ : \\ \forall\ i \in I \ : \\ \quad F_i = F\big|_{\Delta_i} \end{array} \right\} =$$

$$= \left\{ F \in \mathcal{M}^m_T(\Delta) \ \middle| \ \begin{array}{l} \forall\ i \in I \ : \\ \quad F\big|_{\Delta_i} \in \mathcal{F}_i \end{array} \right\}$$

Then, in order to obtain (7.208), and therefore (7.87), it only remains to prove that

$$(7.241) \qquad \mathcal{F}_i = \mathcal{F}\big|_{\Delta_i} \ , \quad i \in I$$

and

$$(7.242) \qquad \mathcal{F} \in \hat{\mathcal{M}}^{m}_{T}(\Delta)^{*} .$$

Now the inclusion '$\supseteq$' in (7.241) follows directly from (7.240). For the inclusion '$\subseteq$' in (7.241) we note that

$$(7.243) \qquad \begin{array}{l} \forall \;\; i \in I, \;\; F_i \in \mathcal{F}_i, \;\; (i,j) \in J : \\ \exists \;\; G = (G_k \mid k \in I) \in \prod_{k \in I} \mathcal{F}_k : \\ \quad *) \;\; G_i = F_i \\ \quad **) \;\; (i,j) \in J_G \end{array}$$

since in view of (7.207), we can take $G_i = F_i$, and then we obtain $G_j \in \mathcal{F}_j$, such that

$$F_i \big|_{\Delta_i \cap \Delta_j} = G_j \big|_{\Delta_i \cap \Delta_j} .$$

Now let us take $i \in I$ and $F_i \in \mathcal{F}_i$. Then (7.243), (7.226) and (7.227) give

$$G = (G_k | k \in I) \in \prod_{k \in I} \mathcal{F}_k$$

such that

$$F_i = G_i, \;\; J_G = J.$$

Thus according to (7.239) and (7.240), we have

$$F_i = G_i = F\big|_{\Delta_i}$$

for a suitable $F \in \mathcal{F}$, and the proof of (7.241) is completed.

In order to show (7.242) we have to prove that

(7.244) $\quad \phi \underset{\neq}{\subset} \mathcal{F} \underset{\neq}{\subset} \mathcal{M}_T^m(\Delta)$

and

(7.245) $\quad \mathcal{F}^{u\ell} \subseteq \mathcal{F}.$

Now $\mathcal{F} \neq \phi$ follows from (7.243), (7.226), (7.227), (7.239) and (7.240). The relation $\mathcal{F} \underset{\neq}{\subset} \mathcal{M}_T^m(\Delta)$ follows from (7.241) and the fact that, for each $i \in I$, we have $\mathcal{F}_i \underset{\neq}{\subset} \mathcal{M}_T^m(\Delta_i)$, since $\mathcal{F}_i \in \sigma'''(\Delta_i) = \hat{\mathcal{M}}_T^m(\Delta_i)^*$. Therefore (7.244) is indeed valid. We turn now to the proof of (7.245).

Let us take therefore any $F^o \in \mathcal{F}^{u\ell}$, then (A.6), (A.7) give

(7.246)
$$\begin{array}{l} \forall\ G \in \mathcal{M}_T^m(\Delta) : \\ \left[\begin{array}{l} \forall\ H \in \mathcal{F} : \\ \quad H \leq_T G \end{array} \right] \Longrightarrow F^o \leq_T G \end{array}$$

We shall show that we have

(7.247) $\quad F^o \in \mathcal{F}$

by using the definition of $\mathcal{F}$ in (7.240), that is, by proving the relations

(7.248) $\quad F^o|_{\Delta_i} \in \mathcal{F}_i \ , \quad i \in I.$

Assume therefore given $i \in I$. Let us use (7.246) and restrict it to Δ_i, in order to show that

(7.249) $\quad F^o|_{\Delta_i} \in \mathcal{F}_i^{u\ell}$

This is sufficient for proving (7.248), since $\mathcal{F}_i$ is a cut, that is $\mathcal{F}_i^{u\ell} = \mathcal{F}_i$, as we have assumed that $\mathcal{F}_i \in \sigma'''(\Delta_i) = \hat{\mathcal{M}}_T^m(\Delta_i)^*$. Now, due to (A.6), (A.7), the relation (7.249) is equivalent with

$$\text{(7.250)} \qquad \begin{array}{l} \forall\ G_i \in \mathcal{M}_T^m(\Delta_i) : \\ \left[\begin{array}{l} \forall\ H_i \in \mathcal{F}_i : \\ \quad H_i \leq_T G_i \end{array} \right] \Longrightarrow F^0|_{\Delta_i} \leq_T G_i \ . \end{array}$$

Therefore, in order to prove (7.250), let us take any $G_i \in \mathcal{M}_T^m(\Delta_i)$ such that

$$\text{(7.251)} \qquad \forall\ H_i \in \mathcal{F}_i : H_i \leq_T G_i .$$

Here it is useful to note that $\mathcal{F}$ is bounded from above in $\mathcal{M}_T^m(\Delta)$, in other words

$$\text{(7.252)} \qquad \begin{array}{l} \exists\ W \in \mathcal{M}_T^m(\Delta) : \\ \forall\ F \in \mathcal{F} : \\ \quad F \leq_T W \end{array}$$

Indeed, following (7.125) - (7.127) and (7.152), we can obtain a finite or countable family of open sets Δ''_μ, with $\mu \in P \subseteq \mathbb{N}$, such that (7.153) and (7.154) hold. Further, in view of (7.126), (7.152), it follows that

$$\begin{array}{l} \forall\ \mu \in P : \\ \exists\ i \in I : \\ \quad \Delta''_\mu \subseteq \Delta_i \end{array}$$

Therefore (7.241) together with (7.112) and (7.113) will give

$$\left[\mathcal{F}\Big|_{\Delta''_\mu} \right]^{u\ell} = \left[\mathcal{F}\Big|_{\Delta_i}\Big|_{\Delta''_\mu} \right]^{u\ell} = \left[\mathcal{F}_i\Big|_{\Delta''_\mu} \right]^{u\ell} = \left[\mathcal{F}_i^{u\ell} \right]\Big|_{\Delta''_\mu} = \mathcal{F}_i\Big|_{\Delta''_\mu}$$

since we assumed that $\mathcal{F}_i \in \hat{\mathcal{M}}_T^m(\Delta_i)^*$. But in view of (7.241)

$$\mathcal{F}_i\Big|_{\Delta''_\mu} = \mathcal{F}\Big|_{\Delta_i}\Big|_{\Delta''_\mu} = \mathcal{F}\Big|_{\Delta''_\mu}$$

Therefore

$$\left[\mathcal{F}\Big|_{\Delta''_\mu}\right]^{u\ell} = \mathcal{F}\Big|_{\Delta''_\mu} \in \hat{\mathcal{M}}_T^m(\Delta''_\mu)^*$$

In particular, (A.20) will imply that

$$\forall\ \mu \in P :$$
$$\exists\ W_\mu \in \mathcal{M}_T^m(\Delta''_\mu)$$
$$\mathcal{F}\Big|_{\Delta''_\mu} \subseteq \langle W_\mu]$$

Now, in view of (7.153), (7.154), we can apply Corollary 7.2 to $(W_\mu | \mu \in P)$ and obtain $W \in \mathcal{M}_T^m(\Delta)$, such that

$$W\Big|_{\Delta''_\mu} = W_\mu, \quad \mu \in P$$

Then owing to (7.153), it is easy to see that W will satisfy (7.252).

Coming back to (7.251), let us take $G \in \mathcal{M}_T^m(\Delta)$ such that

$$(7.253) \qquad G\Big|_{\Delta_i} = G_i\ , \quad G\Big|_{\Delta\setminus\bar{\Delta}_i} = W$$

which is possible according to Corollary 7.3 above. Then (7.253), (7.251), (7.241) and (7.252) give

$$\forall\ H \in \mathcal{F} : H \leq_T G$$

thus (7.246) results in

$$F^o \leq_T G$$

which restricted to Δ_i, will imply (7.250), if (7.253) is taken into account. In this way (7.249), (7.248), (7.247) and finally (7.245) will follow.

In view of (7.244) and (7.245) we completed the second, alternative proof of the fact that (σ''',ρ''') satisfies (7.86) and (7.87).

Remark 7.5

It is well known, Kaneko, that one of the *major shortcomings* of the L Schwartz distributions $\mathcal{D}'(\Omega)$ is that they do *not* constitute a *flabby sheaf*. In other words, given an open set $\Delta \subset \Omega$ and a distribution $F \in \mathcal{D}'(\Delta)$, it may well happen that the relation, see (7.93)

$$F = G|_{\Delta}$$

will *not* hold for any distribution $G \in \mathcal{D}'(\Omega)$. This means that, in general, distributions defined on an open set Δ *cannot be extended* to distributions defined on a larger open set Ω. And the failure in this extension is rather simple. For instance, if $F \in \mathcal{D}'(\Delta)$ is defined by a function

$$f \in \mathcal{L}^1_{loc}(\Delta)\backslash\mathcal{L}^1(\Delta)$$

then F cannot in general be extended to any larger open set Ω, large enough to satisfy

$$\bar{\Delta} \subset \Omega$$

where $\bar{\Delta}$ is the closure of Δ in $\mathbb{R}^n$. A more attentive look at the failure of $\mathcal{D}'(\Omega)$ in being flabby shows that the reasons for that can be traced back at least to *continuous* functions. Indeed, as is well known, locally, that is on compacts, every distribution is the finite order derivative of a continuous function. However the continuous functions in $\mathcal{C}^0(\Omega)$ do equally *fail* to constitute a flabby sheaf, since every *unbounded* continuous function on Δ *fails* to be extendable to a continuous function on Ω, in case for instance

$$\bar{\Delta} \subset \Omega .$$

Therefore the problem arises of finding *flabby sheaves* which contain at least the continuous functions $C^0(\Omega)$. Now, in view of (7.173), $C^0_{nd}(\Omega)$ is such a flabby sheaf. Moreover, in view of Remark 3.1, it is obvious that $C^0_{nd}(\Omega)$ can be seen as the *smallest flabby sheaf* which contains $C^0(\Omega)$. The precise statement on this *minimality* property is given at the end of this Remark. In a similar way, for $\ell \in \bar{\mathbb{N}}$, $C^\ell_{nd}(\Omega)$ can be seen as the *smallest flabby sheaf* containing $C^\ell(\Omega)$.

Indeed, given an open set $\Delta \subset \Omega$, let us consider

$$\Gamma = \Omega \cap \partial\Delta$$

where $\partial\Delta$ denotes the frontier of Δ in $\mathbb{R}^n$. Then Γ is a *closed, nowhere dense* set in Ω. And in case

$$\Gamma \neq \phi$$

let us choose, see (3.15), (3.16)

$$\gamma \in C^\infty(\Omega)$$

such that

$$\Gamma = \{x \in \Omega \mid \gamma(x) = 0\} .$$

If we define now $h : \Delta \longrightarrow \mathbb{R}$ by

$$h(x) = 1/\gamma(x), \quad x \in \Delta$$

then obviously

$$h \in C^\ell(\Delta), \quad \ell \in \bar{\mathbb{N}}$$

however, since h is *unbounded* in a neighbourhood of *every* $x \in \Gamma$, there *cannot* be any continuous function $k \in C^0(\Omega)$, such that

$$h = k|_{\Delta} .$$

On the other hand, if we define $k : \Omega \longrightarrow \mathbb{R}$ by

$$k(x) = \left| \begin{array}{ll} h(x) & \text{if } x \in \Delta \\ 0 & \text{if } x \in \Omega \backslash \Delta \end{array} \right.$$

then obviously

$$k \in \mathcal{C}^{\ell}_{nd}(\Omega), \quad \ell \in \bar{\mathbb{N}}$$

and furthermore we have the restriction property

$$h = k|_{\Delta} .$$

Conversely, given any closed, nowhere dense set $\Gamma \subset \Omega$, if we take

$$\Delta = \Omega \backslash \Gamma$$

then the open set $\Delta \subset \Omega$ will have the property

$$\Gamma \subsetneq \partial\Delta .$$

Therefore, in view of the above argument, in the definition of $\mathcal{C}^{\ell}_{nd}(\Omega)$, $\ell \in \bar{\mathbb{N}}$, see (3.1), we have to consider *arbitrary* closed nowhere dense subsets $\Gamma \subset \Omega$, in case we want to have the *flabbiness* property, see (7.93)

$$\begin{array}{ll} \forall & f \in \mathcal{C}^{\ell}(\Delta) \\ \exists & g \in \mathcal{C}^{\ell}_{nd}(\Omega) : \\ & \quad f = g|_{\Delta} \end{array}$$

valid for any particular $\ell \in \bar{\mathbb{N}}$.

The above heuristic argument on the minimality of $\mathcal{C}^0_{nd}(\Omega)$ as a flabby sheaf containing $\mathcal{C}^0(\Omega)$ can be made precise as follows. Suppose given any *flabby sheaf of sections* (σ,ρ) over Ω, see (7.81)-(7.87), which satisfies the next *'dense definition'* property

$$\text{(DD)}\qquad \begin{array}{l} \forall\ \Delta_1 \subseteq \Delta_2 \subseteq \Omega,\ \Delta_1,\Delta_2 \text{ open},\ \Delta_1 \text{ dense in } \Delta_2 : \\ \forall\ f,g \in \sigma(\Delta_2) : \\ \rho_{\Delta_1,\Delta_2} f = \rho_{\Delta_1,\Delta_2} g \Longrightarrow f = g \end{array}$$

Further, suppose given a *sheaf morphism*, see (7.88)-(7.90)

$$\mathcal{C}^0(\Delta) \xrightarrow{\ i_\Delta\ } \sigma(\Delta),\ \Delta \subseteq \Omega,\ \ \Delta \text{ open}$$

Then we can always construct a sheaf morphism

$$\mathcal{M}^0(\Delta) \xrightarrow{\ j_\Delta\ } \sigma(\Delta),\ \Delta \subseteq \Omega,\ \ \Delta \text{ open}$$

such that

$$j_\Delta\Big|_{\mathcal{C}^0(\Delta)} = i_\Delta$$

where this restriction is well defined in view of the embedding, see (3.11)

$$\mathcal{C}^0(\Delta) \xrightarrow{\ \text{inj}\ } \mathcal{M}^0(\Delta)$$

The construction of j_Δ proceeds in an obvious manner by using (7.40) and (7.36), where Ω is replaced with Δ. In this case, we obtain the embedding

$$\mathcal{M}^0(\Delta) \ni U \longmapsto \mu u \in \mathcal{C}^0(\Delta\backslash\bar{\Gamma})$$

where for a given U, one can choose any $u \in U$. Now we can apply $i_{\Delta\backslash\bar{\Gamma}}$ to μu and obtain

$$i_{\Delta\backslash\bar{\Gamma}}(\mu u) \in \sigma(\Delta\backslash\bar{\Gamma})$$

However, σ is a flabby sheaf, therefore in view of (7.93), we have the extension property

$$\exists \ V \in \sigma(\Delta) \ : \quad i_{\Delta \backslash \bar{\Gamma}}(\mu u) = V|_{\Delta \backslash \bar{\Gamma}}$$

But now, owing to the dense definition property (DD), V above is unique. Therefore, we can define

$$j_\Delta(U) = V$$

and it is easy to see that, in this way, the restriction of j_Δ to $\mathcal{C}^0(\Delta)$ will indeed be i_Δ, since

$$U \in \mathcal{C}^0(\Delta) \Longrightarrow \mu u = U, \quad \bar{\Gamma} = \phi$$

We can conclude: $\mathcal{M}^0(\Omega)$ is the *smallest flabby sheaf* which contains $\mathcal{C}^0(\Omega)$ and satisfies the dense definition property (DD).

7.3 Further Examples and Counterexamples

7.3.1 First we show that (7.73) *cannot* hold, namely, we have, see (7.75)

$$(7.254) \qquad (\mathcal{C}^0_A(\Omega)/\sim_A)^{\wedge} \underset{\neq}{\subset} \hat{\mathcal{M}}^0(\Omega)$$

and therefore

$$(7.255) \qquad (\mathcal{C}^0_A(\Omega)/\sim_A)^{\wedge *} \underset{\neq}{\subset} \hat{\mathcal{M}}^0(\Omega)^* .$$

Indeed, let us assume that we have equality in (7.254). Then in view of (7.61) and (A.40), we obtain the commutative diagram

$$(7.256)\qquad \begin{array}{ccc} \mathcal{C}^0_A(\Omega)/\sim_A & \xrightarrow{\ \ \mathrm{id}\ \ } & \mathcal{M}^0(\Omega) \\ \downarrow & & \downarrow \\ (\mathcal{C}^0_A(\Omega)/\sim_A)^{\wedge} & \xrightarrow[\ \ \hat{\mathrm{id}}\ \ ]{} & \hat{\mathcal{M}}^0(\Omega) \end{array}$$

with $\hat{\mathrm{id}}$ being an order isomorphism. Let us take $f \in \mathcal{C}(\Omega)$ which is unbounded from above on Ω. Then, in view of (3.2) and (3.4), we can consider $F \in \mathcal{M}^0(\Omega)$ being the $\sim$ equivalence class of f. Further, owing to (3.7), we obtain $\langle F] \in \hat{\mathcal{M}}^0(\Omega)$. Now (7.256) gives a cut $A \subseteq \mathcal{C}^0_A(\Omega)/\sim_A$, such that

$$(7.257)\qquad \hat{\mathrm{id}}\ (A) = \langle F]$$

that is

$$(7.258)\qquad A^{u\ell} = \langle F]$$

where u, ℓ and $\langle\ ]$ are defined in $\mathcal{M}^0(\Omega)$. But obviously $\langle F] \underset{\neq}{\subset} \mathcal{M}^0(\Omega)$, since $F \in \mathcal{M}^0(\Omega)$. Therefore (7.256) and (7.257) give

$$A \underset{\neq}{\subset} \mathcal{C}^0_A(\Omega)/\sim_A .$$

Since A is a cut in $\mathcal{C}^0_A(\Omega)/\sim_A$, the relation (A.8) yields $G \in \mathcal{C}^0_A(\Omega)/\sim_A$, such that

$$A \subseteq \langle G]'$$

which clearly implies

$$(7.259)\qquad G \in A^{u'} \subseteq A^{u}$$

where u' and $\langle\ ]'$ are defined in $\mathcal{C}^0_A(\Omega)/\sim_A$. The relations (7.258) and (7.259) yield

$$\langle F] = A^{u\ell} \subseteq \langle G]$$

which means that, in the sense of (3.5), we have

$$(7.260) \quad F \leq G.$$

If we take now any $g \in G$, then (7.260) implies that $g \in \mathcal{C}^0(\Omega\backslash\Gamma)$ and

$$(7.261) \quad f \leq g \quad \text{on} \quad \Omega\backslash\Gamma$$

for a suitable closed, nowhere dense subset $\Gamma \subsetneq \Omega$. However, according to (7.57), the function g is bounded on Ω. On the other hand $f \in \mathcal{C}^0(\Omega)$ and f is unbounded from above on Ω. In this way (7.261) is obviously contradicted, and therefore the proof of (7.254) and (7.255) is completed.

7.3.2 Now we present a construction employing certain Cantor type sets and we shall use it in several subsequent examples or counterexamples, one of which will be related to (7.74).

Let $\Omega = (0,1) \subset \mathbb{R}$. Given $0 < a < 1$, we define

$$(7.262) \quad \Gamma = [0,1] \setminus \bigcup_{1 \leq j < \infty} I_j$$

where the pairwise disjoint open sets $I_1, I_2, I_3, \ldots,$ are constructed according to the following recursive procedure. First, $I_1 = (\frac{1}{2} - \frac{a}{4}, \frac{1}{2} + \frac{a}{4})$ is the open interval of length $\frac{a}{2}$ in the middle of the closed interval $[0,1]$. Then I_2 is the union of two open intervals, each of length $\frac{a}{8}$ and situated at the middle of the respective closed intervals in $[0,1]\backslash I_1$. Further I_3 is the union of four open intervals, each of length $\frac{a}{32}$ and situated at the middle of the respective closed intervals in $[0,1] \setminus (I_1 \cup I_2)$, etc. In this way we obtain

$$(7.263) \quad \begin{aligned} & I_1, I_2, I_3, \ldots \text{ are pairwise disjoint} \\ & \operatorname{mes}(I_j) = \frac{a}{2^j}, \quad 1 \leq j < \infty \\ & \operatorname{mes}\Big(\bigcup_{1 \leq j < \infty} I_j\Big) = a, \quad \operatorname{mes}(\Gamma) = 1 - a. \end{aligned}$$

In addition Γ is closed and nowhere dense in $\bar{\Omega} = [0,1] \subset \mathbb{R}$.

Let us define $f_j : [0,1] \longrightarrow \mathbb{R}$, with $1 \leq j < \infty$, as follows

(7.264) $$f_j = j\, \chi_{I_j}$$

where χ_A denotes the characteristic function of the set A. Further, let us define $f : [0,1] \longrightarrow \mathbb{R}$ by

(7.265) $$f = \sum_{1 \leq j < \infty} f_j$$

then obviously

(7.266) $$f \in \mathcal{C}^0(\bar{\Omega} \backslash \Gamma).$$

We prove now that, with the notation in (7.30) - (7.33), we have

(7.267) $$\bar{f} = f, \quad \bar{\Gamma} = \Gamma$$

in other words f is *discontinuous* at each $x \in \Gamma$ and it *cannot* be extended to a continuous function on any larger set than $\bar{\Omega} \backslash \Gamma$. Indeed, given $x \in \Gamma$ and $\epsilon > 0$, we obviously have

(7.268) $$(x - \epsilon, x + \epsilon) \cap I_j \neq \phi$$

for infinitely many $1 \leq j < \infty$, since (7.262) holds and Γ is nowhere dense in $[0,1]$. Then (7.268), (7.264) and (7.265) give

$$\limsup_{y \to x} f(y) = \infty$$

and the proof of (7.267) is completed.

7.3.3 We can return now to Remark 7.2 in Subsection 7.1 and give an alternative counterexample illustrating the fact that the embedding (7.40), that is

$$(7.269) \qquad \mathcal{M}^0(\Omega) \ni U \longmapsto \mu\, u \in \text{Mes}\ (\Omega)$$

where $u \in U$, is *not* increasing. Indeed, let us take $f, g \in \mathcal{C}^0_{nd}(\Omega)$, where f is given in (7.266), while

$$(7.270) \qquad g(x) = 1/2 \quad \text{for} \quad x \in \Omega.$$

Then obviously

$$g \leq f \quad \text{on} \quad \Omega \setminus \Gamma$$

therefore

$$(7.271) \qquad G \leq F \quad \text{in} \quad \mathcal{M}^0(\Omega)$$

where F and G are the $\sim$ equivalence classes of f and g respectively, see (3.3) - (3.5). However, inspite of (7.271), we have

$$(7.272) \qquad \mu\, g \not\leq \mu\, f \quad \text{in} \quad \text{Mes}\ (\Omega)$$

since (7.267) and (7.270) yield

$$\mu\, f = 0 < \frac{1}{2} = \mu\, g \quad \text{on} \quad \Gamma, \quad \text{and} \quad \text{mes}\ (\Gamma) = 1 - \alpha > 0.$$

Clearly (7.271) and (7.272) prove (7.269).

7.3.4 Curiously, the embedding (7.269) has the following *inverse monotonicity* property. If $U, V \in \mathcal{M}^0(\Omega)$ and $u \in U$, $v \in V$, then

$$(7.273) \qquad \mu\, u \leq \mu\, v \quad \text{in} \quad \text{Mes}\ (\Omega) \Longrightarrow U \leq V \quad \text{in} \quad \mathcal{M}^0(\Omega).$$

Assume indeed that $U \not\leq V$ in $\mathcal{M}^0(\Omega)$. But (7.38) gives

$$\mu\, u \in U, \quad \mu\, v \in V$$

therefore, in view of (3.5) there exists a nonvoid open subset $\Delta \subsetneq \Omega$, such that

$$\mu v < \mu u \quad \text{on} \quad \Delta.$$

However, in that case we cannot have $\mu u \leq \mu v$ in $\mathrm{Mes}(\Omega)$. Thus the proof of (7.273) is completed.

7.3.5 Now we construct an increasing sequence of piecewise continuous functions $g_j \in \mathcal{C}^0_{nd}(\Omega)$, with $j \in \mathbb{N}$, such that

$$(7.274) \qquad (g_j \mid j \in \mathbb{N}) \text{ is unbounded in } \mathrm{Mes}(\Omega)$$

$$(7.275) \qquad (G_j \mid j \in \mathbb{N}) \text{ is bounded in } \mathcal{M}^0(\Omega)$$

where for $j \in \mathbb{N}$, we denote by G_j the $\sim$ equivalence class of g_j, see (3.3). Indeed, let us take

$$(7.276) \qquad g_j = \left| \begin{array}{ll} \sum_{1 \leq i \leq j} f_i & \text{on} \quad \bigcup_{1 \leq i \leq j} I_i \\ j/2 & \text{on} \quad \Omega \setminus \bigcup_{1 \leq i \leq j} I_i \end{array} \right. .$$

Then we have

$$(7.277) \qquad g_1 \leq g_2 \leq \ldots \leq g_j \leq \ldots \quad \text{on} \quad \Omega$$

since given $1 \leq j < \infty$, the relation (7.276) implies

$$g_{j+1} = g_j \quad \text{on} \quad \bigcup_{1 \leq i \leq j} I_i$$

$$g_{j+1} = (j+1)/2 > j/2 = g_j \quad \text{on} \quad \Omega \setminus \bigcup_{1 \leq i \leq j+1} I_i$$

and finally, if we also consider (7.264) then we obtain

$$g_{j+1} = f_{j+1} = j + 1 > j/2 = g_j \quad \text{on} \quad I_{j+1}$$

as $I_1, I_2, \ldots, I_j, I_{j+1}, \ldots$ are pair wise disjoint. We show now that

(7.278) $\quad g_j \leq f \quad \text{on} \quad \Omega \backslash \Gamma, \quad \text{for} \quad 1 \leq j < \infty.$

Indeed, if $x \in \Omega \backslash \Gamma$ then (7.262) yields $1 \leq i < \infty$, such that $x \in I_i$. Then (7.276) and (7.264) give

$$g_j(x) = \left| \begin{array}{ll} i & \text{if} \quad i \leq j \\ j/2 & \text{if} \quad j < i \end{array} \right.$$

while (7.265) and (7.264) result in

$$f(x) = i$$

thus the proof of (7.278), and therefore of (7.275) completed. Further we note that (7.276) results in

(7.279) $\quad g_j = j/2 \quad \text{on} \quad \Gamma, \quad \text{for} \quad 1 \leq j < \infty.$

Assume now that (7.274) does not hold. Then

(7.280) $\quad g_j \leq g \quad \text{on} \quad \Omega \backslash \Sigma, \quad \text{for} \quad 1 \leq j < \infty$

for a suitable $g \in \text{Mes}\ (\Omega)$ and $\Sigma \subseteq \Omega$, with $\text{mes}\ (\Sigma) = 0$. However, since $\text{mes}\ (\Gamma) > 0$, see (7.263), the relations (7.279) and (7.280) contradict each other, and thus the proof of (7.274) is completed.

7.3.6 Finally, we give a *partial negative* answer to (7.74), by proving (7.77). Let

(7.281) $\quad \mathcal{C}_{pc}\ (0,1)$

be the set of all piecewise constant functions $f : \Omega \longrightarrow \mathbb{R}$ which have a finite number of discontinuities. We define on $\mathcal{C}_{pc}\ (0,1)$ the equivalence relation $\sim_{pc}$ by

$$(7.282)\quad f \sim_{pc} g \iff \left[\begin{array}{l} \exists \;\; F \subseteq \Omega \;\; \text{finite :} \\ \;\;\; f = g \;\; \text{on} \;\; \Omega \setminus F \end{array}\right].$$

Then obviously $\mathcal{C}_{pc}\ (0,1) \subseteq \mathcal{C}^0_{nd}\ (0,1)$ and $\sim_{pc}$ is the equivalence $\sim$ restricted to $\mathcal{C}_{pc}\ (0,1)$, see (3.1) and (3.3). Therefore, see (3.4), we have the order isomorphical embedding

$$(7.283)\quad \mathcal{C}_{pc}\ (0,1)/\sim_{pc} \longrightarrow \mathcal{M}^0\ (0,1)\ .$$

However, we also have $\mathcal{C}_{pc}\ (0,1) \subseteq \mathcal{C}^0_z\ (0,1)$, see (7.3). Thus in view of (7.6), the embedding in (7.283) can be written in the stronger form of the order isomorphical embedding

$$(7.284)\quad \mathcal{C}_{pc}\ (0,1)/\sim_{pc} \longrightarrow \mathcal{N}^0\ (0,1).$$

In this way, in view of (7.14), we obtain the order isomorphical embedding

$$(7.285)\quad \mathcal{C}_{pc}\ (0,1)/\sim_{pc} \xrightarrow{\ \pi\ } \text{Mes}\ (0,1).$$

Let us now consider any given mapping

$$(7.286)\quad \varphi : \mathcal{M}^0\ (0,1) \longrightarrow \text{Mes}\ (0,1)$$

and associate with it the mapping, see (A.49)

$$(7.287)\quad \bar{\varphi} : \hat{\mathcal{M}}^0\ (0,1) \longrightarrow \hat{\text{Mes}}\ (0,1)$$

defined for each $A \in \hat{\mathcal{M}}^0\ (0,1)$ as follows

$$(7.288)\quad \bar{\varphi}(A) = \sup_{\hat{\text{Mes}}\ (0,1)} \varphi(A)$$

Then (A.50) implies that $\bar{\varphi}$ is increasing.

In the sense of the embeddings in (7.283) and (7.285), let us assume now that

$$(7.289) \qquad \varphi \Big|_{\mathcal{C}_{pc}(0,1)/\sim_{pc}} = \pi$$

Then we show that

$$(7.290) \qquad \bar{\varphi} : \hat{\mathcal{M}}^0 (0,1) \longrightarrow \hat{\mathrm{Mes}}\,(0,1) \quad \text{is } \textit{not} \text{ injective.}$$

Moreover, we can find $F \in \mathcal{M}^0 (0,1)$ such that

$$(7.291) \qquad \bar{\varphi}(<F]) = +\infty \in \hat{\mathrm{Mes}}\,(0,1) \setminus \mathrm{Mes}\,(0,1).$$

Then, noting that

$$(7.292) \qquad \mathcal{M}^0 (0,1) \in \hat{\mathcal{M}}^0 (0,1), \quad \mathcal{M}^0 (0,1) \neq <F], \quad \bar{\varphi}(\mathcal{M}^0 (0,1)) = +\infty$$

it is clear that (7.290) follows from (7.291) and (7.292). Therefore, it only remains to prove (7.291). Let us take g_j, with $1 \leq j < \infty$, given by (7.276). Further, for $1 \leq j < \infty$, let $G_j \in \mathcal{M}^0 (0,1)$ be the $\sim$ equivalence class of g_j. Finally, let f be given by (7.265) and let $F \in \mathcal{M}^0 (0,1)$ be the $\sim$ equivalence class of f.

Obviously

$$(7.293) \qquad g_1, g_2, \ldots, g_j, \ldots \in \mathcal{C}_{pc} (0,1), \quad f \in \mathcal{C}^0_{nd} (0,1).$$

Moreover (7.277) and (7.278) give

$$(7.294) \qquad G_j \in <F], \quad 1 \leq j < \infty.$$

Therefore (7.288) and (7.294) result in $\hat{\mathrm{Mes}}\,(0,1)$ in the inequality

$$(7.295) \qquad \sup_{\hat{\mathrm{Mes}}\,(0,1)} \{\varphi(G_j) \mid 1 \leq j < \infty\} \leq \bar{\varphi}\,(<F])$$

since $\varphi(G_j) \leq \bar{\varphi}(<G_j])$ even when φ is not increasing. On the other hand (7.289) and (7.293) imply that

$$(7.296) \qquad \varphi(G_j) = \pi(G_j), \quad 1 \leq j < \infty.$$

But there exists $\Sigma \subsetneq (0,1)$, with $\text{mes}(\Sigma) = 0$, such that

$$(7.297) \qquad \pi(G_j) = g_j \quad \text{on} \quad (0,1)\backslash\Sigma, \quad \text{for} \quad 1 \leq j < \infty.$$

And then (7.296), (7.297) and (7.274) give

$$\sup_{\hat{M}\text{es}\,(0,1)} \{\varphi(G_j) \mid 1 \leq j < \infty\} = +\infty$$

which together with (7.295) completes the proof of (7.291), and therefore of (7.290).

7.3.7 An alternative proof of (7.77), and thus a *partial negative* answer to (7.74) can be obtained as follows. Let us start again with (7.286) - (7.288). However, instead of (7.289), let us assume the alternative condition

$$(7.298) \qquad \varphi \Big|_{\mathcal{C}^0[0,1]} = \text{id}$$

recalling that $\mathcal{C}^0[0,1] \underset{\neq}{\subset} \mathcal{M}^0(0,1)$.

We shall then show that (7.290) is again valid. Indeed, let $F \in \mathcal{M}^0(0,1)$ be the $\sim$ equivalence class of f in (7.265). Then it suffices to prove that (7.291) will again hold. For that purpose, let us fix j, with $1 \leq j < \infty$, and then take g_j in (7.276). We recall that $G_j \in \mathcal{M}^0(0,1)$, which is the $\sim$ equivalence class of g_j, will satisfy (7.295). Now, let $S_j \subset [0,1]$ be the set of jump points of g_j. Further let $h \in \mathcal{C}^0[0,1]$ and $H \in \mathcal{M}^0(0,1)$ be the $\sim$ equivalence class of h, and let us assume that in $\mathcal{M}^0(0,1)$ we have

$$H \leq G_j \,.$$

Then obviously

$$(7.299) \quad h(x) \leq g_j(x), \quad x \in (0,1) \backslash S_j \ .$$

Therefore, by an argument similar to that which led to (7.295), we obtain

$$(7.300) \quad \varphi(H) \leq \bar{\varphi}(<H]) \leq \bar{\varphi}(<F]) \ .$$

But owing to (7.298), we have in Mes (0,1)

$$(7.301) \quad \varphi(H) = h \ .$$

Furthermore

$$(7.302) \quad g_j = \sup \left\{ k \in \mathcal{C}^0[0,1] \ \middle| \ \begin{array}{c} \forall\ x \in (0,1) \backslash S_j : \\ k(x) \leq g_j(x) \end{array} \right\}$$

where sup is taken in Mes (0,1). In this way (7.299) - (7.302) give

$$(7.303) \quad g_j \leq \bar{\varphi}(<F]) \quad \text{in Mes } (0,1) \ .$$

Since j, with $1 \leq j < \infty$, is arbitrary in (7.303), and owing to (7.274), it follows that indeed (7.291) is still valid. In this way, under the alternative assumption in (7.298), we again obtain (7.290).

7.3.8 The essence of the above mechanism which leads to the failure in (7.290) is quite simple. The function f defined in (7.265) is clearly an element of Mes (0,1) as it is continuous on $\Omega \backslash \Gamma$ and vanishes on Γ, with Γ being closed, and in fact, nowhere dense as well. The crucial point is that, through (7.276), we can associate with f the sequences in (7.274) and (7.275). Indeed, if we define the extended real valued function $g : \Omega \longrightarrow [0,+\infty]$ by

$$g(x) = \sup \{g_j(x) \mid j \in \mathbb{N}\}, \quad x \in \Omega$$

then we are led to the following two properties

(7.304) $\quad g \sim f \quad \text{in} \quad \mathcal{M}^0(0,1)$

since $g = f$ on $\Omega\setminus\Gamma$, but on the other hand $g = +\infty$ on Γ, and $\text{mes}(\Gamma) > 0$, therefore, noting that $\widehat{\text{Mes}}(0,1)^* = \text{Mes}(0,1)$, it follows that

(7.305) $\quad g \in \widehat{\text{Mes}}(0,1)\setminus\text{Mes}(0,1)$.

7.3.9 The same mechanism can help us in showing that the Nakano & Shimogaki result, see (7.55)

(7.306) $\quad \hat{\mathcal{C}}^0(\bar{\Omega})^* \quad \text{and} \quad \mathcal{C}^0_{\mathbb{Q}}(\bar{\Omega})/\sim_{\mathbb{Q}}$ are order isomorphical

fails to hold if $\bar{\Omega}$ is replaced by Ω. Indeed, we shall prove that the embedding

(7.307) $\quad \mathcal{C}^0(0,1) \xrightarrow{\ \psi\ } \mathcal{C}^0_{\mathbb{Q}}(0,1)/\sim_{\mathbb{Q}}$

does *not* extend to an order isomorphism

(7.308) $\quad \hat{\mathcal{C}}^0(0,1) \xrightarrow{\ \hat{\psi}\ } (\mathcal{C}^0_{\mathbb{Q}}(0,1)/\sim_{\mathbb{Q}})\hat{}$

where $\mathcal{C}^0_{\mathbb{Q}}(0,1)$ and $\sim_{\mathbb{Q}}$ are given by the obvious extensions of the definitions (7.51) and (7.53) respectively, with the boundedness property *) in (7.51) dropped. For that purpose, let us again take f in (7.265) and let F be its $\sim_{\mathbb{Q}}$ equivalence class in $\mathcal{C}^0_{\mathbb{Q}}(0,1)/\sim_{\mathbb{Q}}$. Then clearly $\langle F]$ is not the largest element in $(\mathcal{C}^0_{\mathbb{Q}}(0,1)/\sim_{\mathbb{Q}})\hat{}$. Therefore, if $\hat{\psi}$ in (7.308) were an order isomorphism, it would follow that

$$\hat{\psi}^{-1}(\langle F]) \in \hat{\mathcal{C}}^0(0,1)^* .$$

In view of (7.26) and (A.20) we would obtain in $\hat{\mathcal{C}}^0(0,1)^*$

$$\hat{\psi}^{-1}(\langle F]) \subseteq \langle g]$$

for a suitable $g \in \mathcal{C}^0(0,1)$.

On the other hand, we can define

$$h_j \in \mathcal{C}^0(0,1), \quad 1 \leq j < \infty$$

in such a way that, for $1 \leq j < \infty$, we have, see (7.264)

$$h_j \leq f_j \quad \text{on} \quad (0,1)$$

and at the same time

$$(7.309) \qquad h_j = j$$

at the midpoints of each of the finite number of intervals which constitute I_j. It follows that in $\mathcal{C}^0_{\mathbb{Q}}(0,1)/\sim_{\mathbb{Q}}$, we have for $1 \leq j < \infty$

$$H_j \leq F_j \leq \sup_{1 \leq k < \infty} F_k = F$$

where H_j and F_j are the $\sim_{\mathbb{Q}}$ equivalence classes of h_j and f_j respectively. But if $\hat{\psi}$ in (7.308) were an order isomorphism, then we would have in $\hat{\mathcal{C}}^0(0,1)$

$$(7.310) \qquad \hat{\psi}^{-1}(<H_j]) \leq \hat{\psi}^{-1}(<F]), \quad 1 \leq j < \infty \ .$$

However, by the definition of H_j, we have in $\hat{\mathcal{C}}^0(0,1)$

$$(7.311) \qquad \hat{\psi}^{-1}(<H_j]) = <h_j], \quad 1 \leq j < \infty \ .$$

Now let $x \in \Gamma$ and $\epsilon > 0$. Then owing to (7.268) and (7.309), we can conclude that the sequence of continuous functions $(h_j \mid 1 \leq j < \infty)$ is unbounded from above on $(0,1) \cap (x-\epsilon, x+\epsilon)$. In particular, it cannot be bounded from above by any element of $\mathcal{C}^0(0,1)$. This however contradicts (7.310) and (7.311).

'... Physicists just write a partial differential equation and get a Nobel Prize for it. After that, the prizeless and much harder task of solving those equations is, of course, left to the mathematicians ...'

Anonymous

PART II. APPLICATIONS TO SPECIFIC CLASSES OF NONLINEAR AND LINEAR PDEs

8. THE CAUCHY PROBLEM FOR NONLINEAR FIRST ORDER SYSTEMS

The aim of this Section is to adapt and apply the general method in Sections 2 - 5, in order to prove the *existence* of solutions for a large class of *Cauchy problems* for first order nonlinear systems. Indeed, the general existence results in Theorems 5.1, 5.2, Corollary 5.1 and Theorem 5.3 did not deal with the presence of initial and/or boundary value conditions, see also the related comments in Section 6. Therefore, in order to solve the *Cauchy problems* considered in this Section, one has to *adapt* the mentioned general existence results. However, as seen next, such an adaptation is rather *easy and straightforward*. Moreover, it can be done in more than one way, as for instance, the *alternative* approach in Section 11 shows it. In this way we come to note another of the *advantages* in solving initial and/or boundary value problems for linear and nonlinear PDEs through the method of order completion. Namely, initial and/or boundary value problems are solved by precisely the same kind of constructions as the free problems. On the other hand, as well known, this is not so when functional analytic methods - in particular, involving distributions, their restrictions to lower dimensional manifolds, or the associated trace operators - are used for the solution of such problems.

For convenience, we shall present the case of two independent variables only, however, as seen easily, the respective solution method as well as results extend in an obvious manner to the case of an *arbitrary* finite number of independent variables.

Let $\Omega = (-a,a) \times (-b,b) \subset \mathbb{R}^2$, for some given $a,b > 0$, be the domain of the independent variables (x,y), and suppose we are given the vector valued functions

$$(8.1) \qquad F : \bar{\Omega} \times \mathbb{R}^{2n} \longrightarrow \mathbb{R}^n, \quad F \in \mathcal{C}^0$$

$$(8.2) \qquad f : [-a,a] \longrightarrow \mathbb{R}^n, \quad f \in \mathcal{C}^1$$

for a certain $n \in \mathbb{N}$, $n \geq 1$. The Cauchy problem we intend to solve is

$$(8.3) \qquad D_y U(x,y) + F(x,y,U(x,y),D_x U(x,y)) = 0, \quad (x,y) \in \Omega$$

$$(8.4) \qquad U(x,0) = f(x), \quad x \in (-a,a)$$

the solution being a suitable vector valued function $U : \Omega \longrightarrow \mathbb{R}^n$.

It will be convenient to denote by $T = T(x,y,D)$ the nonlinear partial differential operator in the left hand term of (8.3).

We note that componentwise, T is a particular case of (2.2), however, it is not of the form (2.3), since F in (8.1) can be an arbitrary, not necessarily polynomial type, nonlinear continuous function. Nevertheless, with the notation in (2.10) adapted for systems of PDEs, we obviously have $R_x = \mathbb{R}^n$, with $x \in \bar{\Omega}$.

8.1 Basic Approximations

As a first step, we establish two uniform local approximation properties which specialize and refine the similar property in Lemma 2.1.

Lemma 8.1

We have

$$(8.5) \qquad \begin{aligned} &\forall \ \epsilon > 0 : \\ &\exists \ \delta > 0 : \\ &\forall \ (x_0, y_0) \in \bar{\Omega} : \\ &\exists \ U = U_{\epsilon, x_0, y_0} : \bar{\Omega} \longrightarrow \mathbb{R}^n, \quad U \in \mathcal{C}^1 : \\ &\forall \ (x,y) \in \bar{\Omega} : \\ &\quad \begin{bmatrix} |x - x_0| < \delta \\ |y - y_0| < \delta \end{bmatrix} \Longrightarrow -\epsilon \leq TU(x,y) \leq 0 \end{aligned}$$

Here and in the sequel, we shall denote by c both a given constant $c \in \mathbb{R}$, as well as the corresponding n-dimensional vector $(c,\ldots,c) \in \mathbb{R}^n$. However, the context will make the difference quite clear.

Furthermore, we also have

$$\forall \ \epsilon > 0 :$$
$$\exists \ \delta > 0 :$$
$$\forall \ x_0 \in [-a,a] :$$

(8.6) $$\exists \ U = U_{\epsilon,x_0} : \bar{\Omega} \longrightarrow \mathbb{R}^n, \quad U \in \mathcal{C}^1 :$$

*) $\forall \ (x,y) \in \bar{\Omega}$:

$$\begin{bmatrix} |x - x_0| < \delta \\ |y| < \delta \end{bmatrix} \Longrightarrow -\epsilon \leq TU(x,y) \leq 0$$

**) $U(x,0) = f(x), \quad x \in [-a,a], \quad |x - x_0| < \delta$

Proof

Let $\| \ \|$ be the ℓ^∞ norm on $\mathbb{R}^n$, and let $M_0 > 0$ be a bound for both $\|f\|$ and $\|f'\|$ on $[-a,a]$. Then we can assume given $M_1 > 0$ which is a bound for $\|F\|$ on $\bar{\Omega} \times [-M_0,M_0]^n \times [-M_0,M_0]^n$.

For $\epsilon > 0$, we can take $\delta_0 > 0$ such that

(8.7) $$\|F(x,y,u,v) - F(x_0,y_0,u_0,v_0)\| \leq \epsilon/2$$

provided that (x,y), $(x_0,y_0) \in \bar{\Omega}$, $u,v,u_0,v_0 \in [-M_0 - M_1, M_0 + M_1]^n$ and

$$|x - x_0|, \ |y - y_0|, \ \|u - u_0\|, \ \|v - v_0\| \leq \delta_0 .$$

Further we can take $\delta_1 > 0$ such that

(8.8) $$\|f(x) - f(x_0)\| \leq \delta_0/2, \quad \|f'(x) - f'(x_0)\| \leq \delta_0$$

for all $x,x_0 \in [-a,a]$, with $|x - x_0| \leq \delta_1$.

Let us now show that (8.5) holds. Assume given $(x_0,y_0) \in \bar{\Omega}$ and let us define

$$U : \bar{\Omega} \longrightarrow \mathbb{R}^n$$

by

$$(8.9) \qquad U(x,y) = -(y - y_0)(F(x_0,y_0,0,0) + \epsilon/2), \quad (x,y) \in \bar{\Omega} .$$

For convenience, let us denote

$$(8.10) \qquad A = -F(x_0,y_0,0,0) - \epsilon/2 \in \mathbb{R}^n$$

then, for $(x,y) \in \Omega$, we obtain from (8.9) the relation

$$(8.11) \qquad \begin{aligned} TU(x,y) &= A + F(x,y,(y - y_0)A,0) = \\ &= F(x,y,(y - y_0)A,0) - F(x_0,y_0,0,0) - \epsilon/2 \end{aligned}$$

Let $\delta_2 > 0$ be such that

$$(8.12) \qquad y \in [-b,b], \ |y - y_0| \leq \delta_2 \Rightarrow \|(y - y_0)A\| \leq \delta_0$$

and note that in view of (8.10) and the bound $\|F\| \leq M_1$, we can assume that δ_2 depends only on δ_0 and not on x_0 or y_0.

Taking now $\delta = \min \{\delta_0, \delta_2\}$ and recalling (8.7), we obtain

$$(8.13) \qquad \|F(x,y,(y - y_0)A,0) - F(x_0,y_0,0,0)\| \leq \epsilon/2$$

for all $(x,y) \in \Omega$, for which $|x - x_0|, |y - y_0| \leq \delta$. And then (8.5) follows from (8.11) - (8.13).

We turn now to proving (8.6). Given $x_0 \in [-a,a]$, let us take

$$(8.14) \qquad B = -F(x_0,0,f(x_0),f'(x_0)) - \epsilon/2 \in \mathbb{R}^n$$

and define

$$U : \bar{\Omega} \longrightarrow \mathbb{R}^n$$

by

$$U(x,y) = yB + f(x), \quad (x,y) \in \bar{\Omega} . \tag{8.15}$$

Obviously

$$U(x,0) = f(x), \quad x \in [-a,a] . \tag{8.16}$$

Furthermore, in view of (8.14) and the bounds $\|F\| \leq M_1$, $\|f\|$, $\|f'\| \leq M_0$, the relation (8.15) gives for $(x,y) \in \bar{\Omega}$

$$\|U(x,y)\| \leq M_0 + M_1 \tag{8.17}$$

provided that $|y| \leq \delta_3$, where $\delta_3 > 0$ does not depend on x_0.

Finally, (8.14) yields for $(x,y) \in \bar{\Omega}$ the relation

$$\begin{aligned} TU(x,y) &= B + F(x,y,yB + f(x),f'(x)) = \\ &= F(x,y,yB + f(x),f'(x)) - F(x_0,0,f(x_0),f'(x_0)) - \epsilon/2 \end{aligned} \tag{8.18}$$

Let then $\delta = \min\{\delta_0,\delta_1,\delta_2,\delta_3,\delta_4\}$, where $\delta_4 > 0$ is such that

$$y \in [-b,b], \ |y| \leq \delta_4 \Longrightarrow \|yB\| \leq \delta_0/2 \tag{8.19}$$

independently of $x_0 \in [-a,a]$, which is possible in view of (8.14) and the bounds on F, f and f'.

Now, if $|x - x_0|$, $|y| \leq \delta$ then (8.7) together with (8.17) - (8.19) yield

$$\|F(x,y,yB + f(x),f'(x)) - F(x_0,0,f(x_0),f'(x_0))\| \leq \epsilon/2$$

which completes the proof of (8.6). ■

We proceed now to establish for the nonlinear Cauchy problem (8.3), (8.4) the approximation result corresponding to Proposition 2.1. Here, however, owing to the presence of the initial value problem (8.4), given for

$$(8.20) \qquad y = 0$$

we have to exhibit a certain care in the choice of the possibly occurring singularity sets $\Gamma_\epsilon \subset \Omega$. Indeed, in general, these closed and nowhere dense subsets Γ_ϵ can result in a rather arbitrary way. In particular, it may as well happen that

$$(8.21) \qquad (-a,a) \times \{0\} \subsetneq \Gamma_\epsilon$$

in other words, the portion of the x-axis contained in $\Omega \subset \mathbb{R}^2$, thus corresponding to (8.4) and (8.20), may turn out to be part of some of the singularity sets Γ_ϵ. This is, however, an obviously inadmissible situation, since in such a case, the respective approximating solution $U_\epsilon \in \mathcal{C}^1(\Omega \backslash \Gamma_\epsilon)$ - which in general, may fail to be defined on the singularity set Γ_ϵ - cannot be requested to satisfy the initial value (8.4).

In view of the above, we introduce the following definition. A *finite initial adapted grid* in Ω is a subset $\Gamma \subset \Omega$ which consists of finitely many horizontal and vertical lines, and in addition, is such that

$$(8.22) \qquad \Gamma \cap ((-a,a) \times \{0\}) \text{ is at most a finite set of points.}$$

It follows therefore that (8.21) cannot happen for any finite initial adapted grid $\Gamma \subset \bar{\Omega}$.

Proposition 8.1

If $f \in \mathcal{C}^1[-a,a]$ then

$$\forall \quad \epsilon > 0 :$$

$$\exists \quad \Gamma_\epsilon \subset \Omega \quad \text{finite initial adapted grid}, \quad U_\epsilon \in C^1(\Omega \backslash \Gamma_\epsilon) :$$

(8.23)

$$*) \quad -\epsilon \leq TU_\epsilon \leq 0 \quad \text{on} \quad \Omega \backslash \Gamma_\epsilon$$

$$**) \quad U_\epsilon(x,0) = f(x), \quad (x,0) \in ((-a,a) \times \{0\}) \backslash \Gamma_\epsilon$$

Proof

Given $\epsilon > 0$, we take $\delta > 0$ in Lemma 8.1 and let $\Gamma_\epsilon \subset \Omega$ be any finite initial adapted grid whose nodes are at a distance of at most δ. Then $\Omega \backslash \Gamma_\epsilon$ consists of a finite number of open rectangles

(8.24) $\quad D_1, \ldots, D_\ell$.

Let (x_k, y_k) be the center of the rectangle D_k, with $1 \leq k \leq \ell$. In case for some $1 \leq k \leq \ell$

(8.25) $\quad D_k \cap ((-a,a) \times \{0\} \neq \phi$

that is, D_k intersects the x-axis, we can assume that such a D_k is symmetric with respect to the x-axis, in which case it follows that $y_k = 0$.

Now we can apply Lemma 8.1 to the open rectangles (8.24), more precisely, to their center points. In particular, given $1 \leq k \leq \ell$, if D_k does not satisfy (8.25), then according to (8.5), we obtain

$$U_{\epsilon,k} \in C^1(\bar{\Omega}, \mathbb{R}^n)$$

such that

(8.26) $\quad -\epsilon \leq T(x,y,D)U_{\epsilon,k}(x,y) \leq 0, \quad (x,y) \in D_k$.

In case D_k satisfies (8.25), then (8.6) gives

$$U_{\epsilon,k} \in \mathcal{C}^1(\bar{\Omega}, \mathbb{R}^n)$$

such that

$$(8.27) \qquad -\epsilon \leq T(x,y)U_{\epsilon,k}(x,y) \leq 0, \quad (x,y) \in D_k$$

$$(8.28) \qquad U_{\epsilon,k}(x,0) = f(x), \quad (x,0) \in D_k$$

Now we can take U_ϵ as coinciding on D_k with $U_{\epsilon,k}$, for $1 \leq k \leq \ell$. ∎

8.2 Construction of Solutions Through Order Completion

In view of the presence of the initial value condition (8.4), we have to adapt the general solution method in Sections 2 - 5, since that solution method was not taking into account initial or boundary value conditions. However, as seen below, such an adaptation is quite straightforward and it operates in the same simple and easy manner as was the case with the general method in Sections 2 - 5. In fact, as shown next, one can quite completely *separate* the problem of the existence of solutions from that of satisfying the initial value conditions, see also Remark 8.2 below. More precisely, it is possible *first* to impose the initial value conditions, and do so in a way which is quite near to their classical meaning. Then, as a *second* step, we can use order completion and obtain the existence of generalized solutions. In other words, in this second step, the order completion method will be used on a set of functions which - as a consequence of the first step - already satisfy the initial value conditions.

In a slight modification of (3.1), as required by (8.22), let us define for $\ell \in \bar{\mathbb{N}}$ the set of functions

$$
(8.29) \qquad \mathcal{C}^{\ell}_{nde}(\Omega) = \left\{ u \;\middle|\; \begin{array}{l} \exists\ \Gamma \subset \Omega \text{ closed, nowhere dense :} \\ \quad *)\ \Gamma \cap ((-a,a) \times \{0\}) \text{ is at most a finite} \\ \qquad \text{set of points} \\ \ **)\ u : \Omega \backslash \Gamma \longrightarrow \mathbb{R}^n \\ ***)\ u \in \mathcal{C}^{\ell}(\Omega \backslash \Gamma) \end{array} \right\}
$$

Further, we shall also need the set of functions

$$
(8.30) \qquad \mathcal{C}^{1}_{e}(-a,a) = \left\{ f \;\middle|\; \begin{array}{l} \exists\ \Gamma_e \subset (-a,a) \text{ at most a finite set} \\ \text{of points} \\ \quad *)\ f : (-a,a) \backslash \Gamma_e \longrightarrow \mathbb{R}^n \\ \ **)\ f \in \mathcal{C}^{1}((-a,a) \backslash \Gamma_e) \end{array} \right\}
$$

Remark 8.1

It is important that in the definition of the class of functions in $\mathcal{C}^{\ell}_{nde}(\Omega)$ we do *not* restrict ourselves to allowing only *singularity* sets $\Gamma \subset \Omega$ which are finite initial adapted grids, that is, consist of finitely many horizontal and vertical lines. Indeed, the solutions of the nonlinear Cauchy problem (8.3), (8.4) may exhibit shocks which need not be subsets of such finite initial adapted grids. Therefore the utility of considering singularity sets which can be given by arbitrary closed, nowhere dense subsets $\Gamma \subset \Omega$. For further comments, see pct-s 3) - 5) in Remark 8.3.

Obviously, in analogy with (4.1), (4.2), the nonlinear partial differential operator T given by (8.3), defines a mapping

$$
(8.31) \qquad \begin{array}{ccc} \mathcal{C}^{1}_{nde}(\Omega) & \xrightarrow{\quad \bar{T} \quad} & \mathcal{C}^{0}_{nde}(\Omega) \times \mathcal{C}^{1}_{e}(-a,a) \\ u & \longmapsto & (Tu, u|_{y=0}) \end{array}
$$

Now, similar to (3.3), let us define on $\mathcal{C}^{0}_{nde}(\Omega) \times \mathcal{C}^{1}_{e}(-a,a)$ the equivalence relation $\sim$, by

$$(8.32) \qquad (u,f) \sim (v,g) \iff \left[\begin{array}{lll} 1) & u = v & \text{on } \Omega\backslash\Gamma \\ 2) & f = g & \text{on } (-a,a)\backslash\Gamma_e \end{array} \right]$$

where $\Gamma \subset \Omega$ is closed, nowhere dense, while $\Gamma_e = \Gamma \cap ((-a,a) \times \{0\})$ is at most a finite set of points, with both Γ and Γ_e possibly depending on u, f, v, g, furthermore $u,v \in \mathcal{C}^0(\Omega\backslash\Gamma)$ and $f,g \in \mathcal{C}^1((-a,a)\backslash\Gamma_e)$. Then, similar to (3.4) and (3.5), we define the quotient space

$$(8.33) \qquad X_0(\Omega) = (\mathcal{C}^0_{nde}(\Omega) \times \mathcal{C}^1_e(-a,a))/\sim$$

and the partial order relation $\leq$ on it, by

$$(8.34) \qquad (U,F) \leq (V,G) \iff \left[\begin{array}{lll} 1) & u \leq v & \text{on } \Omega\backslash\Gamma \\ 2) & f = g & \text{on } (-a,a)\backslash\Gamma_e \end{array} \right]$$

where u, v, f and g are representants of U, V, F and G respectively, while Γ and Γ_e are as above.

Remark 8.2

Here we should stop for a moment and note the seemingly peculiar condition 2) in the definition of the partial order $\leq$ in (8.34). Indeed, at first sight it may appear that a more natural version of this condition would rather be the following one

$$(8.35) \qquad f \leq g \quad \text{on} \quad (-a,a)\backslash\Gamma_e \, .$$

However, as seen in Theorem 8.1 below, from the point of view of the *existence* of generalized solutions for the nonlinear Cauchy problem (8.3) and (8.4), the condition (8.35) would lead to an unnecessary complication in the partial order $\leq$ in (8.34) which we can avoid by the *simple* and much more *natural* device above, namely, condition 2) in (8.34). It follows that the partial order $\leq$ in (8.34) does already include the initial value condition (8.4). In other words, when solving the nonlinear PDE in (8.3) under the initial value condition (8.4), we can afford to deal with the initial value condition (8.4) in a *nearly classical way*, that is, asking that the usual equality $f = g$ of functions in condition 2) in (8.34) should hold on $(-a,a)$, except perhaps for a finite number of points. In this way, just as in Sections 2 - 5, the generalized solution procedure based on order completion will only be applied to the nonlinear PDE itself, taken free of any other conditions, that is, as it stands in (8.3). This is, therefore, the meaning of the above statement that one can *separate* the problem of the existence of solutions from that of satisfying the initial value conditions. The abstract mechanism underlying the possibility of this separation can be seen in Sections 9 and 10. However, as seen in Section 11, the separation between the problem of the existence of solutions and that of satisfying initial value conditions can be achieved in an *alternative* way as well. We refer here in particular to the setup in (11.7) - (11.13).

Finally, it is clear that such a separation can be performed also in the presence of boundary value conditions, which for instance, can be lumped together with initial value conditions, in which case the existence of solutions can again be dealt with as a step apart and on its own and in terms of a suitable order completion alone.

Now, in view of (8.31), and similar to (4.1) - (4.4), let us proceed as follows.

We define an equivalence relation $\sim_T$ on $\mathcal{C}^1_{nde}(\Omega)$ by, see (8.32)

$$(8.36) \qquad u \sim_T v \iff (Tu, u|_{y=0}) \sim (Tv, v|_{y=0})$$

and then define the quotient space

$$(8.37) \qquad X_T(\Omega) = \mathcal{C}^1_{nde}(\Omega)/\sim_T .$$

Then, as in (4.1), (4.5), the mapping (8.31) can be extended to an embedding

$$(8.38) \qquad \tilde{T} : X_T(\Omega) \xrightarrow{\text{inj}} X_0(\Omega)$$

by defining $\tilde{T}(U) = (V,F)$, where (V,F) is the $\sim$ equivalence class of any $(Tu, u|_{y=0})$, with u in the $\sim_T$ equivalence class of U. In this way the commutative diagram follows

$$
(8.39) \qquad \begin{array}{ccc}
\mathcal{C}^1_{nde}(\Omega) & \xrightarrow{\quad \bar{T} \quad} & \mathcal{C}^0_{nde}(\Omega) \times \mathcal{C}^1_e(-a,a) \\
\downarrow & & \downarrow \\
X_T(\Omega) & \xrightarrow[\tilde{T}]{\quad inj \quad} & X_0(\Omega)
\end{array}
$$

where $\tilde{T}$ is injective, while the vertical arrows are the respective canonical mappings onto quotient spaces.

Finally, similar to (4.7), we define the partial order $\leq_T$ on $X_T(\Omega)$ by, see (8.34)

$$
(8.40) \qquad U \leq_T V \Longleftrightarrow \tilde{T}U \leq \tilde{T}V .
$$

We note that, again, as in Sections 2 - 5, in particular (4.7), it is the order structure $\leq$ on the range space $X_0(\Omega)$ of the operator $\tilde{T}$ which defines the order structure $\leq_T$ on the domain $X_T(\Omega)$ of that operator.

The immediate effect of this construction is given in

Lemma 8.2

The mapping

$$
(8.41) \qquad \tilde{T} : X_T(\Omega) \longrightarrow X_0(\Omega)
$$

is an order isomorphical embedding. ■

Coming back to $X_0(\Omega)$, we note the following embedding result which is similar to that in Lemma 3.1. Let us consider on $\mathcal{C}^0(\Omega) \times \mathcal{C}^1(-a,a)$ the partial order $\leq$ given by

$$(8.42) \qquad (u,f) \le (v,g) \iff \begin{bmatrix} 1) & u \le v & \text{on} & \Omega \\ 2) & f = g & \text{on} & (-a,a) \end{bmatrix}$$

then we obtain

Lemma 8.3

The embedding

$$(8.43) \qquad \begin{array}{ccc} \mathcal{C}^0(\Omega) \times \mathcal{C}^1(-a,a) & \xrightarrow{\quad i \quad} & X_0(\Omega) \\ (u,f) & \longmapsto & (U,F) \end{array}$$

is an order isomorphical embedding, where (U,F) is the $\sim$ equivalence class of $(u,f) \in \mathcal{C}^0(\Omega) \times \mathcal{C}^1(-a,a) \subseteq \mathcal{C}^0_{\text{nde}}(\Omega) \times \mathcal{C}^1_{\text{e}}(-a,a)$. ■

In view of the fact that we shall apply to $X_0(\Omega)$ the order completion method in the Appendix, it is useful to note that $(X_0(\Omega),\le)$ has no maximum or minimum elements. Similarly, we have

Lemma 8.4

$(X_T(\Omega),\le_T)$ is without maximum or minimum elements.

Proof

Let us take any $U \in X_T(\Omega)$. Then obviously $\tilde{T}U = (V,F) \in X_0(\Omega)$. Let $u \in U$, as well as $v \in V$ and $f \in F$. Then there exists $\Gamma \subset \Omega$ closed, nowhere dense, such that $\Gamma_e = \Gamma \cap ((-a,a) \times \{0\})$ is at most a finite set of points, while $u \in \mathcal{C}^1(\Omega\setminus\Gamma)$, $v \in \mathcal{C}^0(\Omega\setminus\Gamma)$ and $f \in \mathcal{C}^1((-a,a)\setminus\Gamma_e)$. Let us consider the nonvoid, open set

$$\Delta = \Omega\setminus(\Gamma \cup ((-a,a) \times \{0\}))$$

which is in fact a finite collection of nonvoid, open rectangles

$$(8.44) \qquad D_1,\dots,D_k \ .$$

Then (8.39) gives

$$(8.45) \qquad Tu = v \quad \text{on} \quad \Delta \ .$$

Let us consider an open rectangle D whose closure $\bar{D}$ is still contained in Δ, that is, in one of the open rectangles $D_1,\dots,D_k$. Then obviously

$$(8.46) \qquad -\infty < m = \inf \{v(x,y) \mid (x,y) \in \bar{D}\} \ .$$

Let us now apply (8.5) to the operator $T_1 = T - m + 1$. Provided that D is small enough, we can find $u' \in \mathcal{C}^1(\Omega)$, such that

$$(8.47) \qquad Tu' \leq m - 1 \quad \text{on} \quad D.$$

By adding to Γ four lines, two parallel with the x-axis, and two with the y-axis, we can obtain D as one of the rectangles in (8.44). Let us denote by Γ_1 this augmented Γ and let us define $u_1 \in \mathcal{C}^1(\Omega \setminus \Gamma_1)$ by

$$(8.48) \qquad u_1 = \left| \begin{array}{ll} u & \text{on} \quad \Omega \setminus (\Gamma_1 \cup D) \\ u' & \text{on} \quad D \end{array} \right.$$

Then (8.45) - (8.48) give

$$Tu_1 < Tu \quad \text{on} \quad D$$

therefore, if $U_1 \in X_T(\Omega)$ is the $\sim_T$ equivalence class of u_1, then we obtain in $X_0(\Omega)$ the strict inequality

$$\tilde{T}U_1 \leq \tilde{T}U, \quad \tilde{T}U_1 \neq \tilde{T}U$$

which in view of (8.40) means that

$$U_1 \leq_T U, \quad U_1 \neq U.$$

■

We are now in a position to apply the order completion method in Appendix to the spaces $(X_T(\Omega), \leq_T)$ and $(X_0(\Omega), \leq)$. Then (8.41) and (A.40) will yield the commutative diagram

$$\text{(8.49)} \quad \begin{array}{ccc} X_T(\Omega) \ni U & \longmapsto & \tilde{T}U \in X_0(\Omega) \\ \downarrow & & \downarrow \\ \hat{X}_T(\Omega) \ni \langle U] & \longmapsto & \hat{T}U = \langle \tilde{T}U] \in \hat{X}_0(\Omega) \end{array}$$

and the mapping

$$\text{(8.50)} \qquad \hat{T} : \hat{X}_T(\Omega) \longrightarrow \hat{X}_0(\Omega)$$

is again an order isomorphical embedding. We note that in (8.49) and (8.50), the rigorous way of notation would require to write $\hat{\tilde{T}}$ instead of simply $\hat{T}$. However, for reasons of convenience, we shall use the latter notation throughout the rest of this Section.

Before presenting in Theorem 8.1 the result on the existence of generalized solutions for the nonlinear Cauchy problem (8.3), (8.4), we need some technical lemmas. First a lemma which is similar to Lemma 5.1.

Lemma 8.5

For every $f \in \mathcal{C}^1[-a,a]$ we have in $X_0(\Omega)$ the relations

$$\text{(8.51)} \qquad i(0,f) = \sup \left\{ i(u,f) \left| \begin{array}{ll} *) & u \in \mathcal{C}^0(\bar{\Omega}) \\ **) & \forall \ (x,y) \in \bar{\Omega} : \\ & \quad u(x,y) < 0 \end{array} \right. \right\} =$$

$$= \inf \left\{ i(v,f) \left| \begin{array}{ll} *) & v \in \mathcal{C}^0(\bar{\Omega}) \\ **) & \forall \ (x,y) \in \bar{\Omega} : \\ & \quad 0 < v(x,y) \end{array} \right. \right\}$$

Proof

Let us start with the relation

$$(\cdot) \qquad i(0,f) = \sup \ldots$$

If $u \in \mathcal{C}^0(\bar{\Omega})$ and $u < 0$ on $\bar{\Omega}$ then (8.34) obviously implies that

$$i(u,f) \leq i(0,f)$$

therefore the inequality '$\geq$' in the above relation $(\cdot)$ holds. For the converse inequality, let us take $(V,G) \in X_0(\Omega)$ such that

$$(8.52) \qquad i(u,f) \leq (V,G)$$

for all $u \in \mathcal{C}^0(\bar{\Omega})$, for which

$$(8.53) \qquad u(x,y) < 0, \quad (x,y) \in \bar{\Omega}\ .$$

Let $v \in V$ and $g \in G$, then there exists a closed, nowhere dense subset $\Gamma \subset \Omega$, such that

$$v \in \mathcal{C}^0(\Omega \backslash \Gamma), \quad g \in \mathcal{C}^1((-a,a) \backslash \Gamma_e)$$

where

$$\Gamma_e = \Gamma \cap ((-a,a) \times \{0\})$$

Further, for all u satisfying (8.53), we have by (8.52) that

$$(8.54) \qquad u \leq v \quad \text{on} \quad \Omega \backslash \Gamma$$

$$(8.55) \qquad f = g \quad \text{on} \quad (-a,a) \backslash \Gamma_e$$

if we take into account (8.34) and the continuity of u and f. In particular, the above holds for the constant $u = -\epsilon$, with arbitrary $\epsilon > 0$. Therefore

$$0 \leq v \quad \text{on} \quad \Omega \backslash \Gamma$$

which together with (8.55) and (8.34) result in

$$i(0,f) \leq (V,G) \ .$$

The proof of the relation

$$i(0,f) = \inf \ \ldots$$

in (8.51) is similar. ■

Lemma 8.6

For every $f \in \mathcal{C}^1[-a,a]$ we have in $\hat{X}_0(\Omega)$ the relations

$$(8.56) \qquad \begin{aligned} <i(0,f)] &= \sup \left\{ \hat{T}A \ \middle| \ \begin{array}{ll} *) & A \in \hat{X}_T(\Omega) \\ **) & \hat{T}A \subseteq <i(0,f)] \end{array} \right\} = \\ &= \inf \left\{ \hat{T}B \ \middle| \ \begin{array}{ll} *) & B \in \hat{X}_T(\Omega) \\ **) & <i(0,f)] \subseteq \hat{T}B \end{array} \right\} \end{aligned}$$

Proof

By Proposition 8.1, as well as (8.37), (8.39) and (8.34), for $\epsilon > 0$, there exists $U_\epsilon \in X_T(\Omega)$, such that, in $X_0(\Omega)$, the inequality holds

$$\tilde{T}U_\epsilon \leq i(0,f)$$

therefore, in $\hat{X}_0(\Omega)$, we obtain

$$\hat{T}<U_\epsilon] \subseteq <i(0,f)]$$

which means that in $\hat{X}_0(\Omega)$, the relation holds

$$(8.57) \qquad \hat{T}\langle U_\epsilon] \subseteq \sup \left\{ \hat{T}A \;\middle|\; \begin{array}{ll} *) & A \in \hat{X}_T(\Omega) \\ **) & \hat{T}A \subseteq \langle i(0,f)] \end{array} \right\} \subseteq \langle i(0,f)]$$

since we can take $A = \langle U_\epsilon] \in \hat{X}_T(\Omega)$.

Let us now take $u \in \mathcal{C}^0(\bar{\Omega})$, such that

$$u(x,y) < 0, \quad (x,y) \in \bar{\Omega}$$

then there exists $\epsilon > 0$, with the property

$$(8.58) \qquad u(x,y) \leq -\epsilon < 0, \quad (x,y) \in \bar{\Omega} .$$

Using again Proposition 8.1, we can take a finite adapted grid $\Gamma_\epsilon \subset \Omega$ and $u_\epsilon \in \mathcal{C}^1(\Omega \backslash \Gamma_\epsilon)$, such that

$$(8.59) \qquad -\epsilon \leq T\, u_\epsilon \leq 0 \quad \text{on} \quad \Omega \backslash \Gamma_\epsilon$$

$$(8.60) \qquad u_\epsilon(x,0) = f(x), \quad (x,0) \in ((-a,a) \times \{0\}) \backslash \Gamma_\epsilon .$$

Let $U_\epsilon \in X_T(\Omega)$ be the $\sim_T$ equivalence class of u_ϵ. Then (8.34) and (8.57) - (8.60) give in $X_0(\Omega)$ the inequality

$$i(u,f) \leq \tilde{T}U_\epsilon \leq i(0,f)$$

which means that in $\hat{X}_0(\Omega)$ we have

$$\langle i(u,f)] \subseteq \hat{T}\langle U_\epsilon] \subseteq \langle i(0,f)] .$$

Now (8.51) will result in the first equality in (8.56), namely

$$\langle i(0,f)] = \sup \left\{ \hat{T}A \;\middle|\; \begin{array}{ll} *) & A \in \hat{X}_T(\Omega) \\ **) & \hat{T}A \subseteq \langle i(0,f)] \end{array} \right\}$$

The second equality in (8.56) follows in a similar way. ■

Now we come to the basic existence and uniqueness result concerning the solutions of the nonlinear Cauchy problem (8.3), (8.4). This result can be seen as a specialization of the general result in Theorem 5.1, since in that latter theorem initial value conditions have not been considered.

Theorem 8.1

Given $f \in \mathcal{C}^1[-a,a]$, there exists a unique solution $A \in \hat{X}_T(\Omega)$ such that

(8.61) $\quad \hat{T}A = \langle i(0,f)]$.

Proof

It follows from (8.56) and the abstract existence result in Theorem 9.1.

■

Remark 8.3

1) In view of (8.36) - (8.40) and (8.49), the meaning of (8.61) is the following. The unique generalized solution

$$A \in \hat{X}_T(\Omega)$$

is the set of all $\sim_T$ equivalence classes $U \in X_T(\Omega)$, where each $u \in U$ has the properties

(8.62) $\quad u \in \mathcal{C}^1(\Omega \backslash \Gamma)$

(8.63) $\quad Tu \leq 0$ on $\Omega \backslash \Gamma$

(8.64) $\quad u(x,0) = f(x), \quad (x,0) \in ((-a,a) \times \{0\}) \backslash \Gamma$

for a certain closed, nowhere dense subset $\Gamma \subset \Omega$ which may depend on u, and which is such that $\Gamma \cap ((-a,a) \times \{0\})$ is at most a finite set of points. In this way, owing to (8.63), the unique solution A contains as well all the classical *subsolutions* of (8.3), which however must satisfy the initial value condition (8.4).

2) The *uniqueness* of the solution A in (8.61) is a simple consequence of the fact that A collects all solutions and subsolutions (8.62) - (8.64).

3) Every usual *shock wave* solution of (8.3), (8.4) gives rise directly to a generalized solution in (8.61). Indeed, if

$$(8.65) \quad \begin{aligned} &\Gamma \subset \Omega \text{ is closed, nowhere dense} \\ &\Gamma \cap ((-a,a) \times \{0\}) \text{ is a finite set of points} \\ &u \in \mathcal{C}^1(\Omega \setminus \Gamma) \end{aligned}$$

and

$$(8.66) \quad \begin{aligned} &Tu = 0 \quad \text{on} \quad \Omega\setminus\Gamma \\ &u(x,0) = f(x), \quad (x,0) \in ((-a,a) \times \{0\})\setminus\Gamma \end{aligned}$$

then letting $U \in X_T(\Omega)$, be the $\sim_T$ equivalence class of u, we obtain

$$(8.67) \quad A = \langle U] \in \hat{X}_T(\Omega)$$

which satisfies (8.61).

4) The above *consistency* result between usual shock wave solutions and the generalized solutions in (8.61) could not be obtained without allowing in (8.29) for the singularity sets Γ which are more general than finite initial adapted grids.

5) Notwithstanding the previous remark, one can note that in the proof of Theorem 8.1 we only used the approximation result in Proposition 8.1. And this result only requires finite initial adapted grids. It follows that the result in Theorem 8.1 on the existence of generalized solutions for nonlinear Cauchy problems (8.3), (8.4) could be obtained even if in (8.29) we would restrict the singularity sets Γ to finite initial adapted grids.

The difference in this case would be that we would no longer have the simple and direct way in (8.65) - (8.67) to obtain the above consistency result. However this need not mean that in such a case the usual shock wave solutions, with arbitrary shocks not necessarily on finite initial adapted grids, would no longer be obtainable by Theorem 8.1. Indeed, what happens in this case is that for general shock wave solutions A corresponding to (8.65), (8.66), the *simple representation* in (8.67) may in general fail. Instead of it we would in general only have the less detailed representation given in (8.72) below, and obtained as follows. For $\ell \in \bar{\mathbb{N}}$, let us denote, see (8.29)

$$(8.68) \qquad \mathcal{C}^{\ell}_{fia}(\Omega) = \left\{ u \;\middle|\; \begin{array}{l} \exists\, \Gamma \subset \Omega \text{ finite initial adapted grid :} \\ \quad *)\ \ u : \Omega \setminus \Gamma \longrightarrow \mathbb{R}^n \\ \ **)\ \ u \in \mathcal{C}^{\ell}(\Omega \setminus \Gamma) \end{array} \right\}$$

then obviously, see (8.29)

$$(8.69) \qquad \mathcal{C}^{\ell}_{fia}(\Omega) \underset{\neq}{\subset} \mathcal{C}^{\ell}_{nde}(\Omega)$$

therefore, the mapping $\bar{T}$ in (8.31) can be restricted to $\mathcal{C}^{1}_{fia}(\Omega)$. In this way, we can restrict the equivalence relation $\sim_T$ in (8.36) to $\mathcal{C}^{1}_{fia}(\Omega)$ and define the quotient space, see (8.37)

$$(8.70) \qquad Y_T(\Omega) = \mathcal{C}^{1}_{fia}(\Omega)/\sim_T$$

and obtain

$$(8.71) \qquad Y_T(\Omega) \subsetneq X_T(\Omega) \ .$$

Clearly, the relations (8.38) - (8.41), (8.49), (8.50) and (8.56), as well as Lemma 8.4 will remain valid if we replace $X_T(\Omega)$ with $Y_T(\Omega)$ and restrict $\bar{T}$ to that latter space. And then, instead of (8.67), we shall only obtain the representation

$$(8.72) \qquad A \in \hat{Y}_T(\Omega)$$

which means that $A \subsetneq Y_T(\Omega)$ and A is a cut in $(Y_T(\Omega), \leq_T)$, *without* having in general the specific and convenient form

$$(8.73) \quad A = \langle U]$$

for a certain $U \in Y_T(\Omega)$. This fact that, even within the restricted framework of (8.68) - (8.71) we can still obtain (8.72), shows the power of the existence result in Theorem 8.1. The question whether within (8.68) - (8.71) one can obtain (8.73) as well is open. It appears that a positive answer to this question may have applicative, in particular numerical, significance related to the approximation of general shock wave solutions by solutions or subsolutions with a much simpler shock structure.

6) It is obvious that the results in this Section, including Theorem 8.1, can be extended in a straightforward way to equations with *any finite* number of variables.

7) The power of Theorem 8.1 is shown, among others, by the fact that it gives solutions to *distributionally unsolvable* equations as well. For example, in Grushin is shown that for certain $f + ig \in \mathcal{C}^\infty(\mathbb{R}^2)$ the equation

$$D_y(u + iv) + iyD_x(u + iv) = f + ig \quad \text{on} \quad \mathbb{R}^2$$

fails to have distributional solutions in any neighbourhood of $(0,0) \in \mathbb{R}^2$. If we rewrite this equation as a system

$$D_y u - yD_x v - f = 0$$
$$D_y v + yD_x u - g = 0$$

we can apply to it Theorem 8.1 and obtain a unique solution on $\mathbb{R}^2$, for any given initial value condition. A similar remark applies to the well known equation in Lewy

$$D_x U + iD_y U - 2i(x + iy)D_z U = f \quad \text{on} \quad \mathbb{R}^3$$

which for a large class of $f \in C^\infty(\mathbb{R}^3)$, is distributionally unsolvable in any neighbourhood of any point in $\mathbb{R}^3$.

9. AN ABSTRACT EXISTENCE RESULT

In order to better understand the underlying order completion related structure of results concerning the existence of solutions of nonlinear PDEs, such as those in Theorems 5.1, 5.2, 5.3, 8.1, Corollary 5.1, or those in subsequent Sections of Part II, we present here a rather general scheme for solving equations - not necessarily PDEs - in partially ordered sets.

Let X be an arbitrary set and let $(Y,\leq)$ be a partially ordered set without maximum or minimum. Given an arbitrary mapping

$$(9.1) \qquad T : X \longrightarrow Y$$

we shall be interested in finding a solution A for the equation

$$(9.2) \qquad T(A) = F$$

with given $F \in Y$. Ideally, we could have a solution $A \in X$, however, as seen for instance in the case of mappings T corresponding to linear or nonlinear PDEs and with X being a space of usual functions, this may easily fail to happen. Therefore, we should expect that a certain *extension*, or in particular, *completion* process involving X, Y and T may have to be introduced in order to be able to find a solution for the equation (9.2). In the sequel, we present a rather abstract way for such an extension, based on order completion.

Let us start by defining the equivalence relation $\sim_T$ on X, by

$$(9.3) \qquad u \sim_T v \iff T(u) = T(v) \in Y$$

whenever $u,v \in X$. Then we introduce the quotient space

$$(9.4) \qquad X_T = X/\sim_T \, .$$

The mapping (9.1) will canonically lead to an *injective* mapping, denoted for simplicity again by T, according to

$$(9.5) \qquad \begin{array}{rcl} T : X_T & \xrightarrow{\text{inj}} & Y \\ U & \longmapsto & T(u) \end{array}$$

where $u \in U$, that is, U is the $\sim_T$ equivalence class in X_T of $u \in X$, while $T(u) \in Y$ is defined by (9.1).

Finally, we define a partial order $\leq_T$ on X_T by a 'pull-back' of the partial order $(Y,\leq)$ through the mapping T in (9.5). In other words, given $U,V \in X_T$, we define

$$(9.6) \qquad U \leq_T V \Longleftrightarrow T(U) \leq T(V) \quad \text{in} \quad Y \ .$$

In view of (9.5) and (9.6), we obtain the *order isomorphical embedding*

$$(9.7) \qquad X_T \xrightarrow{\ T\ } Y \ .$$

For the rest of this Section, we shall make the assumption

(9.8) $(X_T,\leq_T)$ has no maximum or minimum.

In view of the above we can consider the order completions $\hat{X}_T$ and $\hat{Y}$, according to the Appendix. Moreover, according to Proposition A.1, we obtain the following commutative diagram of order isomorphical embeddings

$$(9.9) \qquad \begin{array}{ccc} X_T & \xrightarrow{\ T\ } & Y \\ \downarrow & & \downarrow \\ \hat{X}_T & \xrightarrow[\ \hat{T}\ ]{} & \hat{Y} \end{array}$$

In particular, for $U \in X_T$ and $A \in \hat{X}_T$ we have in $\hat{Y}$

$$(9.10) \qquad \hat{T}<U] = <TU]$$

$$(9.11) \qquad \hat{T}A = (TA)^{u\ell} = \sup \{<TU] \mid U \in A\} .$$

Returning to the problem of solving the equation (9.2), we reformulate it as follows: given $F \in \hat{Y}$, find a necessary and sufficient condition for the existence of $A \in \hat{X}_T$, such that

$$(9.12) \qquad \hat{T}A = F \quad \text{in} \quad \hat{Y} .$$

First we note that, given $F \in \hat{Y}$, (9.9) implies the inclusions

$$(9.13) \qquad \begin{aligned} &\sup_{\hat{Y}} \left\{\hat{T}U \;\middle|\; \begin{array}{ll} *) & U \in \hat{X}_T \\ **) & \hat{T}U \subseteq F \end{array}\right\} \subseteq \hat{T} \sup_{\hat{X}_T} \left\{U \;\middle|\; \begin{array}{ll} *) & U \in \hat{X}_T \\ **) & \hat{T}U \subseteq F \end{array}\right\} \subseteq \\ &\subseteq \hat{T} \inf_{\hat{X}_T} \left\{V \;\middle|\; \begin{array}{ll} *) & V \in \hat{X}_T \\ **) & F \subseteq \hat{T}V \end{array}\right\} \subseteq \inf_{\hat{Y}} \left\{\hat{T}V \;\middle|\; \begin{array}{ll} *) & V \in \hat{X}_T \\ **) & F \subseteq \hat{T}V \end{array}\right\} \end{aligned}$$

as well as

$$(9.14) \qquad \sup_{\hat{Y}} \left\{\hat{T}U \;\middle|\; \begin{array}{ll} *) & U \in \hat{X}_T \\ **) & \hat{T}U \subseteq F \end{array}\right\} \subseteq F \subseteq \inf_{\hat{Y}} \left\{\hat{T}V \;\middle|\; \begin{array}{ll} *) & V \in \hat{X}_T \\ **) & F \subseteq \hat{T}V \end{array}\right\} .$$

Moreover

$$(9.15) \qquad \{U \in \hat{X}_T \mid \hat{T}U \subseteq F\} \neq \phi$$

since $U = \phi \in \hat{X}_T$ and according to (9.11), we obtain then $\hat{T}U = \hat{T}\phi = \phi \subseteq F$.

In spite of the above useful properties in (9.9) and (9.13)-(9.15), the solution of the equation (9.12) is nevertheless nontrivial. First of all, in the order isomorphical embedding in (9.9), namely

$$\hat{X}_T \xrightarrow{\quad \hat{T} \quad} \hat{Y}$$

the mapping $\hat{T}$ need *not* be surjective, furthermore, this mapping will in general *not* preserve infima or suprema, see in this respect Example 9.1 at the end of this Section. Fortunately however, equation (9.12) can be solved under rather general conditions, as presented next.

The *abstract existence* result, which solves the equation in (9.12) is presented now.

Theorem 9.1

Given $F \in \hat{Y}$, the equation

$$(9.16) \qquad \hat{T}A = F$$

has a solution $A \in \hat{X}_T$, if and only if

$$(9.17) \qquad \sup_{\hat{Y}} \left\{ \hat{T}U \;\middle|\; \begin{array}{ll} *) & U \in \hat{X}_T \\ **) & \hat{T}U \subseteq F \end{array} \right\} = \inf_{\hat{Y}} \left\{ \hat{T}V \;\middle|\; \begin{array}{ll} *) & V \in \hat{X}_T \\ **) & F \subseteq \hat{T}V \end{array} \right\}$$

and, in view of (9.7), this solution is unique, whenever it exists.

In that case, the unique solution $A \in \hat{X}_T$ is given by

$$(9.18) \qquad A = \sup_{\hat{X}_T} \{U \in \hat{X}_T \mid \hat{T}U \subseteq F\} = \inf_{\hat{X}_T} \{V \in \hat{X}_T \mid F \subseteq \hat{T}V\}$$

moreover, see also (9.15)

$$(9.19) \qquad \{V \in \hat{X}_T \mid F \subseteq \hat{T}V\} \neq \phi.$$

Proof

Assume (9.16), then obviously

$$F = \hat{T}A \subseteq \sup_{\hat{Y}} \left\{ \hat{T}U \;\middle|\; \begin{array}{ll} *) & U \in \hat{X}_T \\ **) & \hat{T}U \subseteq F \end{array} \right\}$$

$$\inf_{\hat{Y}} \left\{ \hat{T}V \;\middle|\; \begin{array}{ll} *) & V \in \hat{X}_T \\ **) & F \subseteq \hat{T}V \end{array} \right\} \subseteq \hat{T}A = F$$

therefore (9.13) collapses into the chain of seven equalities

$$F = \hat{T}A = \sup_{\hat{Y}} \ldots = \hat{T} \sup_{\hat{X}_T} \ldots = \hat{T} \inf_{\hat{X}_T} \ldots = \inf_{\hat{Y}} \ldots = \hat{T}A = F .$$

In particular, we obtain (9.17). Further, the injectivity of $\hat{T}$ will give (9.18). Finally, (9.19) is obvious, since we can take $V = A$.

Conversely, let us assume (9.17). Then (9.13) collapses into the chain of three equalities

$$\sup_{\hat{Y}} \ldots = \hat{T} \sup_{\hat{X}_T} \ldots = \hat{T} \inf_{\hat{X}_T} \ldots = \inf_{\hat{Y}} \ldots .$$

In view of (9.14), we can add the equality on the right

$$\sup_{\hat{Y}} \ldots = \hat{T} \sup_{\hat{X}_T} \ldots = \hat{T} \inf_{\hat{X}_T} \ldots = \inf_{\hat{Y}} \ldots = F .$$

Therefore the injectivity of $\hat{T}$ will give (9.16) and (9.18). As above, (9.19) will also follow. ■

Remark 9.1

1) The essence of the solution method for (9.12), in particular, of the proof of Theorem 9.1, can be summarized as follows. Given $F \in \hat{Y}$, let us associate with it the subset of X_T defined by

(9.20) $\quad A = \{U \in X_T \mid \langle TU] \subseteq F\} .$

Now we note that, according to (A.36) and (A.37), the mapping $\hat{T}$ is defined not only on $\hat{X}_T$, but also on the larger set $\mathcal{P}(X_T)$. In particular, we can apply $\hat{T}$ to A in (9.20), and obtain

(9.21) $$\hat{T}A = \sup_{\hat{Y}} \{<TU] \mid U \in A\} .$$

Therefore, the equality

(9.22) $$\hat{T}A = F$$

holds in $\hat{Y}$, if and only if

(9.23) $$\sup_{\hat{Y}} \left\{<TU] \;\middle|\; \begin{array}{ll} *) & U \in X_T \\ **) & <TU] \subseteq F \end{array} \right\} = F .$$

And we note that in (9.23), the inclusion '$\subseteq$' always holds.

However, even if (9.22) - or equivalently (9.23) - holds, this does not yet give a solution to our problem (9.12), unless we show that we also have

(9.24) $$A \in \hat{X}_T$$

that is, A is not only a subset of X_T, but it is as well a *cut* in X_T, see (A.11) and (A.12). This is then the way we are led to condition (9.17).

2) We shall note that condition (9.1) expresses a *local* property in terms of Y, since it is related to the given, fixed $F \in \hat{Y}$. In other words, if it holds for a certain $F \in \hat{Y}$, that need not imply that it will hold for every other $G \in \hat{Y}$. In fact, in Theorem 9.2 below, it is precisely this *global* problem in terms of Y which is being addressed.

3) Here, as a consequence of the above, it is useful to note the following. The existence results in Theorems 5.1, 5.2 or 8.1 as well as in Corollary 5.1 are *not* merely pure and simple order theoretical results. Indeed, on the one hand, they can be included in the general scheme of

solving equations of type (9.12). On the other hand, the ways the respective necessary and sufficient conditions (9.17) or (9.23) which make this general scheme work in the case of the Theorems 5.1, 5.2, 8.1 and Corollary 5.1, do happen to essentially involve *additional specific arguments* intimately connected with properties of the continuous PDEs, linear or nonlinear, single equations or systems, under consideration. Here we refer in particular to such properties as those in (2.5), (2.8), (2.15), (2.22), (5.18), (5.37), (5.45), (8.23), (8.51) or (8.56). This is in contradistinction with the way Freudenthal's 'spectral theorem' is applied, for instance. Indeed, this latter theorem is a purely order theoretical result, both in its formulation and proof. Nevertheless, it happens to imply directly the spectral theory of Hilbert space operators or the Radon-Nikodym theorem in measure theory.

A second *abstract existence* result, this time related to the *global solvability* of the equation in (9.12), that is, to the set of all right hand terms $F \in Y$ for which it has a solution $A \in \hat{X}_T$, is given in

Theorem 9.2

The following are equivalent

(9.25) $\quad \hat{T}(\hat{X}_T) \supseteq Y$

(9.26) $\quad \hat{T}(\hat{X}_T) = \hat{Y}$.

In each of these cases $\hat{T}$ is an order isomorphism between $\hat{X}_T$ and $\hat{Y}$.

Proof

Obviously (9.26) implies (9.25). Let us then assume (9.25). It follows that

$$\begin{aligned} & \forall\ G \in Y : \\ (9.27) \quad & \exists\ B \in \hat{X}_T : \\ & \quad \hat{T}B = G \end{aligned}$$

Let us take now $F \in \hat{Y}$, then (A.23) gives in $\hat{Y}$ the relation

$$F = \sup \left\{ <G] \;\middle|\; \begin{array}{ll} *) & G \in Y \\ **) & <G] \subseteq F \end{array} \right\} .$$

But owing to (9.27), we obviously have in $\hat{Y}$ the inclusion

$$\sup \left\{ <G] \;\middle|\; \begin{array}{ll} *) & G \in Y \\ **) & <G] \subseteq F \end{array} \right\} \subseteq \sup \left\{ \hat{T}B \;\middle|\; \begin{array}{ll} *) & B \in \hat{X}_T \\ **) & T\hat{B} \subseteq F \end{array} \right\} .$$

On the other hand, it is clear that in $\hat{Y}$ we have the inclusion

$$\sup \left\{ \hat{T}B \;\middle|\; \begin{array}{ll} *) & B \in \hat{X}_T \\ **) & \hat{T}B \subseteq F \end{array} \right\} \subseteq \inf \left\{ \hat{T}C \;\middle|\; \begin{array}{ll} *) & C \in \hat{X}_T \\ **) & F \subseteq \hat{T}C \end{array} \right\} .$$

Further, in view of (9.27), we obtain in $\hat{Y}$ the inclusion

$$\inf \left\{ \hat{T}C \;\middle|\; \begin{array}{ll} *) & C \in \hat{X}_T \\ **) & F \subseteq \hat{T}C \end{array} \right\} \subseteq \inf \left\{ <H] \;\middle|\; \begin{array}{ll} *) & H \in Y \\ **) & F \subseteq <H] \end{array} \right\} .$$

Finally, (A.23) gives in $\hat{Y}$ the relation

$$\inf \left\{ <H] \;\middle|\; \begin{array}{ll} *) & H \in Y \\ **) & F \subseteq <H] \end{array} \right\} = F .$$

Now, the above relations together will yield (9.17). Therefore Theorem 9.1 gives $A \in \hat{X}_T$ such that $F = \hat{T}A$. In this way the proof of (9.26) is completed. Then, in view of (9.9), it is obvious that $\hat{T}$ is an order isomorphism between $\hat{X}_T$ and $\hat{Y}$. ■

The nontriviality of Theorem 9.1 is shown, among others by

Example 9.1

Let us take (9.1) in the particular case of

$$X = \mathcal{C}^0[0,1], \quad (Y,\leq) = (\mathcal{B}[0,1],\leq)$$

where $\mathcal{B}[0,1]$ is the set of all the bounded and real valued functions on $[0,1]$, while $\leq$ is the usual partial order among real valued functions. Further, let us assume that

$$T : X \longrightarrow Y$$

is the inclusion mapping between $\mathcal{C}^0[0,1]$ and $\mathcal{B}[0,1]$.

Then (9.3), (9.4) will clearly give

$$X_T = X$$

while in view of (9.6), it follows that the partial order $\leq_T$ on $X_T = X$ is in fact the restriction to $\mathcal{C}^0[0,1]$ of the partial order $\leq$ on $\mathcal{B}[0,1]$. In this way, the condition (9.8) is obviously satisfied.

Now, in view of the fact that in our case $(Y,\leq)$ is Dedekind order complete, it follows that

$$\hat{Y} = Y \cup \{\inf Y, \sup Y\}.$$

Therefore, the order isomorphical embedding, see (9.9)

$$\hat{X}_T = \hat{X} \xrightarrow{\hat{T}} \hat{Y}$$

will in our case *not* be surjective, since

$$\mathcal{C}^0[0,1]^{\hat{}*} \subseteq \text{Mes}\,[0,1] \underset{\neq}{\subset} \mathcal{B}[0,1] .$$

Furthermore, our particular $\hat{T}$ will *not* preserve infima or suprema.

We shall complete this Section by first presenting two dual results which further detail the abstract existence and uniqueness result in Theorem 9.1.

Proposition 9.1

Given $F \in \hat{Y}$, the equation

$$(9.28) \qquad \hat{T}A = F$$

has a solution $A \in \hat{X}_T$, if and only if

$$(9.29) \qquad \sup_{\hat{Y}} \left\{ \hat{T}U \;\middle|\; \begin{array}{ll} *) & U \in \hat{X}_T \\ **) & \hat{T}U \subseteq F \end{array} \right\} = F$$

as well as

$$(9.30) \qquad \sup_{\hat{Y}} \left\{ \hat{T}U \;\middle|\; \begin{array}{ll} *) & U \in \hat{X}_T \\ **) & \hat{T}U \subseteq F \end{array} \right\} = \hat{T} \sup_{\hat{X}_T} \left\{ U \;\middle|\; \begin{array}{ll} *) & U \in \hat{X}_T \\ **) & \hat{T}U \subseteq F \end{array} \right\}$$

Proof

It is clear that (9.29) and (9.30) imply (9.28), since we can take

$$A = \sup_{\hat{X}_T} \left\{ U \;\middle|\; \begin{array}{ll} *) & U \in \hat{X}_T \\ **) & \hat{T}U \subseteq F \end{array} \right\}$$

For the converse implication we note that, according to the proof of Theorem 9.1, the chain of inclusions in (9.13) collapses into seven equalities from which (9.29) and (9.30) follow easily. ■

In a similar manner, we can obtain

Proposition 9.2

Given $F \in \hat{Y}$, the equation

$$(9.31) \qquad \hat{T}A = F$$

has a solution $A \in \hat{X}_T$, if and only if

$$(9.32) \qquad \inf_{\hat{Y}} \left\{ \hat{T}V \;\middle|\; \begin{matrix} *) & V \in \hat{X}_T \\ **) & F \subseteq \hat{T}V \end{matrix} \right\} = F$$

as well as

$$(9.33) \qquad \inf_{\hat{Y}} \left\{ \hat{T}V \;\middle|\; \begin{matrix} *) & V \in \hat{X}_T \\ **) & F \subseteq \hat{T}V \end{matrix} \right\} = \hat{T} \inf_{\hat{X}_T} \left\{ V \;\middle|\; \begin{matrix} *) & V \in \hat{X}_T \\ **) & F \subseteq \hat{T}V \end{matrix} \right\}$$

■

Finally (9.20) - (9.24) yield

Proposition 9.3

Given $F \in \hat{Y}$, if

$$(9.34) \qquad A = \{U \in X_T \mid \langle TU] \subseteq F\} \text{ is a nonvoid cut in } X_T$$

then

$$(9.35) \qquad \hat{T}A = F$$

if and only if

$$(9.36) \qquad \sup_{\hat{Y}} \left\{ \langle TU] \;\middle|\; \begin{matrix} *) & U \in X_T \\ **) & \langle TU] \subseteq F \end{matrix} \right\} = F$$

■

10. PDEs WITH SUFFICIENTLY MANY SMOOTH SOLUTIONS

This Section brings in certain PDE related, and thus more specific structures than those in Section 9, however, the resulting setup still shows that the underlying idea of the basic existence results in Theorems 5.1, 5.2 and Corollary 5.1 is rather abstract, and therefore, it is particularly general. Indeed, the setup in this Section can, among others, quite easily include very general *systems* of continuous nonlinear PDEs, as seen in Remark 10.1. These systems can, in fact, include *infinitely* many equations and unknown functions.

However, at least as important is the fact that the setup in this Section can naturally include *initial* and/or *boundary value* problems. Indeed, one example of (10.1) below we have already seen in (8.31), upon which the Cauchy problem for nonlinear first order systems in an arbitrary number of independent variables was solved. Further examples we shall encounter at the end of this Section.

One of the *major advantages* of the method of solving nonlinear PDEs through order completion is that, according to (A.79), the order completion of a Cartesian product, such as in (10.1), (10.2), is precisely the Cartesian product of order completions. This then offers a powerful *decoupling* effect when we go from T in (10.1) to $\hat{T}$ in (10.10), and this brings with it a significant simplification.

A few applications presented at the end of this Section can give an idea about the power of the abstract method presented here, and in particular, in Theorem 10.1.

First we start with a somewhat more abstract result which will then be applied to specific classes of PDEs.

Let Ω_λ, with $\lambda \in \Lambda$, be $\mathcal{C}^\infty$-smooth manifolds, where Λ is an arbitrary nonvoid set. Given a set X, let

$$(10.1) \qquad T : X \longrightarrow \prod_{\lambda \in \Lambda} \mathcal{C}^0(\Omega_\lambda)$$

be an arbitrary mapping. On each $\mathcal{C}^0(\Omega_\lambda)$, with $\lambda \in \Lambda$, we consider the usual partial order defined for real valued functions, see (3.10). Then, on

$$(10.2) \qquad Y = \prod_{\lambda \in \Lambda} \mathcal{C}^0(\Omega_\lambda)$$

we can consider the product order $\leq$ defined in (A.77). In this case we can apply the abstract scheme in (9.1), (9.3) - (9.7) and obtain the order isomorphical embedding

$$(10.3) \qquad T : X_T \longrightarrow Y$$

between $(X_T, \leq_T)$ and $(Y, \leq)$.

Our basic assumption concerning the mapping T in (10.1) is that it should have solutions for *smooth* right hand terms, that is

$$(10.4) \qquad \begin{array}{l} \forall \ f = (f_\lambda | \lambda \in \Lambda) \in \prod_{\lambda \in \Lambda} \mathcal{C}^\infty(\Omega_\lambda) : \\ \exists \ u \in X : \\ \quad Tu = f \end{array}$$

As seen in the sequel, this condition (10.4) is an extension of the well known P(D)-*convexity* condition, familiar in the case of linear constant coefficient partial differential operators $P(D)$.

Remark 10.1

The abstract scheme in (10.1), and thus in (10.2), (10.3) and (10.10), can easily contain *finite or infinite systems* of nonlinear continuous PDEs. Indeed, let us assume that

$$\Omega_\lambda = \Omega, \quad \lambda \in \Lambda$$

for a certain C^∞-smooth manifold Ω. In this case (10.2) becomes

$$Y = (\mathcal{C}^0(\Omega))^\Lambda \ .$$

Let M be an arbitrary index set and let

$$X = \prod_{\mu \in M} \mathcal{C}^{m_\mu}(\Omega)$$

where $m_\mu \in \mathbb{N}$, with $\mu \in M$. Finally, let

$$T : X \longrightarrow Y$$

be of the form

$$TU = (T_\lambda(U_\mu | \mu \in M) | \lambda \in \Lambda)$$

for $U = (U_\mu | \mu \in M) \in X$. Here we assume that, for each $\lambda \in M$, we have

$$T_\lambda : X \longrightarrow \mathcal{C}^0(\Omega)$$

and T_λ is a nonlinear continuous partial differential operator of order at most m_μ, on the functions $U_\mu \in \mathcal{C}^{m_\mu}(\Omega)$, with $\mu \in M$. It should be noted that both Λ and M can be arbitrary infinite sets.

In (2.2) we obviously have the most simple case when $\Lambda = M = \{1\}$. On the other hand, in (8.3) we have $\Lambda = M = \{1,\ldots,n\}$.

Lemma 10.1

$(X_T, \leq_T)$ is without maximum or minimum elements.

Proof

Let us assume that U is the smallest element in $(X_T, \leq_T)$ and let

$$(10.5) \qquad TU = f = (f_\lambda | \lambda \in \Lambda) \in \prod_{\lambda\in\Lambda} \mathcal{C}^0(\Omega_\lambda) \ .$$

Let $\lambda \in \Lambda$ be given, and let $(\alpha_i | i \in I)$ be a locally finite, compactly supported $\mathcal{C}^\infty$-smooth partition of unity on M_λ, De Rham. Let us take $c_i \in \mathbb{R}$, with $i \in I$, such that

$$c_i \leq f_\lambda - 1 \quad \text{on} \quad \text{supp } \alpha_i \ .$$

But obviously

$$g_\lambda = \sum_{i\in I} c_i \ \alpha_i \in \mathcal{C}^\infty(\Omega_\lambda)$$

and

$$(10.6) \qquad g_\lambda \leq f_\lambda - 1 \quad \text{on} \quad \Omega_\lambda \ .$$

It follows that

$$g = (g_\lambda | \lambda \in \Lambda) \in \prod_{\lambda\in\Lambda} \mathcal{C}^\infty(\Omega_\lambda)$$

therefore, according to (10.4), we have

$$(10.7) \qquad Tv = g$$

for a certain $v \in X$. Let $V \in X_T$ be the $\sim_T$ equivalence class of v. Then (10.7) gives

$$TV = g$$

therefore, according to (10.5) and (10.6), we obtain

$$(10.8) \qquad V \leq_T U \ .$$

However, (10.3) and (10.6) imply that

(10.9) $\quad V \neq U$.

Now (10.8) and (10.9) contradict the assumption that U is the smallest element in $(X_T, \leq_T)$. Similarly one can prove that $(X_T, \leq_T)$ is without a largest element.

∎

In view of (10.3), Lemma 10.1 and (9.9) as well as (A.79), we obtain the commutative diagram of order isomorphical embeddings

$$\text{(10.10)} \qquad \begin{array}{ccc} X_T & \xrightarrow{\quad T \quad} & \prod_{\lambda \in \Lambda} \mathcal{C}^0(\Omega_\lambda) \\ \downarrow & & \downarrow \\ \hat{X}_T^* & \xrightarrow[\quad \hat{T} \quad]{} & \prod_{\lambda \in \Lambda} \mathcal{C}^0(\Omega_\lambda)^{\hat{}\,*} \end{array}$$

Indeed (10.2), (10.3), Lemma 10.1, and (9.9) give the commutative diagram of order isomorphical embeddings

$$\text{(10.11)} \qquad \begin{array}{ccc} X_T & \xrightarrow{\quad T \quad} & Y = \prod_{\lambda \in \Lambda} \mathcal{C}^0(\Omega_\lambda) \\ \downarrow & & \downarrow \\ \hat{X}_T & \xrightarrow[\quad \hat{T} \quad]{} & \hat{Y} = \left(\prod_{\lambda \in \Lambda} \mathcal{C}^0(\Omega_\lambda) \right)^{\hat{}} \end{array}$$

However, in view of (A.36), in particular, of (9.11), it is obvious that

(10.12) $\quad \hat{T}(\min \hat{X}_T) = \min \hat{Y}$

since

$$\hat{T}\phi = \phi$$

Therefore (10.10) follows from (A.79) by restricting $\hat{T}$ in (10.11) to the order isomorphical embedding

$$(10.13) \qquad \hat{X}^*_T \xrightarrow{\ \hat{T}\ } \hat{Y}^*$$

A remarkable fact is that the assumption in (10.4) leads to the following *much stronger* result.

Theorem 10.1

Suppose that the manifolds Ω_λ, with $\lambda \in \Lambda$, admit locally finite smooth partitions of unity, Abraham et. al.

In case (10.4) holds then

$$(10.14) \qquad \begin{array}{l} \forall \;\; F \in \prod_{\lambda\in\Lambda} \mathcal{C}^0(\Omega_\lambda)^{\hat{}\,*} \quad : \\ \exists!\;\; A \in \hat{X}^*_T \;: \\ \quad \hat{T}A = F \end{array}$$

in other words, the mapping, see (10.10)

$$(10.15) \qquad \hat{X}^*_T \xrightarrow{\ \hat{T}\ } \prod_{\lambda\in\Lambda} \mathcal{C}^0(\Omega_\lambda)^{\hat{}\,*}$$

is an order isomorphism.

Proof

We shall apply Theorem 9.2, and for that purpose, we prove the inclusion

$$(10.16) \qquad \hat{T}(\hat{X}_T) \supseteq \prod_{\lambda\in\Lambda} \mathcal{C}^0(\Omega_\lambda) \;.$$

Therefore, let us take any

$$f = (f_\lambda \mid \lambda \in \Lambda) \in \prod_{\lambda\in\Lambda} \mathcal{C}^0(\Omega_\lambda) \;.$$

Then according to Theorem 9.1, in order show that

$$\hat{T}A = f$$

for a suitable $A \in \hat{X}_T$, it suffices to show that in, see (10.2)

$$\hat{Y} = \left[\left(\prod_{\lambda \in \Lambda} \mathcal{C}^0(\Omega_\lambda) \right] \right.^{\wedge}$$

we have

$$(10.17) \quad \sup_{\hat{Y}} \left\{ \hat{T}U \;\middle|\; \begin{array}{l} *)\ U \in \hat{X}_T \\ **)\ \hat{T}U \subseteq \langle f] \end{array} \right\} = \inf_{\hat{Y}} \left\{ \hat{T}V \;\middle|\; \begin{array}{l} *)\ V \in \hat{X}_T \\ **)\ \langle f] \subseteq \hat{T}V \end{array} \right\} .$$

Let us take $\lambda \in \Lambda$ and a locally finite smooth partition of unity $(\alpha_i \mid i \in I)$ of Ω_λ. Given $\epsilon > 0$, it follows that there exists $h_{\epsilon i} \in \mathcal{C}^\infty(\Omega_\lambda)$, with $i \in I$, such that, for $i \in I$, we have

$$f_\lambda - \epsilon \leq h_{\epsilon i} \leq f_\lambda \quad \text{on} \quad \text{supp} \ \alpha_i$$

However, it is clear that

$$g_{\lambda\epsilon} = \sum_{i \in I} \alpha_i \, h_{\epsilon i} \in \mathcal{C}^\infty(\Omega_\lambda)$$

and

$$f_\lambda - \epsilon \leq g_{\lambda\epsilon} \leq f_\lambda \quad \text{on} \quad \Omega_\lambda$$

In this way, we obtain

$$g_\epsilon = (g_{\lambda\epsilon} \mid \lambda \in \Lambda) \in \prod_{\lambda \in \Lambda} \mathcal{C}^\infty(\Omega_\lambda)$$

which satisfies, see (A.77)

$$(10.18)\qquad f - \epsilon \leq g_\epsilon \leq f$$

Now, in view of (10.4), we can find $u_\epsilon \in X$, such that $Tu_\epsilon = g_\epsilon$, therefore (10.18) will give

$$f - \epsilon \leq Tu_\epsilon \leq f$$

Obviously, this means that

$$\sup_{\hat{Y}} \{\hat{T}<u_\epsilon] \mid \epsilon > 0\} = <f]$$

therefore

$$\sup_{\hat{Y}} \left\{\hat{T}U \;\middle|\; \begin{array}{ll} *) & U \in \hat{X}_T \\ **) & \hat{T}U \subseteq <f] \end{array} \right\} = <f]$$

In a similar way one can obtain

$$\inf_{\hat{Y}} \left\{\hat{T}V \;\middle|\; \begin{array}{ll} *) & V \in \hat{X}_T \\ **) & <f] \subseteq \hat{T}V \end{array} \right\} = <f]$$

which means that (10.17) is indeed valid.

Then according to (9.26) it follows that

$$(10.19)\qquad \begin{array}{l} \forall\ F \in (\prod_{\lambda\in\Lambda} \mathcal{C}^0(\Omega_\lambda))^\wedge : \\ \exists\ A \in \hat{X}_T : \\ \quad \hat{T}A = F \end{array}$$

Now (10.14) will follow easily from (10.19) by taking into account Proposition A.2 in the Appendix. Indeed, we simply restrict F in (10.19) to

$$F \in \Big(\prod_{\lambda\in\Lambda} \mathcal{C}^{0}(\Omega_\lambda) \Big)^{\wedge *}$$

and recall (10.13) or (10.10). ∎

At this point, we can present a *nontrivial* application of the rather abstract existence of solution result in Theorem 10.1 above. Let

$$(10.20) \qquad P(D) = \sum_{\substack{p\in\mathbb{N}^n \\ |p|\leq m}} c_p \, D^p$$

a *linear constant coefficient* partial differential operator on an open subset $\Omega \subseteq \mathbb{R}^n$. Further, let us suppose that Ω is $P(D)$-*convex*, in other words, we have satisfied, Hörmander

$$(10.21) \qquad P(D)\mathcal{C}^{\infty}(\Omega) = \mathcal{C}^{\infty}(\Omega)$$

Then we can apply the above general procedure in (10.1) - (10.3), with the respective particularization given by

$$(10.22) \qquad X = \mathcal{C}^{\infty}(\Omega), \quad Y = \mathcal{C}^{0}(\Omega)$$

$$(10.23) \qquad T = P(D)$$

corresponding to an index set Λ containing one single λ, and with respective manifold $\Omega_\lambda = \Omega$.

Now, in view of (10.21), it is obvious that the particular case of (10.4) corresponding to (10.22) and (10.23) is satisfied. Therefore, we can apply Theorem 10.1 and obtain the following *nontrivial* result, see Remark 10.2.

Proposition 10.1

Let Ω be a $P(D)$-convex open subset of $\mathbb{R}^n$. Then

$$(10.24) \qquad \hat{P}(D)\hat{\mathcal{C}}^{\infty}_{P(D)}(\Omega) = \hat{\mathcal{C}}^{0}(\Omega)$$

More precisely

$$\begin{array}{ll} & \forall\ F \in \hat{\mathcal{C}}^0(\Omega) : \\ (10.25) & \exists !\ U \in \hat{\mathcal{C}}^{\infty}_{P(D)}(\Omega) : \\ & \quad \hat{P}(D)U = F \end{array}$$

Proof

Since (10.22), (10.23) corresponds to an index set Λ with one single element, the Cartesian products in Theorem 10.1 are trivial. This applies, in particular, to Proposition A.2 in the Appendix. Therefore, one can revert to the general MacNeille order completion $\hat{\ }$, and there is no need for its restricted version $\hat{\ }^*$. ■

Remark 10.2

Let us assume that Ω is $P(D)$-convex but *not* strictly $P(D)$-convex, Hörmander. Then

$$(10.26) \quad P(D)\mathcal{C}^{\infty}(\Omega) = \mathcal{C}^{\infty}(\Omega)$$

however

$$(10.27) \quad P(D)\mathcal{D}'(\Omega) \underset{\neq}{\subset} \mathcal{D}'(\Omega)$$

Now, in view of Proposition 10.1, we can nevertheless add to (10.26) and (10.27) the following property

$$(10.28) \quad \hat{P}(D)\hat{\mathcal{C}}^{\infty}_{P(D)}(\Omega) = \hat{\mathcal{C}}^0(\Omega) \underset{\neq}{\supset} \mathcal{C}^0(\Omega)$$

which is *nontrivial*, owing to the size of $\hat{\mathcal{C}}^0(\Omega)$ when compared with $\mathcal{C}^{\infty}(\Omega)$.

Indeed, if we recall (7.67), it follows that $\hat{\mathcal{C}}^0(\bar{\Omega})$, for Ω bounded, contains all bounded functions $f : \Omega \longrightarrow \mathbb{R}$ which are *discontinuous* on closed, nowhere dense subsets $\Gamma \subset \Omega$. Further, according to the characterization in Dilworth, the functions in $\hat{C}^0(\bar{\Omega})$ are *normal lower semicontinuous*, while in view of the characterization in Nakano & Shimogaki, they are bounded functions whose set of *discontinuities* is of first category Baire.

■

In the case of *compact* manifolds, a *stronger* version of the result in Theorem 10.1 can be obtained. Suppose that for $\lambda \in \Lambda$

$$(10.29) \quad \Omega_\lambda \text{ is a compact submanifold in an Euclidean space}$$

Further, instead of (10.4), let us make the *weaker* assumption, according to which

$$(10.30) \quad \begin{aligned} & \forall\ f = (f_\lambda \mid \lambda \in \Lambda) \in \prod_{\lambda \in \Lambda} \mathcal{C}^\infty(\Omega_\lambda),\ \ f_\lambda \text{ polynomial :} \\ & \exists\ u \in X\ ; \\ & \quad Tu = f \end{aligned}$$

Then we obtain the *existence and uniqueness* result in

Theorem 10.2

Under the assumption (10.29) and (10.30) it follows that

$$(10.31) \quad \begin{aligned} & \forall\ F \in \prod_{\lambda \in \Lambda} \widehat{\mathcal{C}^0(\Omega)}^{\,*} \\ & \exists !\ A \in \hat{X}^* : \\ & \quad \hat{T}A = F \end{aligned}$$

Proof

Similar to that of Theorem 10.1, except that the existence of the functions $g_{\lambda\epsilon} \in \mathcal{C}^\infty(\Omega_\lambda)$ will this time follow from the theorem of Weierstrass. In particular, these functions $g_{\lambda\epsilon}$ can now be polynomials.

■

As mentioned, one of the major advantages of the order completion method in solving nonlinear PDEs is in the ease and flexibility we encounter in connection with the *various ways* we can deal with initial and/or boundary value problems. Two such ways are presented now. Let $\Omega \subseteq \mathbb{R}^n$ be a bounded open subset and let

$$(10.32) \quad P : \mathcal{C}^m(\bar{\Omega}) \longrightarrow \mathcal{C}^0(\bar{\Omega})$$

be an m-th order continuous nonlinear partial differential operator of the general form (2.1), (2.2). Denoting by $\partial\Omega$ the boundary of Ω, one way of dealing with *boundary value* problems associated with P is as follows. By defining

$$(10.33) \quad \begin{aligned} T : \mathcal{C}^m(\bar{\Omega}) &\longrightarrow \mathcal{C}^0(\bar{\Omega}) \times \mathcal{C}^0(\partial\Omega) \\ u &\longmapsto (Pu, u|_{\partial\Omega}) \end{aligned}$$

we arrive at a particular case of (10.1) and (10.2). Therefore, Theorem 10.1 will result in

Proposition 10.2

Suppose that

$$(10.34) \quad \begin{aligned} &\forall\ f \in \mathcal{C}^\infty(\bar{\Omega}),\quad g \in \mathcal{C}^\infty(\partial\Omega) \\ &\exists\ u \in \mathcal{C}^m(\bar{\Omega}) : \\ &\quad Pu = f \quad \text{on} \quad \bar{\Omega} \\ &\quad u|_{\partial\Omega} = g \end{aligned}$$

Then

$$(10.35) \quad \begin{aligned} &\forall\ F \in \mathcal{C}^0(\bar{\Omega})^{\hat{}*} \times \mathcal{C}^0(\partial\Omega)^{\hat{}*} : \\ &\exists\ A \in \hat{X}_T^* = (\mathcal{C}^m(\bar{\Omega})/\approx_T)^{\hat{}*} : \\ &\quad \hat{T}A = F \end{aligned}$$

■

The relevance of the result in Proposition 10.2 is commented upon in Remark 10.3 below. Meanwhile, we present a *nontrivial* consequence of that proposition in the case of the *Dirichlet* problem for the *Poisson* equation on bounded domains with C^∞-smooth boundaries. These boundaries are regular with respect to the Laplacian Δ, Gilbarg & Trudinger [p 25].

Proposition 10.3

For every bounded open set $\Omega \subset \mathbb{R}^n$ with a regular boundary $\partial\Omega$, we have

$$(10.36)\qquad \begin{aligned} &\forall\ F \in \widehat{C^0(\bar{\Omega})}^*,\quad \Phi \in \widehat{C^0(\partial\Omega)}^* : \\ &\exists!\ U \in \widehat{C^2_T(\bar{\Omega})}^* : \\ &\qquad \Delta U = F \quad \text{in} \quad \hat{C}^0(\bar{\Omega}) \\ &\qquad U\big|_{\partial\Omega} = \Phi \quad \text{in} \quad \hat{C}^0(\partial\Omega) \end{aligned}$$

where T is given by

$$(10.37)\qquad C^2(\bar{\Omega}) \ni u \longmapsto (\Delta u, u\big|_{\partial\Delta}) \in C^0(\bar{\Omega}) \times C^0(\partial\Omega)$$

Proof

It is well known, see pct. 1) in Remark 10.3 next, that the restriction of the mapping in (10.37) given by

$$(10.38)\qquad \begin{aligned} C^\infty(\bar{\Omega}) &\xrightarrow{\quad T \quad} C^\infty(\bar{\Omega}) \times C^\infty(\partial\Omega) \\ u &\longmapsto (\Delta u, u\big|_{\partial\Omega}) \end{aligned}$$

is a bijection. In this way (10.34) is satisfied. Therefore (10.36) follows from (10.35).

■

Remark 10.3

1) It is known, Gilbarg & Trudinger [p 55], that under the conditions in Proposition 10.3, and for every bounded and locally Hölder continuous $f : \Omega \longrightarrow \mathbb{R}$, the Dirichlet problem for the Poisson equation

$$(10.39)\qquad \begin{aligned} &\Delta u = f \quad \text{on} \quad \Omega \\ &u\big|_{\partial\Omega} = \varphi \end{aligned}$$

has a *unique solution*

(10.40) $\quad u \in \mathcal{C}^2(\Omega) \cap \mathcal{C}^0(\bar{\Omega})$

for each boundary value

(10.41) $\quad \varphi \in \mathcal{C}^0(\partial\Omega)$

The $\mathcal{C}^\infty$- smooth version in (10.38) can be found, for instance, in Treves.

2) In view of Remark 10.2, it follows that the *existence and uniqueness* result in Proposition 10.3 holds for a class of right hand terms F and boundary values Φ which is *significantly larger* than those in (10.39). In particular, both F and Φ can be *discontinuous* functions.

3) The result in Proposition 10.2 is not limited to the Poisson equation. Indeed, Dirichlet problems for general nonlinear elliptic equations can be treated in the same manner. ■

A second way of dealing with boundary value problems is presented now. For convenience, we shall again consider the Dirichlet problem for the Poisson equation, although this second way is applicable in more general cases.

Let us take

(10.42) $\quad Y = \mathcal{C}^0(\bar{\Omega})$

with the usual partial order. Further, suppose given $\varphi \in \mathcal{C}^0(\partial\Omega)$ and let us define

(10.43) $\quad X = \{u \in \mathcal{C}^2(\Omega) \cap \mathcal{C}(\bar{\Omega}) \mid u|_{\partial\Omega} = \varphi\}$

It is known, Gilbarg & Trudinger [p 5], that the mapping

(10.44)
$$\begin{array}{ccc} X & \xrightarrow{\ \Delta\ } & Y \\ u & \longmapsto & \Delta u \end{array}$$

is *injective*. Therefore, we can define on X the partial order $\leq_\Delta$ by

$$(10.45) \quad u \leq_\Delta v \iff (\Delta u \leq \Delta v \text{ on } \bar{\Omega})$$

In this way, we are in a particular case of (10.1) - (10.3), with

$$(10.46) \quad T = \Delta, \quad X_T = X$$

Furthermore, in view of (10.38), we have as well

$$(10.47) \quad \begin{array}{l} \forall \ f \in \mathcal{C}^\infty(\bar{\Omega}) \ : \\ \exists! \ u \in X \ : \\ \quad \Delta u = f \ \text{ on } \ \bar{\Omega} \end{array}$$

therefore condition (10.4) is satisfied. In this way, we can apply Theorem 10.1 and obtain

Proposition 10.4

Suppose $\Omega \subset \mathbb{R}^n$ is bounded and open, with $\partial\Omega$ being regular. Then

$$(10.48) \quad \begin{array}{l} \forall \ F \in \mathcal{C}^0(\bar{\Omega})^{\hat{}*} \ : \\ \exists! \ A \in X^{\hat{}*} \ : \\ \quad \Delta A = F \end{array}$$

■

Remark 10.4

1) Again, as with Proposition 10.2, the result in Proposition 10.4 is not limited to the Poisson equation.

2) The construction of X in (10.43) can be extended to *more general* types of boundary values, such as for instance $\varphi \in \mathcal{D}'(\partial\Omega)$. Details in this regard can be found in Estrada & Kanwal and in Egorov, see also Example 12.1 below.

11. NONLINEAR SYSTEMS WITH MEASURES AS INITIAL DATA

11.1 Extension of an Existence Result

We shall start with semilinear hyperbolic systems in two independent variables of the form

$$(11.1) \qquad (D_t + CD_x)U(t,x) = A(U(t,x)), \quad t \geq 0, \quad x \in \mathbb{R}$$

where

$$U : [0,\infty) \times \mathbb{R} \longrightarrow \mathbb{R}^n$$

is the unknown function, C is a diagonal $n \times n$ constant matrix, while $A : \mathbb{R}^n \longrightarrow \mathbb{R}^n$ is $\mathcal{C}^\infty$-smooth. We shall associate with (11.1) the initial value problem

$$(11.2) \qquad U(0,x) = u(x), \quad x \in \mathbb{R}$$

where

$$u : \mathbb{R} \longrightarrow \mathbb{R}^n$$

is given. It will be convenient to write (11.2) in the form

$$(11.3) \qquad RU = u$$

where R denotes the *restriction* operator

$$(11.4) \qquad RU = U\big|_{t = 0}$$

which is obviously well defined for every continuous function $U : [0,\infty) \times \mathbb{R} \longrightarrow \mathbb{R}^n$.

A well known example of (11.1) is the Carleman system

$$(11.5)\qquad \begin{aligned} (D_t + D_x)U_1 &= U_2^2 - U_1^2 \\ (D_t - D_x)U_2 &= U_1^2 - U_2^2 \end{aligned}$$

where

$$U = (U_1, U_2) : [0,\infty) \times \mathbb{R} \longrightarrow \mathbb{R}^2$$

is the unknown function.

It is known, Illner, Platkovski & Illner, Oberguggenberger [3], that (11.5) has a unique solution $U \in \mathcal{C}^\infty([0,\infty) \times \mathbb{R}, \mathbb{R}^2_+)$ for every initial value $u \in \mathcal{C}^\infty(\mathbb{R},\mathbb{R}^2_+) \cap \mathcal{L}^1(\mathbb{R},\mathbb{R}^2_+)$, where we denoted as usual

$$\mathbb{R}^n_+ = \{(x_1,\dots,x_n) \in \mathbb{R}^n \mid x_1,\dots,x_n \geq 0\} .$$

This convex positive cone in $\mathbb{R}^n$ generates a partial order $\leq$ on $\mathbb{R}^n$ defined by

$$(x_1,\dots,x_n) \leq (y_1,\dots,y_n) \iff \begin{bmatrix} x_1 \leq y_1 \\ \dots\dots\dots \\ x_n \leq y_n \end{bmatrix} .$$

The same solution property holds for the general system (11.1) when A is linear and nonnegative, that is, A is a $n \times n$ matrix with nonnegative entries.

It is however important to note that, as shown for instance in Oberguggenberger [2] and in the literature cited there, systems of the form (11.1) which are more general than the cases just mentioned can have global $\mathcal{C}^\infty$-smooth solutions.

In view of the above, in this Section we shall make the *standing assumption*, see (11.9) below, that the nonlinear Cauchy problem (11.1), (11.2) has a unique solution $U \in \mathcal{C}^\infty([0,\infty) \times \mathbb{R}, \mathbb{R}^n_+)$ for every initial value $u \in \mathcal{C}^\infty(\mathbb{R},\mathbb{R}^n_+) \cap \mathcal{L}^1(\mathbb{R},\mathbb{R}^n_+)$.

Our aim is to show in Theorem 11.1 that a unique generalized solution for the nonlinear Cauchy problem (11.1), (11.2) will still exist if we enlarge the initial values, allowing them to belong to

$$(11.6) \qquad \mathcal{FM}(\mathbb{R},\mathbb{R}^n_+)$$

which is the space of nonnegative vector valued measures on $\mathbb{R}$ having finite total mass.

For that purpose, we shall set up an appropriate scheme in order to be able to make use of the general existence theory presented earlier in Section 9. Let us therefore take

$$(11.7) \qquad Y = (\mathcal{C}^\infty(\mathbb{R},\mathbb{R}^n_+) \cap \mathcal{L}^1(\mathbb{R},\mathbb{R}^n_+))\backslash\{0\}$$

and

$$(11.8) \qquad X = \left\{ u \in \mathcal{C}^\infty([0,\infty) \times \mathbb{R}, \mathbb{R}^n_+) \;\middle|\; \begin{array}{ll} *) & u \text{ is a solution of } (11.1) \text{ on } [0,\infty) \times \mathbb{R} \\ **) & \forall\, t \in [0,\infty) : \\ & \quad u(t,\cdot) \in \mathcal{L}^1(\mathbb{R},\mathbb{R}^n_+) \end{array} \right\} \backslash \{0\}$$

Then, according to our above assumption on the unique solution of the nonlinear Cauchy problem (11.1), (11.2), it follows that the restriction operator R in (11.4), when applied on X in (11.8), will give the *bijective* mapping

$$(11.9) \qquad R : X \xrightarrow{\text{bij}} Y \;.$$

We also note that we have the natural embedding

$$(11.10) \qquad Y \xrightarrow{i_0} \mathcal{FM}(\mathbb{R},\mathbb{R}^n_+)$$

defined by the injective mapping

$$(11.11) \qquad \mathcal{C}^\infty(\mathbb{R},\mathbb{R}^n_+) \cap \mathcal{L}^1(\mathbb{R},\mathbb{R}^n_+) \ni f \longmapsto \mu_f \in \mathcal{FM}(\mathbb{R},\mathbb{R}^n_+)$$

where

$$(11.12) \quad \mu_f(a,b) = \int_a^b f(x)dx, \quad (a,b) \subseteq \mathbb{R} .$$

Therefore, according to the earlier general method, see for instance Sections 9, 10, we should start by defining a partial order on Y, then pull it back through R to a partial order on X, turning thus R into an order isomorphism, and then finally, extend R to an order isomorphism between the respective order completions of X and Y. Following this procedure, and based on (11.10), we intend to prove that

$$(11.13) \quad \mathcal{FM}(\mathbb{R},\mathbb{R}^n_+) \subseteq \hat{Y}$$

thus obtaining unique generalized solutions for the nonlinear Cauchy problem (11.1), (11.2) for initial values given by nonnegative measures.

It is useful to note here that (11.1) is a particular case of (8.3). Nevertheless, the set up in (11.7) - (11.3) used in this Section in order to solve (11.1) is quite *different* from that in (8.39) used in solving (8.3). This difference is, however, but a matter of convenience, and it only shows the significant *flexibility* of the method of solving nonlinear PDEs through the method of order completion.

However, it appears that we have to exhibit a certain care in the choice of a useful partial order on Y, which is the first step in the procedure mentioned above. Indeed, we have to make sure that the chosen partial order on Y will give (11.13).

Now we could consider on Y the partial order induced by $\mathcal{C}^0(\mathbb{R},\mathbb{R}^n)$, namely

$$(11.14) \quad f \leq g \Longleftrightarrow \left[\begin{array}{l} \forall \ x \in \mathbb{R} : \\ \quad f(x) \leq g(x) \end{array} \right]$$

where $f,g \in \mathcal{C}^\infty(\mathbb{R},\mathbb{R}^n_+) \cap \mathcal{L}^1(\mathbb{R},\mathbb{R}^n_+)$. We note that this partial order on Y can be induced in a second way as well, namely, through the embedding (11.10) - (11.12), and by the following partial order on $\mathcal{FM}(\mathbb{R},\mathbb{R}^n_+)$

$$(11.15) \qquad \mu \le \nu \Longleftrightarrow \left[\begin{array}{c} \forall\ \varphi \in \mathcal{C}^0_0(\mathbb{R},\mathbb{R}_+) : \\ \int_{\mathbb{R}} \varphi d\mu \le \int_{\mathbb{R}} \varphi d\nu \end{array} \right] .$$

However, the partial order (11.14), that is (11.15), is not suitable since with it, Y is *not* order dense in $\mathcal{FM}(\mathbb{R},\mathbb{R}^n_+)$. In particular (11.13) will not hold. Indeed, the difficulty with the partial order (11.14), (11.15) is that it is *too strong*. In other words, there are many elements in $\mathcal{FM}(\mathbb{R},\mathbb{R}^n_+)$ which are *not comparable* in this partial order, see Remark 11.1 below.

Therefore, we consider on $\mathcal{FM}(\mathbb{R},\mathbb{R}^n_+)$ another *weaker* partial order, given as follows. Let us define the mapping

$$(11.16) \qquad \mathcal{FM}(\mathbb{R},\mathbb{R}^n_+) \ni \mu \longmapsto F_\mu : \mathbb{R} \longrightarrow \mathbb{R}^n_+$$

by

$$(11.17) \qquad F_\mu(x) = \mu(-\infty,x], \quad x \in \mathbb{R} .$$

It is clear that F_μ is monotonous, continuous from the right and bounded. It also satisfies

$$(11.18) \qquad \lim_{x\to -\infty} F_\mu(x) = 0, \quad \lim_{x\to \infty} F_\mu(x) = \int_{\mathbb{R}} d\mu .$$

We note that in (11.12), the function f is called the *density* of the measure μ_f, while in (11.17), the function F_μ is called the *distribution* of the measure μ.

We shall define now on $\mathcal{FM}(\mathbb{R},\mathbb{R}_+^n)$ the partial order $\leq_D$ by

$$(11.19) \qquad \mu \leq_D \nu \Longleftrightarrow \left[\begin{array}{l} \forall\ x \in \mathbb{R} : \\ \quad F_\mu(x) \leq F_\mu(x) \end{array} \right] .$$

Then for $\mu,\nu \in \mathcal{FM}(\mathbb{R},\mathbb{R}_+^n)$ we obviously have

$$(11.20) \qquad \mu \leq \nu \Longrightarrow \mu \leq_D \nu$$

however

$$(11.21) \qquad \mu \leq_D \nu \not\Longrightarrow \mu \leq \nu$$

see for details pct. 2) in Remark 11.1. In this way $\leq_D$ is indeed *weaker* than $\leq$.

Now we come to see in the next two propositions the way we can secure the property (11.13), and do so under a yet more precise form. The proofs are given at the end of this Subsection.

Proposition 11.1

$(\mathcal{FM}(\mathbb{R},\mathbb{R}_+^n),\ \leq_D)$ is Dedekind order complete. ■

In view of (11.10), the result in Proposition 11.1 implies that in case (11.13) holds in the sense of the partial order $\leq_D$, then we shall actually have, see (7.26)

$$(11.22) \qquad \mathcal{FM}(\mathbb{R},\mathbb{R}_+^n)\setminus\{0\} = \hat{Y}^*$$

where 0, which denotes the identically null measure, has to be subtracted, since obviously it is the minimum element of $\mathcal{FM}(\mathbb{R},\mathbb{R}_+^n)$. This relation holds indeed, as seen in

Proposition 11.2

In the sense of the partial order $\leq_D$, we have

$$(11.23) \quad \mathcal{FM}(\mathbb{R},\mathbb{R}^n_+)\setminus\{0\} = \hat{Z}^*$$

both for

$$(11.24) \quad Z = \mathcal{C}^\infty(\mathbb{R},\mathbb{R}^n_+) \cap \mathcal{L}^1(\mathbb{R},\mathbb{R}^n_+)\setminus\{0\}$$

and for

$$(11.25) \quad Z = \mathcal{C}^\infty(\mathbb{R},\mathbb{R}^n_+) \cap \mathcal{L}^1(\mathbb{R},\mathbb{R}^n_+) \cap \mathcal{L}^\infty(\mathbb{R},\mathbb{R}^n_+)\setminus\{0\}$$ ■

In this way (11.22) follows indeed from (11.23) and (11.24).

Now, based on (11.9), let us pull back the partial order $\leq_D$ on Y through R, and define a partial order $\leq_R$ on X by

$$(11.26) \quad u \leq_R v \Longleftrightarrow Ru \leq_D Rv$$

whenever $u,v \in X$. In this way we obtain from (11.9) and (11.26), the *order isomorphism*

$$(11.27) \quad R : X \longrightarrow Y$$

which according to Proposition A.1, we can extend to the *order isomorphism*

$$(11.28) \quad \hat{R} : \hat{X} \longrightarrow \hat{Y}$$

In view of (11.22), we obtain

Theorem 11.1

Given any $\mu \in \mathcal{FM}(\mathbb{R},\mathbb{R}^n_+)$, there exists a *unique* $U \in \hat{X}$, see (11.8), such that

(11.29) $\hat{R}U = \mu.$ ■

The meaning of (11.29) will be clarified in the next Subsection.

Here we only note that owing to (11.27), (11.28), (11.10), (11.22) and (A.39) we obtain commutative diagram of order isomorphical embeddings

$$
(11.30)\qquad
\begin{array}{ccccc}
X & \xrightarrow[\text{bij}]{R} & Y & \xrightarrow[\text{inj}]{i_0} & \mathcal{FM}(\mathbb{R},\mathbb{R}^n_+) \\
\downarrow & & \downarrow & & \downarrow \\
\hat{X} & \xrightarrow[\text{bij}]{\hat{R}} & \hat{Y} & \xrightarrow[\text{bij}]{\hat{i}_0} & \mathcal{FM}(\mathbb{R},\mathbb{R}^n_+) \cup \{\infty\}
\end{array}
$$

where R, $\hat{R}$ and $\hat{i}_0$ are in fact order isomorphisms. It follows that the solution $U \in \hat{X}$ in (11.29) can be written in the form

(11.31) $U = (\hat{R}^{-1} \circ \hat{i}_0^{-1})\mu, \quad \mu \in \mathcal{FM}(\mathbb{R},\mathbb{R}^n_+)$

Remark 11.1

1) For simplicity, we shall consider $n = 1$, and show that

(11.32) $\mathcal{C}^0(\mathbb{R},\mathbb{R}_+) \cap \mathcal{L}^1(\mathbb{R},\mathbb{R}_+)$

is *not* order dense in

(11.33) $\mathcal{FM}(\mathbb{R},\mathbb{R}_+)$

in the sense of the partial order $\leq$ in (11.14), (11.15). Actually we show the stronger property, namely that

(11.34) $f \nleq \delta \nleq g$

for every $f \in (\mathcal{C}^0(\mathbb{R},\mathbb{R}_+) \cap \mathcal{L}^1(\mathbb{R},\mathbb{R}_+))\setminus\{0\}$, $g \in \mathcal{C}^0(\mathbb{R},\mathbb{R}_+) \cap \mathcal{L}^1(\mathbb{R},\mathbb{R}_+)$, where $\delta \in \mathcal{FM}(\mathbb{R},\mathbb{R}_+)$ denotes the Dirac measure concentrated at $x = 0 \in \mathbb{R}$.

Indeed, according to (11.12), (11.15), the inequality $f \leq \delta$ would mean that we have

$$(11.35) \qquad \int_{\mathbb{R}} f(x)\varphi(x)dx \leq \varphi(0), \quad \varphi \in \mathcal{C}^0_0(\mathbb{R},\mathbb{R}_+) \ .$$

However, we can obviously assume that

$$\varphi(0) = 0$$
$$f\varphi \neq 0$$

in which case (11.35) cannot hold. On the other hand, the inequality $\delta \leq g$ will in a similar way give

$$(11.36) \qquad \varphi(0) \leq \int_{\mathbb{R}} g(x)\varphi(x)dx, \quad \varphi \in \mathcal{C}^0_0(\mathbb{R},\mathbb{R}_+) \ .$$

If we choose φ so that

$$g(0) + 2 \leq \varphi(0), \quad \int_{\mathbb{R}} \varphi(x)dx = 1, \quad \text{supp } \varphi \subseteq [-a,a]$$

for a suitable $a > 0$, for which we also have

$$g(x) \leq g(0) + 1, \quad x \in [-a,a]$$

then (11.36) implies the contradiction

$$\varphi(0) \leq \int_{\mathbb{R}} g(x)\varphi(x)dx \leq g(0) + 1 < g(0) + 2 \leq \varphi(0) \ .$$

In this way (11.34) is indeed proved.

2) Let $\delta_a, \delta_b \in \mathcal{FM}(\mathbb{R},\mathbb{R}_+)$ be the Dirac measures concentrated at $a,b \in \mathbb{R}$ respectively. If $a < b$ then (11.17), (11.19) give

$$\delta_b \leq_D \delta_a$$

but in view of (11.15) we obviously have

$$\delta_b \nleq \delta_a \nleq \delta_b \ .$$

Proof of Proposition 11.1

Let $\mu_i \in \mathcal{FM}(\mathbb{R},\mathbb{R}^n_+)$, with $i \in I$, be an arbitrary family of measures which is bounded from above, that is, for a certain $\mu \in \mathcal{FM}(\mathbb{R},\mathbb{R}^n_+)$, we have

$$\mu_i \leq_D \mu, \quad i \in I$$

Then (11.19) gives

$$(11.37) \qquad F_{\mu_i}(x) \leq F_\mu(x), \quad i \in I, \quad x \in \mathbb{R}$$

Therefore, we can define the function $\tilde{F} : \mathbb{R} \longrightarrow \mathbb{R}^n$ by

$$(11.38) \qquad \tilde{F}(x) = \sup_{\mathbb{R}^n_+} \{F_{\mu_i}(x) \mid i \in I\}, \quad x \in \mathbb{R}$$

In view of (11.37), (11.38) we obtain

$$(11.39) \qquad \tilde{F}(x) \leq F_\mu(x), \quad x \in \mathbb{R}$$

We show now that $\tilde{F}$ is increasing. Indeed, assume given $x,y \in \mathbb{R}$, $x \leq y$. Then (11.17) yields

$$F_{\mu_i}(x) \leq F_{\mu_i}(y), \quad i \in I$$

hence, according to (11.38), we obtain

$$F_{\mu_i}(x) \leq F_{\mu_i}(y) \leq \tilde{F}(y), \quad i \in I$$

therefore, by using (11.38) again, it follows that

(11.40) $\quad \tilde{F}(x) \le \tilde{F}(y)$

Furthermore, the relations (11.39) and (11.18) imply that $\tilde{F}$ is bounded on $\mathbb{R}$, and

(11.41) $$\lim_{x \to -\infty} \tilde{F}(x) = 0$$

Now let $F : \mathbb{R} \longrightarrow \mathbb{R}^n_+$ be the right-continuous function associated to $\tilde{F}$. Then obviously F is well defined, since $\tilde{F}$ is increasing and bounded. Moreover in view of (11.39), (11.40), (11.41) it is easy to see that F is also bounded, increasing and satisfies

$$\lim_{x \to -\infty} F(x) = 0$$

It follows that, for a certain $\nu \in \mathcal{FM}(\mathbb{R},\mathbb{R}^n_+)$, we have with the notation in (11.16)

(11.42) $\quad F = F_\nu$

Indeed, the *vector* valued nonnegative finite total mass measures $\lambda \in \mathcal{FM}(\mathbb{R},\mathbb{R}^n_+)$ are in one to one correspondence with n-*tuples* of *real* valued nonnegative finite total mass measures

$$\lambda_1,\dots,\lambda_n \in \mathcal{FM}(\mathbb{R},\mathbb{R}_+)$$

according to

(11.43) $\quad \lambda(-\infty,x] = (\lambda_1(-\infty,x],\dots,\lambda_n(-\infty,x]), \quad x \in \mathbb{R}$

However, it is well known that (11.42) holds for $n = 1$. Therefore, in view of (11.43), it follows that (11.42) will in particular hold for all $n \ge 1$.

In this way it is clear that

$$\nu = \sup_{i\in I} \mu_i \quad \text{in} \quad (\mathcal{FM}(\mathbb{R},\mathbb{R}^n_+),\ \leq_D)$$

The existence in $\mathcal{FM}(\mathbb{R},\mathbb{R}^n_+)$ of the infimum of every family $(\mu_i \mid i \in I)$ can be proved in a similar way. We note that automatically, every such family will have 0 as a lower bound. ■

Proof of Proposition 11.2

Let us first take Z given in (11.24) and in view of the embedding (11.11), let us consider it partially ordered by $\leq_D$. In view of Proposition 11.1, it follows that

$$\mathcal{FM}(\mathbb{R},\mathbb{R}^n_+) \cup \{\infty\}$$

is order complete. Therefore, with Z in (11.24), we have

$$\hat{Z} \subseteq \mathcal{FM}(\mathbb{R},\mathbb{R}^n_+) \cup \{\infty\}$$

In particular, we obtain the order isomorphical embedding

$$(11.44) \qquad \hat{Z}^* \subseteq \mathcal{FM}(\mathbb{R},\mathbb{R}^n_+)\setminus\{0\}$$

Now, in order to prove (11.23) for Z in (11.24), it only remains to establish the inclusion '$\supseteq$' in (11.44). Let us take $\mu \in \mathcal{FM}(\mathbb{R},\mathbb{R}^n_+)\setminus\{0\}$. Then Lemma 11.1 below gives

$$(11.45) \qquad f_\epsilon,\ g_\epsilon \in (\mathcal{C}^\infty(\mathbb{R},\mathbb{R}^n_+) \cap \mathcal{L}^1(\mathbb{R},\mathbb{R}^n_+) \cap \mathcal{L}^\infty(\mathbb{R},\mathbb{R}^n_+))\setminus\{0\}, \quad \epsilon > 0$$

such that in $(\mathcal{FM}(\mathbb{R},\mathbb{R}^n_+),\ \leq_D)$ we have

$$(11.46) \qquad \mu = \sup_{\epsilon>0} f_\epsilon = \inf_{\epsilon>0} g_\epsilon$$

Let us now take

$$(11.47) \qquad H = \{h \in (\mathcal{C}^\infty(\mathbb{R},\mathbb{R}^n_+) \cap \mathcal{L}^1(\mathbb{R},\mathbb{R}^n_+) \cap \mathcal{L}^\infty(\mathbb{R},\mathbb{R}^n_+))\setminus\{0\} \left| \begin{array}{c} \forall\ \epsilon > 0 : \\ h \leq_D g_\epsilon \end{array} \right.\}$$

Then in $(\mathcal{FM}(\mathbb{R},\mathbb{R}^n_+), \leq_D)$ we have

$$(11.48) \qquad \mu = \sup H$$

Indeed, (11.46), (11.47) will obviously give

$$\sup H \leq_D \inf_{\epsilon>0} g_\epsilon = \mu$$

on the other hand, they also imply that

$$f_\epsilon \in H, \quad \epsilon > 0$$

which means that

$$\mu = \sup_\epsilon f_\epsilon \leq \sup H$$

In this way, the proof of (11.48), and with it of (11.23), (11.24) as well, are completed. Finally, in view of (11.45), it is clear that the above proof can be applied to Z in (11.25) as well, and then we can again obtain (11.23). ■

Lemma 11.1

Given $\mu \in \mathcal{FM}(\mathbb{R},\mathbb{R}^n_+)$, there exist sequences

$$(11.49) \qquad f_\epsilon, g_\epsilon \in \mathcal{C}^\infty(\mathbb{R},\mathbb{R}^n_+) \cap \mathcal{L}^1(\mathbb{R},\mathbb{R}^n_+) \cap \mathcal{L}^\infty(\mathbb{R},\mathbb{R}^n_+), \quad \epsilon > 0$$

such that in $(\mathcal{FM}(\mathbb{R},\mathbb{R}^n_+), \leq_D)$ we have

$$(11.50) \qquad \mu = \sup_{\epsilon>0} f_\epsilon = \inf_{\epsilon>0} g_\epsilon$$

Proof

Take $\varphi : \mathbb{R} \longrightarrow \mathbb{R}$, with $\varphi \in \mathcal{D}(\mathbb{R})$, $\varphi \geq 0$, supp $\varphi \subsetneq (0,1)$ and

$$\int_{\mathbb{R}} \varphi(x)dx = 1$$

and for $\epsilon > 0$, define $\varphi_\epsilon : \mathbb{R} \longrightarrow \mathbb{R}$ by $\varphi_\epsilon(x) = \varphi(x/\epsilon)/\epsilon$, with $x \in \mathbb{R}$.

Let us now define $f_\epsilon : \mathbb{R} \longrightarrow \mathbb{R}^n$ by

$$(11.51) \qquad f_\epsilon(x) = \int_{\mathbb{R}} \varphi_\epsilon(x - y)d\mu(y), \quad \epsilon > 0, \quad x \in \mathbb{R}$$

Then in view of (11.12), (11.17), and by denoting for simplicity

$$F_{\mu_{f_\epsilon}} = F_\epsilon, \quad \epsilon > 0$$

we obtain

$$F_\epsilon(x) = \int_{-\infty}^{x} \int_{\mathbb{R}} \varphi_\epsilon(z - y)d\mu(y)dz, \quad \epsilon > 0, \quad x \in \mathbb{R}$$

hence

$$(11.52) \qquad F_\epsilon(x) = \int_{\mathbb{R}} \psi_{\epsilon,x}(y)d\mu(y), \quad \epsilon > 0, \quad x \in \mathbb{R}$$

where

$$(11.53) \qquad \psi_{\epsilon,x}(y) = \int_{-\infty}^{(x-y)/\epsilon} \varphi(z)dz, \quad \epsilon > 0, \quad x,y \in \mathbb{R}$$

Therefore

$$\psi_{\epsilon,x}(y) = \begin{vmatrix} 1 & \text{if} & y \leq x - \epsilon \\ 0 & \text{if} & y \geq x \end{vmatrix}$$

thus denoting by $\chi_{(-\infty,x)}$ the characteristic function of the interval $(-\infty,x) \subset \mathbb{R}$, we obtain

$$\psi_{\epsilon,x} \leq \chi_{(\infty,x)} , \quad \epsilon > 0, \quad x \in \mathbb{R}$$

and

$$\lim_{\epsilon \downarrow 0} \psi_{\epsilon,x}(y) = \chi_{(-\infty,x)}(y), \quad x,y \in \mathbb{R}$$

In view of Lebesgue's dominated convergence theorem, it follows from (11.53) that

$$(11.54) \qquad \lim_{\epsilon \downarrow 0} F_{\epsilon}(x) = \mu(-\infty,x), \quad x \in \mathbb{R}$$

We prove now that (11.54) can in fact be written as

$$(11.55) \qquad \sup_{\mathbb{R}^n} \{F_{\epsilon}(x) \,|\, \epsilon > 0\} = \mu(-\infty,x), \quad x \in \mathbb{R}$$

Indeed, for a fixed $x \in \mathbb{R}$ and for $0 < \delta \leq \epsilon$, we obtain from (11.53) the relations

$$\psi_{\epsilon,x}(y) = \psi_{\delta,x}(y) = 0, \quad y \in [x,\infty)$$

and

$$\psi_{\epsilon,x}(y) = \int_{-\infty}^{(x-y)/\epsilon} \varphi(z)dz \leq \int_{-\infty}^{(x-y)/\delta} \varphi(z)dz = \psi_{\delta,x}(y), \quad y \in (-\infty,x]$$

Therefore (11.52) yields

$$(11.56) \qquad F_{\epsilon}(x) \leq F_{\delta}(x), \quad 0 < \delta \leq \epsilon, \quad x \in \mathbb{R}$$

In this way (11.55) follows from (11.54) and (11.56). Now in view of (11.55), we can apply an argument similar with that in the proof of Proposition 11.1 and obtain that

$$\mu = \sup_{\epsilon>0} f_\epsilon$$

In order to obtain the other half of (11.50), we can proceed as above, except that we shall require the condition

$$\mathrm{supp}\ \varphi \subset (-1,0)$$

and then let

$$g_\epsilon(x) = \int_{\mathbb{R}} \varphi_\epsilon(x - y) d\mu(y), \quad \epsilon > 0, \quad x \in \mathbb{R}$$

■

11.2 The Meaning of the Generalized Solutions

Let us return to the meaning of the existence and uniqueness result in (11.29) in Theorem 11.1, obtained under the assumption in (11.9), valid throughout this Section.

First we note that, in view of *) in (11.8), the following is immediate. Given $\mu \in \mathcal{FM}(\mathbb{R},\mathbb{R}^n_+)$, then the resulting unique $U \in \hat{X}$ in (11.29) will clearly satisfy, see (11.8) and (A.12)

$$U \subseteq \mathcal{C}^\infty([0,\infty) \times \mathbb{R}, \mathbb{R}^n_+)$$

Therefore *) in (11.8) yields

$$(11.57) \qquad \begin{array}{l} \forall \ u \in Y : \\ \quad u \text{ is a classical solution of (11.1), (11.2)} \end{array}$$

In particular, the *Carleman system* (11.5) can be solved with *nonnegative measures* as initial values. We shall see later in this Subsection the more specific interpretation of such solutions for (11.1), or for that matter (11.5).

Let us first address the issue of the *coherence* between the generalized solution and the classical one, when the latter exists as well. In the *linear* case it can be shown that the generalized and classical solutions do in fact *coincide*. For that purpose, let us recall a classical existence result which is a special case of the semilinear theory presented in Demengel & Rauch.

Proposition 11.3

Suppose that the smooth function A in (11.1) is linear and it is given by a constant matrix with nonnegative elements. Given any $\mu \in \mathcal{FM}(\mathbb{R},\mathbb{R}_+^n)$, there exists a unique

$$(11.58) \qquad u \in \mathcal{L}^1_{loc}([0,\infty),\ \mathcal{FM}(\mathbb{R},\mathbb{R}_+^n)) \cap \mathcal{C}^0([0,\infty),\ \mathcal{FM}(\mathbb{R},\mathbb{R}_+^n)_{w^*})$$

which solves (11.1), (11.2) in the sense of distributions. Here $\mathcal{FM}(\mathbb{R},\mathbb{R}_+^n)_{w^*}$ denotes $\mathcal{FM}(\mathbb{R},\mathbb{R}_+^n)$ equipped with the weak star topology.

Proof

The idea of the proof is as follows. One can rewrite (11.1), (11.2) as an integral equation and then solve it by a contraction fixed point argument in the Banach space $\mathcal{L}^1([0,T],\ \mathcal{FM}(\mathbb{R},\mathbb{R}_+^n))$, where $T > 0$ will only depend on A in (11.1) and on the initial value μ in (11.2). It follows that the solution can be extended up to $T = \infty$. ∎

Now let Z be the space of all mappings from $[0,\infty)$ to $\mathcal{FM}(\mathbb{R},\mathbb{R}_+^n)$, that is

$$(11.59) \qquad Z = \mathcal{FM}(\mathbb{R},\mathbb{R}_+^n)^{[0,\infty)}$$

According to (A.75), we can equip Z with the product order denoted again by $\leq_D$, and given by, see (11.19)

$$(11.60) \qquad (Z, \leq_D) = (\mathcal{FM}(\mathbb{R},\mathbb{R}^n_+), \leq_D)^{[0,\infty)}$$

In other words, according to (A.77), we obtain for $U,V \in Z$

$$(11.61) \qquad U \leq_D V \iff \left[\begin{array}{l} \forall\ t \in [0,\infty) : \\ \quad U(t,\cdot) \leq_D V(t,\cdot) \end{array} \right]$$

Then in view of Proposition 11.1 and Proposition A.2, it is clear that

$$(11.62) \qquad (Z, \leq_D) \text{ is Dedekind order complete}$$

Now, in the *linear* case dealt with in Proposition 11.3, let us define the *solution operator*

$$(11.63) \qquad S : \mathcal{FM}(\mathbb{R},\mathbb{R}^n_+) \longrightarrow Z$$

by

$$(11.64) \qquad \mathcal{FM}(\mathbb{R},\mathbb{R}^n_+) \ni \mu \longmapsto S\mu = u$$

where u is given in (11.58).

Proposition 11.4

Under the conditions in Proposition 11.3 on A in (11.1), the solution operator, see (11.63)

$$(11.65) \qquad S : \mathcal{FM}(\mathbb{R},\mathbb{R}^n_+) \longrightarrow Z$$

is an *order isomorphical embedding.*

Proof

We have to show that for $\mu,\nu \in \mathcal{FM}(\mathbb{R},\mathbb{R}^n_+)$

$$(11.66) \quad \mu \leq_D \nu \Longleftrightarrow \left[\begin{array}{l} \forall \;\; t \in [0,\infty) : \\ \quad (S\mu)(t,\cdot) \leq_D (S\nu)(t,\cdot) \end{array} \right]$$

Let us start by noting that the implication '$\Longleftarrow$' is obvious, since in view of (11.2), we have $\mu = (S\mu)(0,\cdot)$, $\nu = (S\nu)(0,\cdot)$. For the converse implication '$\Longrightarrow$' we note that, see (11.16)

$$(11.67) \quad F_\mu(x) = (H*\mu)(x), \; F_\nu(x) = (H*\nu)(x), \quad x \in \mathbb{R}$$

where H is the right continuous version of the Heaviside function H, i.e. $H(0) = 1$. Since both

$$U = S\mu, \quad V = S\nu$$

are solutions of (11.1), (11.2), and since all coefficients in (11.1) are constant and thus the convolution commutes with the operations in (11.1), we obtain

$$(11.68) \quad \begin{array}{l} (D_t + CD_x)(H*U) = A(H*U) \\ R(H*U) = H*\mu \end{array}$$

as well as

$$(11.69) \quad \begin{array}{l} (D_t + CD_x)(H*V) = A(H*V) \\ R(H*V) = H*\nu \end{array}$$

where, as in (11.67), the convolution is taken in the x-variable alone.

However, under the given conditions on A, it is easy to see that the solution operator considered as a mapping

$$(11.70) \quad S : \mathcal{L}^\infty(\mathbb{R},\mathbb{R}^n) \longrightarrow \mathcal{C}^0([0,\infty), \mathcal{L}^\infty(\mathbb{R},\mathbb{R}^n))$$

is increasing, when we consider the usual order on $\mathcal{L}^\infty(\mathbb{R},\mathbb{R}^n)$, namely, defined for $f,g \in \mathcal{L}^\infty(\mathbb{R},\mathbb{R}^n)$ by

$$f \leq g \Longleftrightarrow (f(x) \leq g(x) \;\text{ for almost all }\; x \in \mathbb{R})$$

Now (11.68) - (11.70) will give

$$(11.71) \quad \left[\begin{array}{l}(H*\mu)(x) \leq (H*\nu)(x) \\ \text{for almost all } x \in \mathbb{R}\end{array}\right] \Longrightarrow \left[\begin{array}{l}(H*U)(t,x) \leq (H*V)(t,x) \\ \text{for all } t \in [0,\infty) \text{ and} \\ \text{almost all } x \in \mathbb{R}\end{array}\right]$$

On the other hand, both $(H*U)(t,\cdot)$ and $(H*V)(t,\cdot)$ are right continuous, hence the right hand term of the implication in (11.71) will in fact hold as well for all $x \in \mathbb{R}$. In this way (11.67), (11.19) and (11.71) will give

$$U(t,\cdot) \leq_D V(t,\cdot), \quad t \in [0,\infty)$$

■

Let us now denote by

$$(11.72) \quad \mathcal{R}(S) \subseteq Z$$

the range in Z of the solution operator (11.65). Then by Proposition 11.4

$$(11.73) \quad S : \mathcal{FM}(\mathbb{R},\mathbb{R}^n_+) \longrightarrow \mathcal{R}(S)$$

is an *order isomorphism*, in particular, in view of Proposition 11.1, it follows that

$$(11.74) \quad (\mathcal{R}(S), \leq_D) \text{ is Dedekind order complete}$$

Further we note that in view of (11.8), (11.58), (11.59) and (11.63) we obtain the inclusion mapping

$$(11.75) \quad X \xrightarrow{\ i\ } \mathcal{R}(S)$$

therefore, the classical existence and uniqueness result in Proposition 11.3 gives, see also (11.30), (11.31)

$$(11.76) \quad i = S \circ i_0 \circ R \text{ is an order isomorphical embedding}$$

Now we can formulate the *coherence* result between generalized and classical solutions.

Proposition 11.5

Under the conditions in Proposition 11.3 on A in (11.1), let $\mu \in \mathcal{FM}(\mathbb{R},\mathbb{R}^n_+)$ and correspondingly, let $U \in \hat{X}$ be given by (11.29). Then, as an element of Z, the generalized solution U *coincides* with the classical solution $S\mu$, that is, see (11.75)

(11.77) $\quad S\mu = \hat{i}(U)$

More precisely, the following commutative diagram of order isomorphical embeddings holds

(11.78)

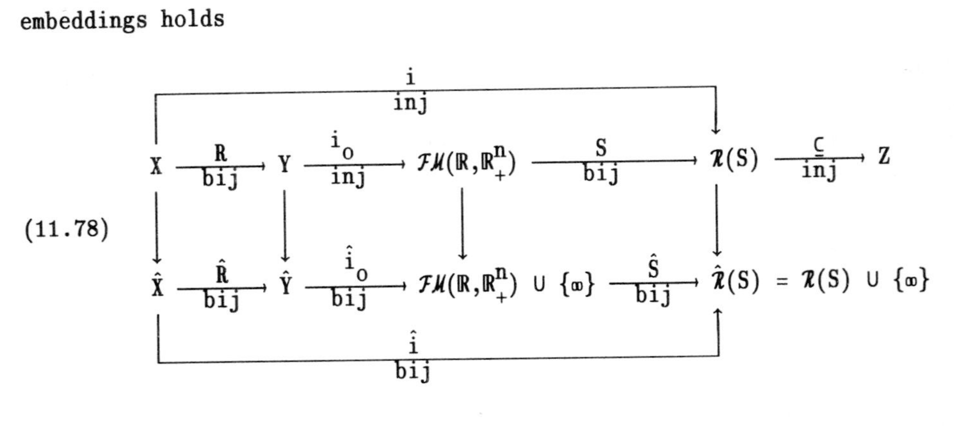

where R, $\hat{R}$, $\hat{i}_0$, S, $\hat{S}$ and $\hat{i}$ are in fact order isomorphisms

Proof

It is easy to see that (11.78) follows from (11.30), (11.73) - (11.76), (A.39) and (A.58). In particular, four of the order isomorphisms are related according to

(11.79) $\quad \hat{i} = \hat{S} \circ \hat{i}_0 \circ \hat{R}$

or equivalently

(11.80) $\quad \hat{S} = \hat{i} \circ \hat{R}^{-1} \circ \hat{i}_0^{-1}$

Now, let $\mu \in \mathcal{FM}(\mathbb{R},\mathbb{R}^n_+)$, then the classical solution is, see (11.58), (11.64)

$$(11.81) \quad u = S\mu$$

while in view of (11.31), the generalized solution is

$$(11.82) \quad U = (\hat{R}^{-1} \circ \hat{i}_0^{-1})\mu$$

However (11.79) - (11.82) will clearly give (11.77). ∎

Following the linear *coherence* result in Proposition 11.5, we now return to the issue of the *meaning* of the *generalized solution* U in (11.29) in the case of the *nonlinear* system (11.1), which contains as a particular instance the Carleman equations (11.5).

Let be given $\mu \in \mathcal{FM}(\mathbb{R},\mathbb{R}^n_+)$, then (11.30) yields

$$(11.83) \quad F = \hat{i}_0^{-1}(\mu) \in \hat{Y}$$

According to (9.18), the generalized solution U in (11.29) is given in our case by

$$(11.84) \quad U = \sup_{\hat{X}} \{V \in \hat{X} \mid \hat{R}V \subseteq F\}$$

On the other hand, given any $f \in F \in \hat{Y}$, then (11.83) and the assumption on classical solutions in (11.9) yield a corresponding classical solution $u \in X$, such that

$$Ru = f$$

In this way, we obtain

$$F = \sup_{\hat{Y}} \left\{ <f] \;\middle|\; \begin{array}{l} *)\ f \in Y \\ **)\ <f] \subseteq F \end{array} \right\} \subseteq \sup_{\hat{Y}} \left\{ <Ru] \;\middle|\; \begin{array}{l} *)\ u \in X \\ **)\ <Ru] \subseteq F \end{array} \right\} \subseteq F$$

which is precisely the condition (9.36) in Proposition 9.3. We also have satisfied, see Proposition A.3 in the Appendix, the condition (9.34), which in our case becomes

(11.85) $\{u \in X \mid \langle Ru] \subsetneq F\}$ is a nonvoid cut in X

Indeed, we can apply (A.94) with $X \xrightarrow{R} \hat{Y}$, instead of $X \xrightarrow{\varphi} Y$.

Therefore, according to Proposition 9.3, it follows that the *generalized solution* U in (11.84) can also be expressed as

(11.86) $U = \{u \in X \mid \langle Ru] \subsetneq F\} = \{u \in X \mid Ru \in F\}$

where F is given in (11.83).

In other words, given $\mu \in \mathcal{FM}(\mathbb{R},\mathbb{R}^n_+)$, then the *generalized solution* $U \in \hat{X}$, see (11.29), of the equation

(11.87) $\hat{R}U = \mu$

is according to (11.86), precisely the set of all *classical solutions* $u \in X$ of the family of equations

(11.88) $Ru = f, \quad f \in \hat{i}_0^{-1}(\mu)$

and in view of the assumption (11.9), the equation $Ru = f$ has a unique solution $u \in X$ for every $f \in \hat{i}_0^{-1}(\mu)$.

Finally, let us indicate one more possible interpretation for the meaning of the *unique* generalized solution U of (11.1), (11.2). This time, however, we shall use *two* operators, and not only one, namely R in (11.9) alone, as we have done it so far. For that purpose, let us consider the spaces

(11.89) $$G = \left\{u \in \mathcal{C}^\infty([0,\infty) \times \mathbb{R}, \mathbb{R}^n_+) \;\middle|\; \begin{array}{l} \forall\ t \in [0,\infty) : \\ \quad u(t,\cdot) \in \mathcal{L}^1(\mathbb{R},\mathbb{R}^n_+) \end{array} \right\} \setminus \{0\}$$

and

(11.90) $\quad H = \mathcal{C}^{\infty}([0,\infty) \times \mathbb{R}, \mathbb{R}^n)$

We define the operator P by

$$(11.91) \qquad \begin{array}{ccc} G & \xrightarrow{\;\;P\;\;} & H \\ u & \longmapsto & (D_t + CD_x)u - A(u) \end{array}$$

and note that in view of (11.9), we can define the extension

$$(11.92) \qquad G \xrightarrow{\;\;R\;\;} Y$$

by the same (11.4). Furthermore, it is clear that

(11.93) $\quad X = \{u \in G \mid Pu = 0\}$

Now we extend the operators P and R to the respective power sets, and do so in the natural set theoretical way, namely

$$(11.94) \qquad \begin{array}{ccc} \mathcal{P}(G) & \xrightarrow{\;\;\tilde{P}\;\;} & \mathcal{P}(H) \\ W & \longmapsto & \{Pw \mid w \in W\} \end{array}$$

and

$$(11.95) \qquad \begin{array}{ccc} \mathcal{P}(G) & \xrightarrow{\;\;\tilde{R}\;\;} & \mathcal{P}(Y) \\ W & \longmapsto & \{Rw \mid w \in W\} \end{array}$$

Finally, we associate with our original semilinear hyperbolic system and its initial value problem (11.1), (11.2) the following two equations: given $\mu \in \mathcal{FM}(\mathbb{R},\mathbb{R}^n_+)$, find $W \in \mathcal{P}(G)$, such that

$$(11.96) \qquad \begin{array}{l} \tilde{P}W = \{0\} \\ \tilde{R}W = \hat{i}_0^{-1}(\mu) \end{array}$$

Proposition 11.6

Given $\mu \in \mathcal{FM}(\mathbb{R},\mathbb{R}^n_+)$, then (11.96) has a *unique* solution $W \in \mathcal{P}(G)$ and it coincides with the *unique* generalized solution $U \in \hat{X}$ of (11.29), that is, see (11.93)

$$(11.97) \qquad W = U \in \hat{X} \subseteq \mathcal{P}(G)$$

Proof

First we show that $U \in \hat{X}$ given in Theorem 11.1 is a solution of (11.96). Indeed, in view of (A.12) and (11.93), we obtain

$$U \in \hat{X} \Longrightarrow U \subseteq X \Longrightarrow \tilde{P}U = \{0\}$$

hence the first equation in (11.96) is satisfied. On the other hand (11.86), (11.83) give

$$(11.98) \qquad U = \{u \in X \mid Ru \in \hat{i}_0^{-1}(\mu)\}$$

thus in view of (11.88) and (11.9)

$$\tilde{R}U = \hat{i}_0^{-1}(\mu)$$

and the second equation in (11.96) is also satisfied.

Now it only remains to prove the uniqueness of the solution of (11.96). For that purpose, let $V \in \mathcal{P}(G)$ be another solution of (11.96). Then $\tilde{P}V = \{0\}$ and (11.93) give

$$V \subseteq X$$

However, we also have

$$(11.99) \qquad \tilde{R}V = \hat{i}_0^{-1}(\mu)$$

which means that

$$Rv \in \hat{i}_0^{-1}(\mu), \quad v \in V$$

hence (11.98) implies that

$$(11.100) \quad V \subseteq U$$

On the other hand, if $u \in U$ then (11.98) gives $Ru \in \hat{i}_0^{-1}(\mu)$. Therefore, according to (11.99), there exists $v \in V$, such that

$$(11.101) \quad Rv = Ru$$

But in view of (11.9), the classical solutions are unique, hence (11.101) implies $u = v$, which means that $u \in V$. In this way the inclusion

$$U \subseteq V$$

was proved, which together with (11.100) leads to $U = V$, and thus proves the uniqueness of the solution of (11.96). ∎

Remark 11.2

The extension method we used in (11.94) and (11.95) is purely set theoretical. Therefore, it is not the extension in (A.36), and which is intimately connected with the MacNeille order completion, see (A.37). However, if we use the extension in (A.36), that is, if we define

$$\begin{array}{ll} \hat{P} : \mathcal{P}(G) \longrightarrow \hat{H} & \hat{R} : \mathcal{P}(G) \longrightarrow \hat{Y} \\ \quad W \longmapsto (P(W))^{u\ell} \quad \text{and} & \quad W \longmapsto (R(G))^{u\ell} \end{array}$$

it appears that we may *lose* the uniqueness of the solution of (11.96). Indeed, U in (11.29) would still be a solution of (11.96), since we would have

$$\hat{P}U = \langle 0] \in \hat{H}, \quad \hat{R}U = (\hat{i}_0^{-1}(\mu))^{u\ell} = \hat{i}_0^{-1}(\mu)$$

However, if as in the proof of Proposition 11.6, $V \in \mathcal{P}(G)$ is another solution of (11.96), we could only conclude that $\hat{P}V = <0]$, which does not appear to mean that $V \subseteq X$ as well.

12. SOLUTION OF PARTIAL DIFFERENTIAL EQUATIONS AND THE COMPLETION OF UNIFORM SPACES

The purpose of this Section is to show that the *order completion* method used in the previous Sections for the solution of general continuous nonlinear partial differential equations can be seen as a parallel and alternative method to the well established *customary functional analytic method* which uses the *completion of uniform spaces*, Bourbaki, Köthe, Pervin, be they normed or mere locally convex vector spaces. We shall also point out the relative advantages of these two methods. However, from the outset, one should note that the customary solution method through the completion of uniform spaces is rather geared to the solution of *linear* partial differential equations where it can easily offer its main advantages. On the other hand, the solution method based on order completion performs equally well *both for linear and nonlinear* partial differential equations.

Furthermore, one advantage of the order completion method is that, as seen in Subsection 7.1, the *generalized solutions* obtained both for linear and nonlinear partial differential equations can be assimilated with *usual measurable functions* on Euclidean spaces. The same, of course, does *not* hold for many of the generalized solutions - even of linear partial differential equations - when they are obtained through the completion of uniform spaces.

The completion of normed or locally convex vector spaces is often used for the solution of *linear* partial differential equations. In fact, many of the results concerning the existence of *generalized* solutions for such equations are obtained in this way. And the advantage is quite obvious. Indeed, one may start with a rather small vector space of sufficiently smooth functions containing classical solutions or merely approximate solutions, and then, by a completion of a suitable uniform structure on them, one may build up large enough spaces of generalized functions, for instance, spaces of distributions or ultradistributions, which may contain generalized solutions.

However, as we shall see later in this Section, in the case of *nonlinear* partial differential equations one may no longer enjoy so easily the convenience of dealing solely with *vector spaces* of functions and with their normed or locally convex uniform topologies.

For clarity, let us start with a well known, rather basic and typical example which illustrates the way the completion of uniform spaces can yield generalized solutions for linear partial differential equations.

Let $\Omega \subset \mathbb{R}^n$ be a bounded open set with a smooth boundary $\partial\Omega$. It is well known, Treves, Gilbarg & Trudinger, that for any given $f \in \mathcal{C}^\infty(\bar{\Omega})$, the equation

$$(12.1) \qquad \Delta u(x) = f(x), \quad x \in \Omega$$

has a unique solution $u \in X$, where

$$(12.2) \qquad X = \{v \in \mathcal{C}^\infty(\bar{\Omega}) \mid v|_{\partial\Omega} = 0\}$$

It follows that

$$(12.3) \qquad X \ni v \longmapsto \|\Delta x\|_{\mathcal{L}^2(\Omega)}$$

defines a norm on the vector space X. Now let

$$(12.4) \qquad Y = \mathcal{C}^\infty(\bar{\Omega})$$

with the topology induced by $\mathcal{L}^2(\Omega)$, then in view of (12.1) - (12.4), it follows that

$$(12.5) \qquad \Delta : X \longrightarrow Y$$

is a *uniformly continuous* linear *bijection*. Therefore, it can be extended in a unique manner to an *isomorphism* of Banach spaces

$$(12.6) \qquad \Delta : \bar{X} \longrightarrow \bar{Y} = \mathcal{L}^2(\Omega)$$

In other words, through the *completion* of the *uniform spaces* X and Y in (12.5) which led us to (12.6), we obtain the quite general *existence and uniqueness* result

$$
(12.7) \quad \begin{array}{l} \forall\ f \in \mathcal{L}^2(\Omega) : \\ \exists!\ u \in \bar{X} : \\ \quad \Delta u = f \end{array}
$$

The power and the simplicity of the above solution method through completion of uniform spaces is quite obvious. Indeed, this is illustrated, among others, by the fact that the space $\bar{Y} = \mathcal{L}^2(\Omega)$ is much larger than the original $Y = \mathcal{C}^\infty(\bar{\Omega})$.

The only problem left with the extended existence result in (12.7) is about the *structure* of the space $\bar{X}$, that is, about the nature of the *generalized solutions* $u \in \bar{X}$.

In the *linear* case, such as for instance in the example above, this problem can be settled through an appropriate study of the partial differential operator involved, which in our case is the Laplacean Δ. And in the case of (12.7), it turns out that

$$
(12.8) \quad \bar{X} = H^2(\Omega) \cap H^1_0(\Omega)
$$

In view of the above, we shall consider the following general setup

$$
(12.9) \quad T : X \xrightarrow{\text{bij}} Y
$$

where X is an arbitrary set, Y is a Hausdorff uniform space and T is a bijection. Then X can be endowed with the initial uniformity with respect to T. In other words, the uniform structure on X will be the 'pull-back' through T of the uniform structure of Y. In this case T becomes uniformly continuous form X to Y. In fact, T will be an *isomorphism* of the uniform spaces X and Y. Therefore, it can be *extended* to an isomorphism of uniform spaces, see below

$$
(12.10) \quad \bar{T} : \bar{X} \longrightarrow \bar{Y}
$$

where $\bar{X}$, $\bar{Y}$ denote the completion of the uniform spaces X, Y respectively. In this way we obtain the commutative diagram of uniformly continuous mappings

(12.11)

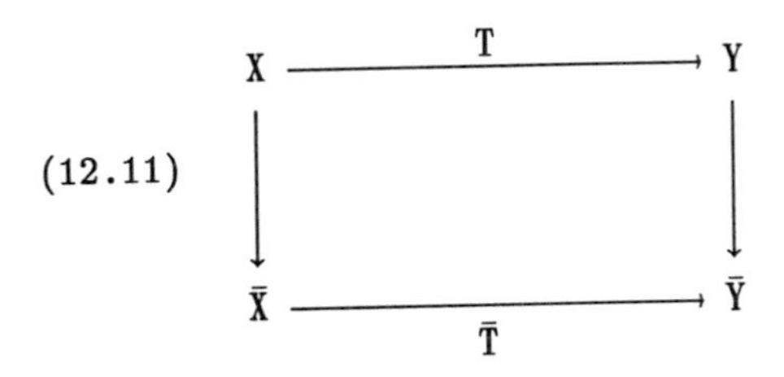

where T and $\bar{T}$ are actually isomorphisms of the respective uniform spaces, while the vertical arrows are the canonical embeddings of the respective uniform spaces into their completions.

In particular, one obtains the following *existence and uniqueness* of generalized solutions

$$\begin{aligned} &\forall \ f \in \bar{Y} : \\ (12.12) \qquad &\exists! \ u \in \bar{X} : \\ &\bar{T}u = f \end{aligned}$$

Usually, X in (12.11) is a rather small space of *classical* functions. On the other hand, the space of *generalized functions* $\bar{X}$ - which may *depend* on T - can be quite large. This fact can obviously facilitate the existence of generalized solutions. Beyond this existence problem, one will have to determine the nature or structure of the elements of $\bar{X}$ which give the generalized solutions.

However, *two advantages* of the solution method in (12.12) through the completion of uniform spaces remain valid regardless of the possible problem of the further clarification of the structure of the elements of these completions. First, the *existence* of the generalized solutions $u \in \bar{X}$ can be asserted without knowing the structure of the elements of $\bar{X}$. Second, these generalized solutions $u \in \bar{X}$ can be conveniently *approximated* by nets of classical functions from X, since X will be *dense* in $\bar{X}$.

In case T in (12.9) is *nonlinear*, X and $\bar{X}$ need no longer be topological vector spaces, such as for instance normed or locally convex. Nevertheless, the existence of solution result in (12.12) will still hold. In such a situation, however, the clarification of the structure of the elements in $\bar{X}$ may become even more difficult, and not less important.

We note therefore that, with the method of *order completion*, the resulting change in approach, when compared with the customary functional analytic handling of nonlinear partial differential equations, is then obvious. Namely, we are no longer looking for generalized solutions in topological vector spaces - spaces usually tied to linear partial differential operators, like for instance the various Sobolev spaces, defined by means of selected sets of partial derivatives. Instead, we tie our spaces of generalized functions to the specific nonlinear partial differential operator under consideration. In this way, the *solvability* becomes a rather easy task, after which we have to deal with the problem of the *regularity* of solutions. And as mentioned, this latter problem centers around the structure of the elements of $\bar{X}$.

We now recall without proofs several basic results from the theory of uniform spaces, Bourbaki, Köthe [1], Pervin. Then we apply them to prove a *basic extension* result of the type (12.11), (12.12). Finally, we give a few applications to linear and nonlinear partial differential equations.

Given a set X and a filter $\mathcal{N}$ on $X \times X$, we call $(X,\mathcal{N})$ a *uniform space*, if and only if the following three conditions are satisfied

$$\Delta_X = \{(x,x) \mid x \in X\} \subseteq N, \quad N \in \mathcal{N} \tag{12.13}$$

$$N^{-1} = \{(y,x) \mid (x,y) \in N\} \in \mathcal{N}, \quad N \in \mathcal{N} \tag{12.14}$$

and

$$\begin{aligned} &\forall\ N \in \mathcal{N} : \\ &\exists\ M \in \mathcal{N} : \\ &\quad M \circ M \subseteq N \end{aligned} \tag{12.15}$$

where for $A, B \subseteq X \times X$ we denote

$$A \circ B = \left\{ (x,z) \;\middle|\; \begin{array}{l} \exists\, y \in Y : \\ \quad (x,y) \in A, \quad (y,z) \in B \end{array} \right\}$$

The sets $N \in \mathcal{N}$ are called *vicinities* of the *diagonal* Δ_X, see (12.13). For $N \in \mathcal{N}$ and $x \in X$ we define the N-*neighbourhood* of x by

$$(12.16) \qquad V_N(x) = \{ y \in X \mid (x,y) \in N \}$$

It follows that $V_N(x)$, with $x \in X$, $N \in \mathcal{N}$ form a *base of neighbourhoods* for a topology on X, called the *associated uniform topology*. Moreover, this topology on X will be Hausdorff, if and only if

$$(12.17) \qquad \bigcap_{N \in \mathcal{N}} N = \Delta_X$$

in which case $(X, \mathcal{N})$ is called a *Hausdorff* uniform space.

Given two uniform spaces $(X, \mathcal{N})$ and $(Y, \mathcal{M})$, a mapping $T : X \longrightarrow Y$ is called *uniformly continuous*, if and only if

$$(12.18) \qquad \begin{array}{l} \forall\, M \in \mathcal{M} : \\ \exists\, N \in \mathcal{N} : \\ \quad (x,y) \in N \Longrightarrow (Tx, Ty) \in M \end{array}$$

The mapping $T : X \longrightarrow Y$ is called an *isomorphism* of the uniform spaces X and Y, if and only if it is bijective and both T and T^{-1} are uniformly continuous.

A net $(x_i)_{i \in I}$ in X is called *Cauchy*, if and only if

$$(12.19) \qquad \begin{array}{l} \forall\, N \in \mathcal{N} : \\ \exists\, i \in I : \\ \forall\, j,k \in I \;\; : \\ \quad j,k \geq i \Longrightarrow (x_j, x_k) \in N \end{array}$$

Finally, a uniform space $(X,\mathcal{N})$ is called *complete*, if and only if every Cauchy net in X converges to an element in X, when X is considered with the associated uniform topology.

The basic result on the *completion of uniform spaces* is given in

Proposition 12.1

For every Hausdorff uniform space $(X,\mathcal{N})$ there exists a *complete* Hausdorff uniform space $(\bar{X},\bar{\mathcal{N}})$, with $X \subseteq \bar{X}$, such that the inclusion mapping

(12.20) $\quad X \xrightarrow{\subseteq} \bar{X}$

is uniformly continuous, and

(12.21) $\quad X$ is dense in $\bar{X}$

where $\bar{X}$ is considered with its associated uniform topology.

Furthermore $(\bar{X},\bar{\mathcal{N}})$ has the following *universal* property: given any complete Hausdorff uniform space $(Y,\mathcal{M})$ and a uniformly continuous mapping $T : X \longrightarrow Y$, there *exist a unique* uniformly continuous extension

(12.22) $\quad \bar{T} : \bar{X} \longrightarrow Y$

In particular $(\bar{X},\bar{\mathcal{N}})$ is *unique* up to isomorphisms of uniform spaces. Finally, if $(Z,\mathcal{L})$ is any Hausdorff uniform space and $S : X \longrightarrow Z$ is uniformly continuous, then there exists a *unique* uniformly continuous extension

(12.23) $\quad \bar{S} : \bar{X} \longrightarrow \bar{Z}$ ∎

Remark 12.1

The complete Hausdorff uniform space $(\bar{X},\bar{\mathcal{N}})$ can be obtained in a *constructive* manner. Indeed, let $\mathcal{B} \subseteq \mathcal{N}$ be any filter base of $\mathcal{N}$, such that

$$N^{-1} = N, \quad N \in \mathcal{B}$$

Let us index the elements of $\mathcal{B}$ according to

$$\mathcal{B} = \{N_i \mid i \in I\}$$

and define a directed partial order on I by

$$i \leq j \iff N_i \supseteq N_j$$

Further, let

$$\tilde{X} = \{(x_i)_{i \in I} \mid (x_i)_{i \in I} \text{ Cauchy net in } (X, \mathcal{N})\}$$

Now we define an equivalence relation $\sim$ on $\tilde{X}$ by

$$(x_i)_{i \in I} \sim (y_i)_{i \in I} \iff \begin{bmatrix} \forall\ i \in I : \\ \exists\ j \in I : \\ \forall\ k \in I, \quad k \geq j : \\ \quad (x_k, y_k) \in N_i \end{bmatrix}$$

Then we can take

$$\bar{X} = \tilde{X}/\sim$$

Finally, the definition of $\bar{\mathcal{N}}$ as well can be obtained in a constructive way, in terms of the above chosen filter base $\mathcal{B}$ for $\mathcal{N}$, see Köthe [1, p 34].

■

Let X be a set, $(Y, \mathcal{M})$ a Hausdorff uniform space and $T : X \longrightarrow Y$ any given mapping. For $M \in \mathcal{M}$, we define

$$(12.24) \qquad N_{T,M} = \{(x, x') \in X \times X \mid (Tx, Tx') \in M\}$$

Further, we define

(12.25) $\mathcal{B}_T = \{N_{T,M} \mid M \in \mathcal{M}\}$

Then $\mathcal{B}_T$ is the filter base of a uniform space $(X,\mathcal{N}_T)$. We call $(X,\mathcal{N}_T)$ the *initial* uniform space on X with respect to T.

In the case $X \subseteq Y$ and T is the inclusion mapping $X \xrightarrow{\subseteq} Y$, the respective initial uniform space on X is called the uniform space *induced* by $(Y,\mathcal{M})$.

As was mentioned, the case of particular interest to us will be that of *bijections* $T : X \longrightarrow Y$, see (12.9), where Y is a uniform space, while X is a set which is going to be endowed with the *initial* uniform space with respect to T. In this case, the basic result is given by

Proposition 12.2

Given a bijection $T : X \longrightarrow Y$ between as set X and a uniform space $(Y,\mathcal{M})$, then

$$T : (X,\mathcal{N}_T) \longrightarrow (Y,\mathcal{M})$$

is an *isomorphism* of uniform spaces and it has a *unique* extension

$$\bar{T} : (\bar{X},\bar{\mathcal{N}}_T) \longrightarrow (\bar{Y},\bar{\mathcal{M}})$$

which is again an isomorphism of uniform spaces. ∎

In view of the above, we can formulate in terms of *completions of uniform spaces* the basic existence and uniqueness property, which to a certain extent, see Remark 12.2 below, parallels the result in Theorem 5.3, a result obtained through *order completion*.

Theorem 12.1

Let X be a set, $(Y,\mathcal{M})$ a Hausdorff uniform space and $T : X \longrightarrow Y$ a bijective mapping. If we consider on X the initial uniform space $(X,\mathcal{N}_T)$ with respect to T, then

$$\forall \quad \bar{y} \in \bar{Y} :$$

$$(12.26) \qquad \exists! \ \bar{x} \in \bar{X} :$$

$$\bar{T}(\bar{x}) = \bar{y}$$

Further, for every $\bar{x} \in \bar{X}$ and net $(\bar{x}_i)_{i\in I}$ in $\bar{X}$ which converges to $\bar{x}$, we obtain in $\bar{Y}$ that the net $(\bar{T}(\bar{x}_i))_{i\in I}$ will converge to $\bar{T}(\bar{x})$.

Finally, since X is dense in $\bar{X}$, for every $\bar{x} \in \bar{X}$ there exist nets $(x_i)_{i\in I}$ in X which converge in $\bar{X}$ to $\bar{x}$. In particular, it follows that $(T(x_i))_{i\in I}$ will converge in $\bar{Y}$ to $\bar{T}(\bar{x})$. ∎

Remark 12.2

1) When compared with Theorem 12.1, it is clear that Theorem 5.3 requires a *weaker hypothesis* than that in (12.9) which is *essential* in the former theorem, see Remark 12.3 below. Indeed, the mapping in Theorem 5.3, see (4.6)

$$(12.27) \qquad \mathcal{M}^m_T(\Omega) \xrightarrow[\text{inj}]{T} \mathcal{M}^0(\Omega)$$

is *only injective*. And as mentioned in (1.9) - (1.13), this is precisely one of the major problems in solving linear or nonlinear partial differential equations. Therefore, the special strength and interest of Theorem 5.3, when compared with Theorem 12.1 for instance, is that it turns the injective, and generally *nonsurjective*, mapping in (12.27) into the *order isomorphism* in (5.82), namely

$$(12.28) \qquad \hat{\mathcal{M}}^m_T(\Omega) \xrightarrow[\text{bij}]{\hat{T}} \hat{\mathcal{M}}^0(\Omega)$$

And it is *particularly important* to mention that, especially in the case of a *nonlinear* T in (12.27), it is not at all easy to obtain a satisfactory description of the *range* set of T, that is of $T(\mathcal{M}^m_T(\Omega))$, based on which we may try to reduce the merely injective setting in (12.27) to a bijective one, such as in (12.9). On the other hand, as seen in Section 4, the merely injective setting in (12.27) follows in a rather immediate manner for large classes of nonlinear T.

2) Along with the above, the interest in the abstract existence and uniqueness result in Theorem 9.2 is similar to that in Theorem 5.3, rather than to that in Theorem 12.1. Namely, Theorem 9.2 gives the *necessary and sufficient condition* for turning the *injective* mapping, see (9.7)

$$(12.29) \qquad X_T \xrightarrow[\text{inj}]{T} Y$$

into the *order isomorphism*, see (9.26)

$$(12.30) \qquad \hat{X}_T \xrightarrow[\text{bij}]{\hat{T}} \hat{Y}$$

3) Nevertheless, the setup in (12.9) or in Theorem 12.1 is relevant in a variety of situations. It applies, for instance, in the case of (10.3) and (10.4), provided that X_T and Y are restricted accordingly, namely, Y is replaced with its subset

$$(12.31) \qquad \prod_{\lambda \in \Lambda} \mathcal{C}^\infty(\Omega_\lambda)$$

while X_T is replaced with the inverse image of (12.31) through T.

Obviously (12.9) applies in the case of (11.9), and as seen in (12.5) or in the examples below, it can be encountered in a variety of problems related to the solution of linear or nonlinear partial differential equations.

■

As a first application of Theorem 12.1, let us return to the problem dealt with in (12.1) - (12.8) and consider it in its more general form of the Dirichlet problem for the Poisson equation with *distributional* boundary values.

Example 12.1

Let $\Omega \subset \mathbb{R}^n$ be open and bounded and with a smooth boundary $\partial\Omega$. Further, let

$$(12.32) \quad X = \mathcal{C}^\infty(\bar{\Omega}), \quad Y = \mathcal{C}^\infty(\bar{\Omega}) \times \mathcal{C}^\infty(\partial\Omega)$$

It is well known, Treves, Gilbarg & Trudinger, that then the mapping

$$(12.33) \quad \begin{array}{ccc} X & \xrightarrow{\ T\ } & Y \\ u & \longmapsto & (\Delta u, u|_{\partial\Omega}) \end{array}$$

is bijective.

Let us now consider on Y the uniform space induced by the product of the strong topologies $\mathcal{D}'(\Omega) \times \mathcal{D}'(\partial\Omega)$. Then we can equip X with the resulting initial uniform space with respect to T. In this case Theorem 12.1 will yield an *isomorphism* of uniform spaces

$$(12.34) \quad \bar{T} : \bar{X} \longrightarrow \bar{Y}$$

However, it is clear that

$$(12.35) \quad \bar{Y} = \mathcal{D}'(\Omega) \times \mathcal{D}'(\partial\Omega)$$

Therefore we conclude that the problem

$$(12.36) \quad \Delta u = f \quad \text{on} \quad \Omega, \quad u|_{\partial\Omega} = g$$

has a *unique generalized solution* $u \in \bar{X}$, for any $f \in \mathcal{D}'(\Omega)$ and $g \in \mathcal{D}'(\partial\Omega)$. Concerning the *structure* of $\bar{X}$, and in particular of its elements $v \in \bar{X}$, we shall see later in this Section that

$$(12.37) \quad H^1(\Omega) \subset \bar{X}$$

Moreover, if $f \in H^{-1}(\Omega)$ and $g \in H^{1/2}(\Omega)$, then the generalized solution $u \in \bar{X}$ of (12.36) *coincides* with the classical solution in $H^1(\Omega)$.

Example 12.2

Let us consider the following version of the semilinear hyperbolic problem in (11.1), (11.2), namely

(12.38) $\quad LU(t,x) = (D_t + CD_x)U(t,x) + A(U(t,x)) = f(t,x), \quad t \geq 0, \quad x \in \mathbb{R}$

(12.39) $\quad RU(t,x) = U(0,x) = g(x), \quad x \in \mathbb{R}$

where C is a diagonal $n \times n$ constant matrix, $A : \mathbb{R}^n \longrightarrow \mathbb{R}^n$ is $\mathcal{C}^\infty$-smooth, with $\nabla_u A(u)$ bounded for $u \in \mathbb{R}^n$. We let

(12.40) $\quad X = \mathcal{C}^\infty([0,\infty) \times \mathbb{R},\mathbb{R}^n), \quad Y = \mathcal{C}^\infty([0,\infty) \times \mathbb{R},\mathbb{R}^n) \times \mathcal{C}^\infty(\mathbb{R},\mathbb{R}^n)$

As seen in Subsection 11.1, under these conditions the mapping

(12.41)
$$\begin{array}{ccc} X & \xrightarrow{\quad T \quad} & Y \\ u & \longmapsto & (Lu,Ru) \end{array}$$

is a *bijection*, see also Oberguggenberger [2, pp 121-127] for this result in the classical existence theory.

Now we equip Y with structure of uniform space induced by $Z = \mathcal{D}'((0,\infty) \times \mathbb{R}) \times \mathcal{D}'(\mathbb{R})$, while X will be considered with the corresponding initial uniform space with respect to T. Then, according to Theorem 12.1, it follows that we obtain the extension

(12.42) $\quad \bar{X} \xrightarrow{\quad \bar{T} \quad} \bar{Y}$

where $\bar{T}$ is an isomorphism of uniform spaces.

Since the corresponding version of (12.35) again applies, that is

(12.43) $\quad \bar{Y} = Z = \mathcal{D}'((0,\infty) \times \mathbb{R}) \times \mathcal{D}'(\mathbb{R})$

we obtain that the *semilinear* problem (12.38), (12.39) has a *unique generalized solution* $u \in \bar{X}$, for every $f \in \mathcal{D}'((0,\infty) \times \mathbb{R})$ and $g \in \mathcal{D}'(\mathbb{R})$.

The structure of such generalized solutions will be discussed later in this Section.

Example 12.3

We consider this time the case of a nonlinear ordinary differential equation which can, nevertheless, illustrate a *nontrivial* result. Let

$$(12.44) \qquad U'(t) + U(t)^3 = f(t), \quad t \in \mathbb{R}$$

It is known that in the setting of distributions the equation (12.44) *cannot* in general be solved for arbitrary $f \in \mathcal{E}'(\mathbb{R})$. Indeed, in order to be able to make sense of (12.44) in $\mathcal{D}'(\mathbb{R})$, we should require that

$$U \in \mathcal{L}^3_{loc}(\mathbb{R})$$

But then U and U' are distributions of order 1. Hence taking $f \in \mathcal{E}'(\mathbb{R})$ a distribution of order at least two would lead to a contradiction.

In order to apply Theorem 12.1 to the equation in (12.44), let Y be the vector space generated by the Dirac distributions on $\mathbb{R}$, that is, the set of all distributions of the form

$$(12.45) \qquad f = \sum_{0 \leq i \leq m} c_i \, \delta(t - t_i)$$

where $n \in \mathbb{N}$, $c_i, t_i \in \mathbb{R}$. Then clearly $Y \subset \mathcal{E}'(\mathbb{R})$, and

$$(12.46) \qquad \bar{Y} = \mathcal{E}'(\mathbb{R})$$

if we equip Y with the structure of uniform space induced by $\mathcal{E}'(\mathbb{R})$.

Now, let X be the space of all piecewise smooth functions $u : \mathbb{R} \longrightarrow \mathbb{R}$ which vanish on a left half line, and which in addition, satisfy

$$(12.47)\qquad u' + u^3 \in Y$$

Clearly, condition (12.47) is equivalent with requiring

$$(12.48)\qquad u'(t) + u(t)^3 = 0$$

at each $t \in \mathbb{R}$ where u is smooth.

Then, it is easy to see that the mapping

$$(12.49)\qquad \begin{array}{ccc} X & \xrightarrow{\ T\ } & Y \\ u & \longmapsto & u' + u^3 \end{array}$$

is *bijective*. Indeed, given $t_0, u_0 \in \mathbb{R}$, the solution to

$$(12.50)\qquad \begin{array}{l} U'(t) + U(t)^3 = 0, \quad t \in [t_0, \infty) \\ U(t_0) = u_0 \end{array}$$

is

$$(12.51)\qquad U(t) = u_0/(1 + 2(t - t_0)u_0^2)^{1/2}, \quad t \in [t_0, \infty)$$

It follows that the solution $U \in X$ to

$$(12.52)\qquad U'(t) + U(t)^3 = c_0\ \delta(t - t_0), \quad t \in \mathbb{R}$$

for any given $t_0, c_0 \in \mathbb{R}$, is

$$(12.53)\qquad U(t) = \left| \begin{array}{ll} 0 \ \text{ if } \ t < t_0 \\ c_0/(1 + 2(t - t_0)c_0^2)^{1/2} & \text{if } \ t \geq t_0 \end{array} \right.$$

In this way, for any f in (12.45), the solution of

$$(12.54)\qquad U'(t) + U(t)^3 = f(t), \quad t \in \mathbb{R}$$

can be obtained by patching up solutions of type (12.53), with suitably placed discontinuities.

In view of (12.49) we can apply Theorem 12.1 to equation (12.44), and owing to (12.46), we obtain the isomorphism of uniform spaces

$$(12.55) \quad \bar{X} \xrightarrow{\bar{T}} \bar{Y} = \mathcal{E}'(\mathbb{R})$$

This means that, unlike within the distributional framework, the equation (12.44) will have a *unique generalized solution* $U \in \bar{X}$, for every $f \in \mathcal{E}'(\mathbb{R})$. ∎

We now turn to the issue of the *coherence* between the generalized solutions given by Theorem 12.1 and the classical solutions, whenever the latter happen to exist. However, here we have to take a more careful look at the *extension* properties of uniformly continuous mappings, properties of which that in Proposition 12.2 has been but a particular case. In this respect, the basic results on the *extension of uniformly continuous* mappings are presented in the next two propositions.

Proposition 12.3

Let $T : S \longrightarrow Y$ be a uniformly continuous mapping from a dense subset S of a uniform space X, where S is considered with the induced structure of uniform space, into a complete Hausdorff uniform space Y. Then there exists a *unique* uniformly continuous extension of T onto the whole of X, that is

$$(12.56) \quad \bar{T} : X \longrightarrow Y$$ ∎

Proposition 12.4

Let S,Q be dense subsets of the complete Hausdorff uniform spaces X,Y respectively. If $T : S \longrightarrow Q$ is an *isomorphism* of the uniform spaces induced by X,Y on S,Q respectively, then there exists a *unique* extension

$$(12.57) \qquad \bar{T} : X \longrightarrow Y$$

which is an *isomorphism* of the uniform spaces X,T. ∎

Remark 12.3

The extension properties in (12.56) and (12.57) cannot be strengthened in general. Indeed, it is possible that a *bijective uniformly continuous* mapping between two uniform spaces X,Y

$$(12.58) \qquad T : X \longrightarrow Y$$

which *fails* to be an isomorphism of X,Y, will only extend to a *uniformly continuous* mapping

$$(12.59) \qquad \bar{T} : \bar{X} \longrightarrow \bar{Y}$$

with $\bar{T}$ being *neither injective, nor surjective.*

This further *highlights* the role of the *particular* framework of *bijective* mappings in (12.9) and of *initial* uniform spaces, see Proposition 12.2, within the solution method based on the *completion of uniform spaces.* Indeed, in general this particular framework *cannot* so easily be relaxed in order to accommodate mappings which are *merely injective*, such as for instance in (12.27), or more abstractly, in (12.29). And as mentioned in pct. 1) in Remark 12.2, the particular nature of the framework in (12.9) may present *disadvantages* especially when solving *nonlinear* partial differential equations.

Examples of the failure of $\bar{T}$ in (12.59) to be injective or surjective can be found in Bourbaki. Here we give a simple illustration of the failure of injectivity in (12.59). Let $X = Y = Z$ be the set of smooth functions $f : \mathbb{R} \longrightarrow \mathbb{R}$ which vanish on some left half line. We consider the bijective mappings

$$(12.60) \qquad X \xrightarrow{\ i\ } Y \xrightarrow{\ D_t\ } Z$$

where i is the identity. Further X and Z are equipped with the uniform space induced by $\mathcal{D}'(\mathbb{R})$, while Y has the initial uniform space with respect to D_t. In that case, it is clear that i is uniformly continuous. However, the extended mapping

$$(12.61) \qquad \bar{X} \xrightarrow{\ \bar{i}\ } \bar{Y} \text{ is not injective}$$

Indeed, let us note that

$$(12.62) \qquad \bar{X} = \mathcal{D}'(\mathbb{R})$$

Now, let us take $c \in \mathbb{R}$, $c \neq 0$, and a net $(f_j)_{j \in I}$ in X, such that $f_j \longrightarrow c$ in $\mathcal{D}'(\mathbb{R})$. Then the topology of $\mathcal{D}'(\mathbb{R})$ gives that $D_t(f_j) \longrightarrow D_t c = 0$ in $\mathcal{D}'(\mathbb{R})$. This however means that $i(f_j) \longrightarrow 0$ in $\bar{Y}$. Thus the uniform continuity of $\bar{i}$ results in $\bar{i}(c) = 0$, although we assumed that $c \neq 0$. ■

Nevertheless, the following *extension* property is valid. And in Remark 12.4, we shall see that, in fact, it amounts to a *regularity* result, when applied in the context of solving partial differential equations.

Theorem 12.2

Suppose given the commutative diagram

$$(12.63) \qquad \begin{array}{ccccc} X & \xrightarrow{\ \subsetneq\ } & X_1 & & \\ T \downarrow & & \downarrow T_1 & & \\ Y & \xrightarrow[\subsetneq]{} & Y_1 & \xrightarrow[\subsetneq]{} & Z \end{array}$$

Here X, Y are sets, X_1, Y_1, Z are Hausdorff uniform spaces, X dense in X_1, with X_1 and Z being complete, while the inclusion $Y_1 \subsetneq Z$ is uniformly continuous. Further, T, T_1 are bijective and T_1 is uniformly continuous. We equip Y with the uniform space induced by Z, while on X we consider the initial uniform space with respect to T.

Then the inclusion $X \subsetneq \bar{X}^T$ *lifts* to an inclusion $X_1 \subsetneq \bar{X}^T$ and we can extend (12.63) to the commutative diagram

(12.64) 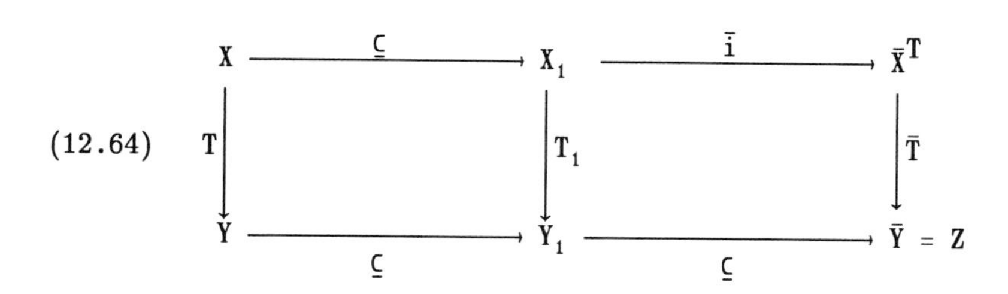

where $\bar{X}^T$ is the completion of X in the initial uniform space with respect to T, $\bar{i}$ is injective and uniformly continuous, while $\bar{T}$ is an isomorphism of the uniform spaces $\bar{X}^T$ and Z.

We note that the inclusions $Y_1 \subsetneq Z$, $Y \subsetneq Z$ are uniformly continuous, however, the inclusions $X \subsetneq X_1$, $Y \subsetneq Y_1$ need *not* be uniformly continuous.

Proof

Let us consider the identity mapping

$$X \xrightarrow{\ i\ } X$$

and equip its domain with the uniform space $(X,\mathcal{N}_1)$ induced by X_1, while its range with the initial uniform space $(X,\mathcal{N}_T)$ with respect to T. Then

$$(12.65) \qquad (X,\mathcal{N}_1) \xrightarrow{\ i\ } (X,\mathcal{N}_T)$$

is a uniformly continuous bijection. An arbitrary vicinity in the uniform space $(X,\mathcal{N}_T)$ is of the form

$$M = \{(x,y) \in X \times X \mid (Tx,Ty) \in N\}$$

where N is a vicinity in the uniform space Z. However, $T_1 : X_1 \longrightarrow Y_1$ and the inclusion $Y_1 \subseteq Z$ are uniformly continuous. Hence, there exists a vicinity L in the uniform space X_1, such that

$$(x,y) \in L \Rightarrow (T_1x,T_1y) \in N$$

But then

$$L \cap (X \times X) \subseteq M$$

and the proof of (12.65) is completed.

Now, according to Proposition 12.3, there exists a unique uniform continuous extension of (12.65), given by

(12.66) $\quad \bar{X}^1 = X_1 \xrightarrow{\quad \bar{i} \quad} \bar{X}^T$

where $\bar{X}^1$ and $\bar{X}^T$ denote the completions of $(X,\mathcal{N}_1)$ and $(X,\mathcal{N}_T)$ respectively. We prove that

(12.67) $\quad \bar{i}$ in (12.66) is injective

Indeed, let $f,g \in X_1$ and assume that

(12.68) $\quad \bar{i}(f) = \bar{i}(g)$

Since X is dense in X_1, we can take two nets $(f_j)_{j\in I}$ and $(g_j)_{j\in I}$ in X, such that

(12.69) $\quad f_j \longrightarrow f$ and $g_j \longrightarrow g$ in X_1

Then (12.66), (12.68) and (12.69) yield

(12.70) $\quad \bar{i}(f_j) \longrightarrow \bar{i}(f) = \bar{i}(g) \longleftarrow \bar{i}(g_j)$ in $\bar{X}^T$

However $f_j, g_j \in X$, with $j \in I$, while (12.66) is an extension of (12.65), therefore

(12.71) $\bar{i}(f_j) = i(f_j) = f_j, \quad \bar{i}(g_j) = i(g_j) = g_j, \quad j \in I$

Now (12.70), (12.71) imply that

$$f_j \longrightarrow \bar{i}(f) = \bar{i}(g) \longleftarrow g_j \quad \text{in} \quad \bar{X}^T$$

hence, by the definition of an initial uniform structure, we obtain that

(12.72) $T(f_j) \longrightarrow \bar{T}(\bar{i}(f)) = \bar{T}(\bar{i}(g)) \longleftarrow T(g_j) \quad \text{in} \quad Z$

On the other hand, the commutativity of (12.63) yields

(12.73) $T(f_j) = T_1(f_j), \quad T(g_j) = T_1(g_j), \quad j \in I$

However (12.69) implies

(12.74) $T_1(f_j) \longrightarrow T_1(f), \quad T_1(g_j) \longrightarrow T_1(g) \quad \text{in} \quad Y_1$

since T_1 is uniformly continuous. But (12.74) and the uniform continuity of the inclusion $Y_1 \subseteq Z$ give

$$T_1(f_j) \longrightarrow T_1(f), \quad T_1(g_j) \longrightarrow T_1(g) \quad \text{in} \quad Z$$

which together with (12.72), (12.73) result in

$$T_1(f) = T_1(g)$$

and then the injectivity of T_1 completes the proof of (12.67).

Finally, we show that

(12.75) $\bar{T}\big|_{X_1} = T_1$

For that purpose, let $f \in X_1$ and choose a net $(f_j)_{j\in I}$ in X, such that $f_j \longrightarrow f$ in X_1. Then similar to the argument above, we obtain

$$\bar{T}(\bar{i}(f)) = \bar{T}(\lim_{\bar{X}^T} f_j) = \lim_{Z} \bar{T}(f_j) = \lim_{Z} T(f_j)$$

while on the other hand

$$T_1(f) = T_1(\lim_{X_1} f_j) = \lim_{Y_1} T_1(f_j) = \lim_{Y_1} T(f_j) = \lim_{Z} T(f_j)$$

hence (12.74) holds. ∎

Remark 12.4

As mentioned, the extension property in Theorem 12.2 is in fact a *regularity* result in the context of partial differential equations. Indeed, since $\bar{T}$ is bijective, we have in particular that for every $y_1 \in Y_1$, there exists a unique $u \in \bar{X}^T$, such that

$$(12.76) \qquad \bar{T}(u) = y_1$$

On the other hand, the bijectivity of T_1 gives a unique $x_1 \in X_1$ which satisfies

$$T_1(x_1) = y_1$$

However, the commutativity of (12.64) yields

$$T_1(x_1) = \bar{T}(\bar{i}(x_1))$$

and then the injectivity of $\bar{T}$ will imply

$$(12.77) \qquad u = \bar{i}(x_1)$$

(13.6)

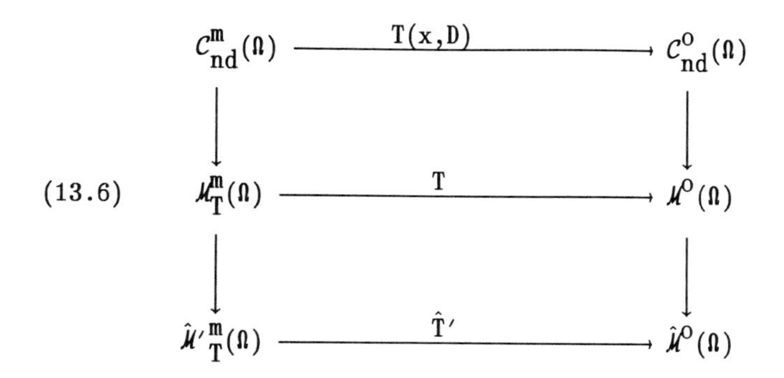

where $\hat{\mathcal{M}}'^m_T(\Omega)$ is the Dedekind order completion of $(\mathcal{M}^m_T(\Omega), \dashv)$, while $\hat{T}'$ is the corresponding mapping $\hat{\varphi}$ in (A.39). Furthermore, $\hat{T}'$ is *increasing*. However, it need *not* be injective or an order isomorphical embedding.

A *difficulty* with diagrams (13.6) is in the fact that, although $\dashv$ is *smaller* than $\leq_T$, it is nevertheless not so easy to establish a relationship between $\hat{\mathcal{M}}^m_T(\Omega)$ and $\hat{\mathcal{M}}'^m_T(\Omega)$, other than that resulting from MacNeille's Theorem, see Appendix, namely, the existence of the order isomorphical embeddings.

(13.7)

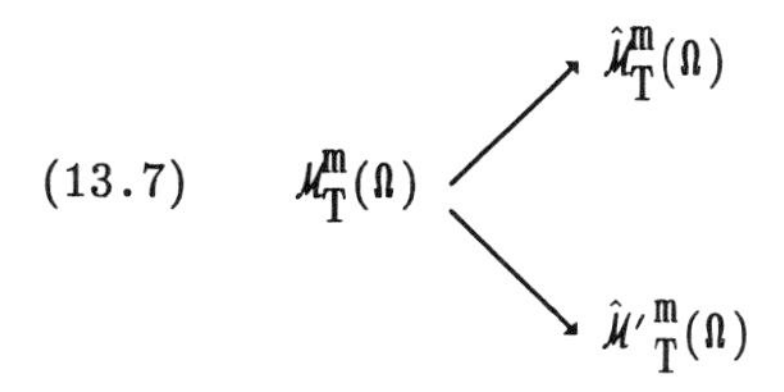

each corresponding to the respective partial order $\leq_T$ and $\dashv$ on $\mathcal{M}^m_T(\Omega)$.

Here, however, we shall set aside for the time being the diagrams (13.6), and we shall be interested in another *class* of commutative diagrams containing as particular cases the diagram (13.1) and avoiding the problem with the difficulty in (13.7) in establishing a connection between $\hat{\mathcal{M}}^m_T(\Omega)$ and $\hat{\mathcal{M}}'^m_T(\Omega)$. The interest in this new class of diagrams, see (13.14) below, becomes obvious in view of the fact that they lead to immediate *extensions* of the existence results presented in Section 5. For brevity, we shall only show the way Theorem 5.1 can be extended in order to obtain the *existence* result in Theorem 13.1 and Corollary 13.1 below. However, similar extensions can be obtained for the other existence results in Section 5.

We note that the interest in the class of diagrams (13.14) and in existence results such as in Theorem 13.1 and Corollary 13.1 becomes obvious in Section 18.

Before we go further, let us recall the way the poset $(\mathcal{M}^m_T(\Omega), \leq_T)$, see (13.1) - (13.3) was defined in (4.3) - (4.8). Namely, one starts with the *equivalence* relation $\sim_T$ on $\mathcal{C}^m_{nd}(\Omega)$ given by

$$u \sim_T v \iff T(x,D)u \sim T(x,D)v$$

Then one defines

$$\mathcal{M}^m_T(\Omega) = \mathcal{C}^m_{nd}(\Omega)/\sim_T$$

and obtains the *injective* mapping

$$\mathcal{M}^m_T(\Omega) \xrightarrow{\quad T \quad} \mathcal{M}^0(\Omega)$$

Finally, one defines the partial order $\leq_T$ on $\mathcal{M}^m_T(\Omega)$ as a 'pull-back' through T of the partial order $\leq$ on $\mathcal{M}^0(\Omega)$, that is

$$U \leq_T V \iff TU \leq TV$$

Now let us suppose that we are given another equivalence relation $\sim_{T,*}$ on $\mathcal{C}^m_{nd}(\Omega)$, such that

$$(13.8) \qquad \begin{aligned} &\forall \ u,v \in \mathcal{C}^m_{nd}(\Omega) : \\ &\quad u \sim_{T,*} v \Longrightarrow u \sim_T v \end{aligned}$$

Then defining

(13.9) $$\mathcal{M}^m_{T,*}(\Omega) = \mathcal{C}^m_{nd}(\Omega)/\sim_{T,*}$$

we obtain the *commutative diagram*

(13.10)

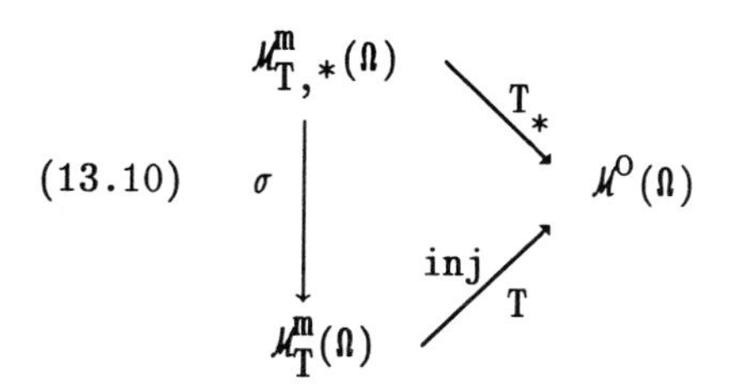

where σ is the *canonical* surjective mapping between the two respective quotient spaces and whose existence is guaranteed by (13.8), namely

$$\mathcal{M}^m_{T,*}(\Omega) \ni U \longmapsto U' = \sigma(U) \in \mathcal{M}^m_T(\Omega)$$

with U' being the $\sim_T$ equivalence class of any $u \in U$, while T_* is the *canonical* quotient mapping corresponding to (3.4), (13.8), (13.9), that is

$$\mathcal{M}^m_{T,*}(\Omega) \ni U \longmapsto T_* U \in \mathcal{M}^0(\Omega)$$

where T_*U is the $\sim$ equivalence class of any $T(x,D)u$, with $u \in U$.

Finally, we define the partial order $\leq_{T,*}$ om $\mathcal{M}^m_{T,*}(\Omega)$ as follows: if $U,V \in \mathcal{M}^m_{T,*}(\Omega)$ then

(13.11) $$U \leq_{T,*} V \iff \left[\begin{array}{l} *)\ U = V \\ \text{or} \\ **)\ T_*U \underset{\neq}{\leq} T_*V \ \text{ in } \mathcal{M}^0(\Omega) \end{array}\right]$$

In this way (13.10) becomes the commutative diagram of *increasing* mappings

$$(13.12) \qquad \begin{array}{ccc} (\mathcal{M}^{m}_{T,*}(\Omega), \leq_{T,*}) & \searrow^{T_*} & \\ \sigma \downarrow & & (\mathcal{M}^{0}(\Omega), \leq) \\ & \nearrow^{T}_{inj} & \\ (\mathcal{M}^{m}_{T}(\Omega), \leq_{T}) & & \end{array}$$

with T being in fact an order isomorphical embedding, see (4.8). Furthermore, it is easy to see that

$$(13.13) \qquad (\mathcal{M}^{m}_{T,*},(\Omega), \leq_{T,*}) \text{ has no minimum or maximum}$$

Finally, it is clear that, in case the equivalence relation $\sim_{T,*}$ is identical with $\sim_T$ then

$$(\mathcal{M}^{m}_{T,*}(\Omega), \leq_{T,*}) = (\mathcal{M}^{m}_{T}(\Omega), \leq_{T})$$

moreover σ becomes the identity mapping, while $T_* = T$.

Remark 13.1

It is important to note that, in general, the partial order $\leq_{T,*}$ defined in (13.11) on $\mathcal{M}^{m}_{T,*}(\Omega)$ is *not* the 'pull-back' through T_* of the partial order $\leq$ on $\mathcal{M}^{0}(\Omega)$. Indeed, let $\dashv$ denote the 'pull-back' of $\leq$ through T_*, then for $U, V \in \mathcal{M}^{m}_{T,*}(\Omega)$ we have

$$U \dashv V \Longleftrightarrow T_* U \leq T_* V$$

However, since T_* need *not* be injective, it is clear that $\dashv$ may *fail* to be antisymmetric, and therefore, $\dashv$ will in general *not* be a partial order. Further, it is obvious that

$$\begin{array}{l} \forall \; U, V \in \mathcal{M}^{m}_{T,*}(\Omega) : \\ \quad U \leq_{T,*} V \Rightarrow U \dashv V \end{array}$$

in other words, as a binary relation, $\leq_{T,*}$ is *smaller* than $\dashv$. ■

Now in view of Proposition A.1, the diagrams (13.1) become a *particular case* of the following class of commutative diagrams

$$\begin{array}{ccc} \mathcal{C}^m_{nd}(\Omega) & \xrightarrow{T(x,D)} & \mathcal{C}^0_{nd}(\Omega) \\ \downarrow & & \downarrow \\ \mathcal{M}^m_{T,*}(\Omega) & \xrightarrow{T_*} & \mathcal{M}^0(\Omega) \\ \downarrow & & \downarrow \\ \hat{\mathcal{M}}^m_{T,*}(\Omega) & \xrightarrow{\hat{T}_*} & \hat{\mathcal{M}}^0(\Omega) \end{array} \qquad (13.14)$$

with both T_* and $\hat{T}_*$ being *increasing* mappings.

As mentioned, the interest in diagrams (13.14) comes from the fact that they allow an extension of the *existence* results in Section 5. However, this time T_* and $\hat{T}_*$ need *not* be injective. For instance, we obtain the following more general version of Theorem 5.1:

Theorem 13.1

For every bounded open set $\Omega \subset \mathbb{R}^n$ we have

$$(13.15) \quad \hat{T}_*(\hat{\mathcal{M}}^m_{T,*}(\Omega)) \supseteq \mathcal{C}^0(\bar{\Omega})$$

Moreover, if $f \in \mathcal{C}^0(\bar{\Omega})$ then

$$(13.16) \quad F = \{U \in \mathcal{M}^m_{T,*}(\Omega) \mid T_*U \leq f \text{ in } \mathcal{M}^0(\Omega)\} \in \hat{\mathcal{M}}^m_{T,*}(\Omega)$$

is the *maximum* global generalized solution of the continuous nonlinear PDE, see (2.1), (2.2)

$$(13.17) \quad T(x,D)U(x) = f(x), \quad x \in \Omega$$

that is, we have

$$(13.18) \quad \hat{T}_* F = f \quad \text{in} \quad \hat{\mathcal{M}}^0(\Omega)$$

and at the same time

$$(13.19) \quad \begin{array}{l} \forall\ F' \in \hat{\mathcal{M}}^m_{T,*}(\Omega) : \\ \quad \hat{T}_* F' = f \quad \text{in} \quad \hat{\mathcal{M}}^0(\Omega) \Longrightarrow F' \leq F \quad \text{in} \quad \hat{\mathcal{M}}^m_{T,*}(\Omega) \end{array}$$

Proof

We follow the proof of Theorem 5.1 with slight adaptations.

We take any $f \in \mathcal{C}^0(\bar{\Omega})$. If $\epsilon > 0$ then (2.22) yields a closed, nowhere dense set $\Gamma_\epsilon \subset \Omega$ and a function $u_\epsilon \in \mathcal{C}^m(\Omega \backslash \Gamma_\epsilon)$, such that

$$(13.20) \quad f + \epsilon/2 \leq T(x,D)u_\epsilon \leq f + \epsilon \quad \text{on} \quad \Omega \backslash \Gamma_\epsilon$$

Let $U_\epsilon \in \mathcal{M}^m_{T,*}(\Omega)$ be the $\sim_{T,*}$ equivalence class generated by u_ϵ, see (13.9). Then (13.12), (13.20) give

$$(13.21) \quad f + \epsilon/2 \leq T_* U_\epsilon \leq f + \epsilon \quad \text{in} \quad \mathcal{M}^0(\Omega)$$

Now we take

$$(13.22) \quad A = \{U_\epsilon \mid \epsilon > 0\} \subseteq \mathcal{M}^m_{T,*}(\Omega)$$

and define, see (A.7)

$$(13.23) \quad F = A^\ell \subseteq \mathcal{M}^m_{T,*}(\Omega)$$

in the sense of the partial order $\leq_{T,*}$.

Then (A.16), (A.12) imply that

(13.24) $\quad F \in \hat{\mathcal{M}}^m_{T,*}(\Omega)$

Moreover

(13.25) $\quad \phi \underset{\neq}{\subset} F \underset{\neq}{\subset} \mathcal{M}^m_{T,*}(\Omega)$

Indeed, (2.15) yields $V \in \mathcal{M}^m_{T,*}(\Omega)$, such that

$$T_* V \leq f \quad \text{in} \quad \mathcal{M}^0(\Omega)$$

hence (13.21), (13.11) imply that in $\mathcal{M}^m_{T,*}(\Omega)$ we have the inequalities

$$V \leq_{T,*} U_\epsilon, \quad \epsilon > 0$$

which in view of (13.22), (13.23) means that $V \in F$. On the other hand (13.21), (13.22) imply that $A \neq \phi$, therefore (A.8) and (13.23) will result in $F \underset{\neq}{\subset} \mathcal{M}^m_{T,*}(\Omega)$, and the proof of (13.25) is completed.

An essential fact we prove now is the relation

(13.26) $\quad F = \{U \in \mathcal{M}^m_{T,*}(\Omega) \mid T_* U \leq f \quad \text{in} \quad \mathcal{M}^0(\Omega)\}$

For the inclusion '$\supseteq$' let U be in the right hand term of (13.26). Then in view of (13.9), (13.10), we obtain

$$T(x,D)u \leq f \quad \text{on} \quad \Omega\backslash\Gamma$$

for $u \in U$ and a suitable closed, nowhere dense subset $\Gamma \subset \Omega$. According to (13.20), it follows that for all $\epsilon > 0$ we have

$$T(x,D)u + \epsilon/2 \leq T(x,D)u_\epsilon \quad \text{on} \quad \Omega\backslash(\Gamma \cup \Gamma_\epsilon)$$

However, according to (13.11), this means that

$$U \leq_{T,*} U_\epsilon, \quad \epsilon > 0$$

and therefore (13.22), (13.23) will imply $U \in F$. For the converse inclusion '$\subseteq$' in (13.26), let $U \in F$ and $u \in U$. Then (13.10), (13.22), (13.23) give for every $\epsilon > 0$ the inequality, see (13.20).

$$(13.27) \quad T(x,D)u \leq T(x,D)u_\epsilon \leq f + \epsilon \quad \text{on} \quad \Omega \setminus \Gamma'_\epsilon$$

where $\Gamma'_\epsilon \subset \Omega$ is closed, nowhere dense. Let us now assume that

$$T_* U \leq f \quad \text{does } not \text{ hold in} \quad \mathcal{M}^0(\Omega)$$

Then, according to (3.5) and (13.10), for every closed, nowhere dense subset $\Gamma \subset \Omega$, there exists $x_0 \in \Omega \setminus \Gamma$, such that

$$T(x_0,D)u(x_0) > f(x_0)$$

But then the continuity properties of $T(x,D)$, u and f will yield $\delta > 0$ and an open subset $\Delta \subseteq \Omega \setminus \Gamma$, with $x_0 \in \Delta$, such that

$$T(x,D)u \geq f + \delta \quad \text{on} \quad \Delta$$

which will clearly contradict (13.27), and hence complete the proof of (13.26).

Now it only remains to show that F given in (13.23) satisfies the equation

$$(13.28) \quad \hat{T}_* F = \langle f] \quad \text{in} \quad \mathcal{M}^0(\Omega)$$

For that we note the inclusion

$$-\epsilon/4 \leq T_\lambda(x,D)P_{x_0}(x) - a_{x_0}m! \leq \frac{\epsilon}{4}, \quad \lambda \in \Lambda$$

$$-\epsilon/4 \leq f(x_0) - f(x) \leq \epsilon/4$$

But for $x, x_0 \in \Omega$, $\lambda \in \Lambda$, we have

$$T_\lambda(x,D)P_{x_0}(x) - f(x) =$$

$$= (T_\lambda(x,D)P_{x_0}(x) - a_{x_0}m!) + (a_{x_0}m! - f(x_0)) + (f(x_0) - f(x))$$

while

$$a_{x_0}m! - f(x_0) = -\frac{\epsilon}{2}$$

and in this way the proof of (14.3) is completed. ■

Based on the above lemma, we obtain the following *simultaneous approximation* version of Proposition 2.1

Proposition 14.1

Let $f \in \mathcal{C}^0(\bar{\Omega})$ then

$$\begin{aligned} &\forall\ \epsilon > 0 : \\ &\exists\ \Gamma_\epsilon \subset \Omega \text{ closed, nowhere dense, } U_\epsilon \in \mathcal{C}^m(\Omega \setminus \Gamma_\epsilon) : \\ &\forall\ \lambda \in \Lambda : \\ &\quad f - \epsilon \leq T_\lambda(x,D)U_\epsilon \leq f \text{ on } \Omega \setminus \Gamma_\epsilon \end{aligned} \qquad (14.5)$$

Proof

It follows from Lemma 14.1 in a manner similar with that in which Proposition 2.1 results from Lemma 2.1.

Remark 14.1 (see Remark 2.1)

1) It is possible to construct Γ_ϵ in (14.5) in such a way that, see (2.7)

$$(14.6) \qquad \text{mes } (\Gamma_\epsilon) = 0$$

2) We can obtain a version of (14.5) in which

$$(14.7) \qquad f \leq T(x,D)U_\epsilon \leq f + \epsilon \quad \text{on} \quad \Omega \backslash \Gamma_\epsilon$$

■

Before proceeding further, we note the existence of the mappings

$$(14.8) \qquad T_\lambda(x,D) : \mathcal{C}^m_{nd}(\Omega) \longrightarrow \mathcal{C}^0_{nd}(\Omega), \quad \lambda \in \Lambda$$

Let us now define the equivalence relation $\sim_\Lambda$ on $\mathcal{C}^m_{nd}(\Omega)$ by, see (4.3)

$$(14.9) \qquad u \sim_\Lambda v \Longleftrightarrow \begin{bmatrix} \forall \ \lambda \in \Lambda : \\ \quad T_\lambda(x,D)u \sim T_\lambda(x,D)v \end{bmatrix}$$

where $u,v \in \mathcal{C}^m_{nd}(\Omega)$, while $\sim$ is the equivalence relation on $\mathcal{C}^0_{nd}(\Omega)$, defined in (3.3). Then we can define, see (4.4)

$$(14.10) \qquad \mathcal{M}^m_\Lambda(\Omega) = \mathcal{C}^m_{nd}(\Omega)/\sim_\Lambda$$

and obtain for each $\lambda \in \Lambda$, the commutative diagram, see (4.6)

$$(14.11) \qquad \begin{array}{ccc} \mathcal{C}^m_{nd}(\Omega) & \xrightarrow{\quad T_\lambda(x,D) \quad} & C^0_{nd}(\Omega) \\ \downarrow & & \downarrow \\ \mathcal{M}^m_\Lambda(\Omega) & \xrightarrow[\quad T_\lambda \quad]{} & \mathcal{M}^0(\Omega) \end{array}$$

where, see (4.5)

$$(14.12)\quad \begin{array}{ccc} \mathcal{M}^m_\Lambda(\Omega) & \xrightarrow{\quad T_\lambda \quad} & \mathcal{M}^0(\Omega) \\ U \ni u & \longmapsto & T_\lambda(x,D)u \in F \end{array}$$

in other words, for given $U \in \mathcal{M}^m_\Lambda(\Omega)$, we obtain $F = T_\lambda(U)$ as the $\sim$ equivalence class in $\mathcal{M}^0(\Omega)$ generated by $T_\lambda(x,D)u$, where $u \in U$ can be taken arbitrarily.

Here it should be mentioned that, unlike in (4.5), (4.6), the mappings T_λ in (14.11), (14.12) will in general *not* be injective.

We define now a partial order $\leq_\Lambda$ on $\mathcal{M}^m_\Lambda(\Omega)$ by, see (4.7)

$$(14.13)\quad U \leq_\Lambda V \Longleftrightarrow \left[\begin{array}{l} \forall\ \lambda \in \Lambda : \\ \quad T_\lambda U \leq T_\lambda V \quad \text{in} \quad \mathcal{M}^0(\Omega) \end{array}\right]$$

Then clearly, each of the mappings

$$(14.14)\quad \mathcal{M}^m_\Lambda(\Omega) \xrightarrow{\quad T_\lambda \quad} \mathcal{M}^0(\Omega), \quad \lambda \in \Lambda$$

will be *increasing*, however, as we already noted, they need *not* be injective as well. Nevertheless, this does not prevent us from applying (A.39) to the increasing mappings (14.14), and thus obtain the commutative diagrams for $\lambda \in \Lambda$, see (14.11)

$$(14.15)\quad \begin{array}{ccc} \mathcal{C}^m_{nd}(\Omega) & \xrightarrow{\quad T_\lambda(x,D) \quad} & \mathcal{C}^0_{nd}(\Omega) \\ \downarrow & & \downarrow \\ \mathcal{M}^m_\Lambda(\Omega) & \xrightarrow{\quad T_\lambda \quad} & \mathcal{M}^0(\Omega) \\ \downarrow & & \downarrow \\ \hat{\mathcal{M}}^m_\Lambda(\Omega) & \xrightarrow{\quad \hat{T}_\lambda \quad} & \hat{\mathcal{M}}^0(\Omega) \end{array}$$

where, in addition to T_λ, the mapping $\hat{T}_\lambda$ as well will be increasing.

Remark 14.2 (see Remark 13.1)

1) It is important to note that, just as in (13.11), see also Remark 13.1, the poset

$$(14.16) \quad (\mathcal{M}^m_\Lambda(\Omega), \leq_\Lambda)$$

is in general *not* the 'pull-back' of the poset $(\mathcal{M}^0(\Omega), \leq)$ through any of the mappings T_λ, with $\lambda \in \Lambda$.

2) In order to be able to obtain $\hat{\mathcal{M}}^m_\Lambda(\Omega)$ in (14.15), we have to know that the poset in (15.16) does not have a minimum or a maximum element. This however follows easily from Lemma 14.1 in a way similar to that in which Lemma 4.2 is proved by using Lemma 2.1.

3) In case the index set Λ contains *one single* element, it is obvious that the construction in the previous part of this Subsection reduces to the corresponding construction in Sections 2-5. ■

Given $f \in \mathcal{C}^0(\bar{\Omega})$, we denote by

$$(14.17) \quad \mathcal{U}^-_f, \ \mathcal{U}^+_f \subseteq \mathcal{M}^m_\Lambda(\Omega)$$

the set of all $U \in \mathcal{M}^m_\Lambda(\Omega)$ which for a suitable $\epsilon > 0$, satisfy the inequality, see (14.5)

$$(14.18) \quad f - \epsilon \leq T_\lambda U \leq f \ \text{ in } \mathcal{M}^0(\Omega), \ \text{ for } \lambda \in \Lambda$$

respectively, see (14.7)

$$(14.19) \quad f \leq T_\lambda U \leq f + \epsilon \ \text{ in } \mathcal{M}^0(\Omega), \ \text{ for } \lambda \in \Lambda$$

Then clearly, we have in $\hat{\mathcal{M}}^m_\Lambda(\Omega)$ the inequality, see (14.13)

$$(14.20) \quad \sup \mathcal{U}^-_f \leq_\Lambda \inf \mathcal{U}^+_f$$

The result on the *simultaneous solvability* of families of nonlinear PDEs is presented in

Theorem 14.1

Let $f \in \mathcal{C}^0(\bar{\Omega})$ and $F \in \hat{\mathcal{M}}^m(\Omega)$, then

(14.21) $\sup \mathcal{U}_f^- \leq_\Lambda F \leq_\Lambda \inf \mathcal{U}_f^+$

will imply

(14.22) $\hat{T}_\lambda F = f, \quad \lambda \in \Lambda$

Proof

Given $\epsilon > 0$, let us $U^- \in \mathcal{U}_f^-$, $U^+ \in \mathcal{U}_f^+$ such that for $\lambda \in \Lambda$, the inequalities hold in $\mathcal{M}^0(\Omega)$

$$f - \epsilon \leq T_\lambda U^- \leq f \leq T_\lambda U^+ \leq f + \epsilon$$

Now (14.21) gives in $\hat{\mathcal{M}}_\Lambda^m(\Omega)$ the inequalities

$$U^- \leq_\Lambda F \leq_\Lambda U^+$$

and then (14.14) and (14.15) yield for $\lambda \in \Lambda$ the following inequalities in $\hat{\mathcal{M}}^0(\Omega)$

(14.23) $f - \epsilon \leq \hat{T}_\lambda U^- \leq \hat{T}F \leq \hat{T}_\lambda U^+ \leq f + \epsilon$

However, as $\epsilon > 0$ is arbitrary, it is clear that (14.23) will imply (14.22). ■

Corollary 14.1

For $f \in \mathcal{C}^0(\bar{\Omega})$ the family of nonlinear PDEs, see (14.1)

$$(14.24) \qquad T_\lambda(x,D)U(x) = f(x), \quad x \in \Omega, \quad \lambda \in \Lambda$$

has a *simultaneous generalized solution*

$$(14.25) \qquad F \in \hat{\mathcal{M}}^m_\Lambda(\Omega)$$

Proof

In view of (14.20), the condition (14.21) can always be satisfied for a certain $F \in \hat{\mathcal{M}}^m_\Lambda(\Omega)$. ■

Remark 14.3

1) The condition of *uniform strength* in (14.2) is clearly satisfied whenever Λ is a *finite* set. Therefore, in view of Corollary 14.1, it follows that *every finite family* of nonlinear PDEs, see (14.1)

$$(14.26) \qquad \begin{array}{l} T_1(x,D)U(x) = f(x), \quad x \in \Omega \\ \vdots \\ T_\ell(x,D)U(x) = f(x), \quad x \in \Omega \end{array}$$

with $f \in \mathcal{C}^0(\bar{\Omega})$ given, has a *simultaneous generalized solution*

$$(14.27) \qquad F \in \hat{\mathcal{M}}^m_{\{1,\ldots,\ell\}}(\Omega)$$

2) We shall now present an explanation of this apparently *paradoxical* result of *simultaneous generalized solutions* which may occur when the index set Λ has *at least two* elements. For that purpose, let us return to the general situation in (14.1), (14.2). Further, let us assume the existence of

$$(14.28) \qquad u_\lambda \in \mathcal{C}^m(\Omega), \quad \lambda \in \Lambda$$

such that

$$(14.29) \qquad T_\lambda(x,D)u_\lambda(x) = f(x), \quad x \in \Omega, \quad \lambda \in \Lambda$$

where $f \in \mathcal{C}^0(\bar{\Omega})$ is given. Here however, as it will usually happen in the case when classical solutions exists, we shall *not* assume that these classical solutions u_λ, with $\lambda \in \Lambda$, are necessarily the same.

For $\lambda \in \Lambda$, let $U_\lambda \in \mathcal{M}_\Lambda^m(\Omega)$ be the $\sim_\Lambda$ equivalence class of u_λ. Then (14.29) and (14.12) give in $\mathcal{M}^0(\Omega)$ the relations

(14.30) $\quad T_\lambda U_\lambda = f, \quad \lambda \in \Lambda$

therefore (14.15) will yield in $\hat{\mathcal{M}}^0(\Omega)$ the relations

(14.31) $\quad \hat{T}_\lambda U_\lambda = f, \quad \lambda \in \Lambda$

However, since the mappings $\hat{T}_\lambda$, with $\lambda \in \Lambda$, are *not* necessarily injective, see (14.14), it follows that the equations (14.31) and (14.22) need *not* imply any particular relationship between F, and the other hand, U_λ, with $\lambda \in \Lambda$.

On the other hand, if Λ has *one single* element, say λ, then as pointed out in pct. 3) in Remark 14.2, we are back to the situation in Sections 2-5. In particular, $\hat{T}_\lambda$ will be *injective*, and in fact, it will be an *order isomorphical embedding*. And then, it follows that the *classical* solution U_λ and the *generalized solution* F are related according to

(14.32) $\quad U_\lambda \in F$

with F being the *unique generalized solution*, see Section 6 for further details.

In view of the above, we can conclude as follows.

The situation when, as in (14.13), we do *not* use a 'pull-back' partial order on the space of generalized solutions, the order completion method for solving continuous nonlinear PDEs can lead to results such as in this Subsection.

This however, does *not* disqualify the use of such partial orders which are not obtained through 'pull-back'. Indeed, the results in Sections 13 and 18 testify in this respect.

Furthermore, as seen in Sections 8-12, and next, in Subsection 14.3, the order completion method for solving continuous nonlinear PDEs can offer genuine and relevant extensions of existence results known so far in the literature.

The overall conclusion is, therefore, that the order completion method is particularly general and powerful with regard to delivering *existence* results. And as seen in Section 9, it can give solutions not only to PDEs. In this way, its proper use in the context of solving linear or nonlinear PDEs has to be accompanied by an appropriate consideration of the circumstances involved.

14.3 Strengthened Approximation and Existence Results with Conditions on Derivatives

Here we shall briefly indicate one possible way to *strengthen* the approximation, and therefore, the *existence* results in Sections 2 and 5, respectively.

Let us consider a *bounded* open subset $\Omega \subset \mathbb{R}^n$ and the nonlinear partial differential operator, see (2.3)

$$(14.33) \quad T(x,D) = D_t^m + \sum_{1 \leq i \leq h} c_i(x) \prod_{1 \leq j \leq k_i} D_t^{p_{ij}} D_y^{q_{ij}}$$

where $x = (t,y) \in \Omega, \quad t \in \mathbb{R}, \quad y \in \mathbb{R}^{n-1}$,

$$p_{ij} \in \mathbb{N}, \quad q_{ij} \in \mathbb{N}^{n-1}, \quad p_{ij} < m, \quad p_{ij} + |q_{ij}| \leq m,$$

while about the coefficients we shall assume this time that

$$(14.34) \quad c_i \in \mathcal{C}^{\ell}(\bar{\Omega})$$

for a certain given $\ell \in \bar{\mathbb{N}} = \mathbb{N} \cup \{\infty\}$. Further, let us assume given a finite subset $S \subseteq \mathbb{N}^n$ of orders of partial differentiation, such that each $s \in S$ satisfies the condition $|s| \leq \ell$. For convenience, we shall assume that for the n-tuple of zeros $(0,\ldots,0) \in \mathbb{N}^n$ we have $(0,\ldots,0) \in S$.

Lemma 14.2 (see Lemma 2.1)

Given $f \in \mathcal{C}^\ell(\bar{\Omega})$, then

$$
(14.35)\qquad
\begin{array}{l}
\forall \ \epsilon > 0 : \\
\exists \ \delta > 0 : \\
\forall \ x_0 \in \Omega : \\
\exists \ P_{x_0} \text{ polynomial in } x \in \mathbb{R}^n : \\
\forall \ x \in \Omega, \ \ \|x - x_0\| \leq \delta, \ \ s \in S : \\
\quad D^s f(x) - \epsilon \leq D^s T(x,D) P_{x_0}(x) \leq D^s f(x)
\end{array}
$$

Proof

Let us assume that each $s \in S$ is of the form $s = (s', s'')$, where $s' \in \mathbb{N}$ and $s'' \in \mathbb{N}^{n-1}$ correspond to partial derivation in t and y respectively. Given now $\epsilon > 0$ and $x_0 = (t_0, y_0) \in \Omega$, let us define the polynomial in $x = (t,y) \in \mathbb{R}^n$, by

$$
P_{x_0}(x) = a_0 (t - t_0)^m + \sum_{s \in S} a_s (t - t_0)^{m+s'} (y - y_0)^{s''}
$$

where the coefficients are such that

$$
(m + s')! s''! a_s = D^s f(x_0) - \epsilon/2, \quad s \in S
$$

Then for $s \in S$ we obtain

$$
D^s T(x,d) P_{x_0}(x) = D^s f(x_0) - \epsilon/2 + r_s(t - t_0, y - y_0)
$$

where r_s are polynomials, with

$$r_s(0,0) = 0$$

In this way, for $x = (t,y) \in \bar{\Omega}$ and $s \in S$, we obtain

$$|D^s T(x,D) P_{x_0}(x) - D^s f(x) + \epsilon/2| \leq c'|t - t_0| + c''|y - y_0|$$

provided that

$$|t - t_0|, \ \|y - y_0\| \leq 1$$

where $c', c'' > 0$ do not depend on $x_0, x \in \Omega$. Then it follows that

$$-\epsilon \leq D^s T(x,D) P_{x_0}(x) - D^s f(x) \leq 0$$

for $x_0, x \in \Omega$, $s \in S$, provided that

$$\|x - x_0\| \leq \delta$$

where $\delta > 0$ does not depend on x_0, x. ∎

Now it is easy to obtain

Proposition 14.2 (see Proposition 2.1)

Let $f \in C^{\ell}(\bar{\Omega})$, then

$$\begin{aligned} &\forall\ \epsilon > 0 : \\ &\exists\ \Gamma_\epsilon \subset \Omega \text{ closed, nowhere dense, } U_\epsilon \in \mathcal{C}^{m+\ell}(\Omega \backslash \Gamma_\epsilon) \\ &\forall\ s \in S : \\ &\quad D^s f - \epsilon \leq D^s T(x,D) U_\epsilon \leq D^s f \ \text{ on } \Omega \backslash \Gamma_\epsilon \end{aligned} \tag{14.36}$$

∎

'... Whatever is thought by us is either conceived through itself, or involves the concept of another ... So one must either proceed to infinity, or all thoughts are resolved into those which are conceived through themselves ... Every idea is analysed perfectly only when it is demonstrated a priori that it is possible ... Since, however, it is not in our power to demonstrate the possibility of things in a perfectly a priori way, that is, to analyse them into God and nothing, it will be sufficient for us to reduce their immense multitude to a few, whose possibility can either be supposed and postulated, or proved by experience ...'

G.W. Leibnitz, Of an Organum
or Ars Magna (c. 1679)

PART III. GROUP INVARIANCE OF GLOBAL GENERALIZED SOLUTIONS OF NONLINEAR PDEs

15. INTRODUCTION

Next, in Part III, based on the *general global existence* results in Sections 5 and 13, as well as on the earlier more particular ones in Rosinger [5-7] and Colombeau [1-3], one obtains a certain degree of fulfillment *for the first time* in the literature, for Lie's original program of group invariance of nonlinear PDEs, started in 1874. In fact, the results concerning the group transformations of large spaces of global generalized solutions can also be seen as *three* different and independent solutions to Hilbert's fifth problem, when this problem is considered in its natural extended sense, see Subsection 16.1, corresponding to non smooth solutions of nonlinear PDEs.

It is useful to note, however, that these three solutions presented in Sections 16-18, are not only different and independent, but also, they exhibit rather *different properties*.

What is *common* to all of them is that they allow the definition of nonlinear group transformations on large spaces of generalized solutions which, as in Sections 16 and 17, will contain the L Schwartz distributions, see Subsections 16.4, 16.5, 17.2, 17.3, 18.3 and 18.4. This alone is a first in the literature, especially when the *simplicity* of the respective definitions is taken into account, see the more detailed comments in Subsection 16.1 and at the end of Subsection 16.5.

The *differences* between the three methods start with the *computation* of symmetry groups of nonlinear PDEs.

Indeed, in this regard, a first main distinction is to be found between the *easier* methods in Sections 16 and 18, and on the other hand, the method in Section 17. The methods in Sections 16, 18, which are based on the existence results in Rosinger [5-7], respectively, those in the earlier Sections 5 and 13 of this work, allow a surprisingly simple and direct way - the way which is precisely the expected *natural* one - for the *computation* of *nonlinear* symmetry groups of large classes of nonlinear PDEs, provided that these groups are *projectable*. More precisely, in the case of such projectable groups, each *classical* symmetry group will automatically be a symmetry group for *global generalized solutions*, see Subsections 16.6 and 18.5.

On the other hand, the method in Section 17 does *not* allow such an easy and natural extension of classical symmetry groups, see Subsection 17.4. The reason for that comes from the particular way Colombeau's algebra of generalized functions is defined. In particular, this way will impose *growth conditions* both on the nonlinear PDEs considered and on their symmetry groups.

A further distinction can be noted between the two methods in Sections 16 and 18, respectively. The method in Section 16 is based on the *global Cauchy-Kovalevskaia theorem*, see Theorem 16.1, which was obtained in Rosinger [5-7] through the use of algebras of generalized functions defined with the help of the *nowhere dense ideals*, see (16.20), (16.23). These algebras, which contain the L Schwartz distributions, can in an easy and natural way be endowed with large classes of nonlinear groups of transformations, see Subsections 16.4, 16.10. On the other hand, in the case of *nonprojectable* groups of transformations, the extension of such group actions to the algebras may involve technical complications, see Subsection 16.8.

In the case of the method in Section 18, the natural extension of group action to the spaces of generalized solutions appears to be easier, see Subsection 18.6. This, then, indicates another *advantage* of the method of solving nonlinear PDEs through Dedekind order completion, the method presented in Parts I and II of this work.

16. GROUP INVARIANCE OF GLOBAL GENERALIZED SOLUTIONS OF NONLINEAR PDEs OBTAINED THROUGH THE ALGEBRAIC METHOD

16.0 Preliminaries

Three results are presented in this Section as a *first in the literature*:

1) Large classes of *nonlinear* Lie groups of transformations are defined on algebras of generalized functions, algebras which contain the L Schwartz distributions. The respective definitions do *not* involve *infinite* dimensional manifolds.

2) The usual projectable Lie groups of symmetry for classical smooth solutions of nonlinear PDEs *extend* directly to symmetry groups for *global generalized solutions* of the respective PDEs. The way more general Lie groups of symmetry can be extended is indicated. In this way a large class of nonlinear Lie groups of symmetry can easily be *computed* for global generalized solutions.

3) Applications are presented for the delta wave solutions of semilinear hyperbolic equations and for Riemann type solvers of the nonlinear shock wave equation.

These results can be seen as one possible solution to Hilbert's fifth problem when considered in its extended version. Two other different and independent solutions are presented in the next two Sections.

16.1 Hilbert's Fifth Problem

In 1900, Hilbert proposed the following as his fifth problem:

> 'How far Lie's concept of continuous groups of transformations is approachable in our investigations without the assumption of the differentiability of the functions?'

see details in Hilbert, or Browder [pp 12-14].

Often, this problem has been interpreted in a *narrow* sense, see for instance Varadarajan [p 42], Yang. Namely, it has been reduced to the manifolds of independent and dependent variables upon which the Lie groups of transformations themselves are defined. And in this context, the question was asked whether these groups and manifolds - which can be *defined by continuity* conditions alone - will in fact necessarily have a *differentiable* structure as well. In other words, let $\mathbf{M}$ be a finite dimensional manifold and G a finite dimensional Lie group, Olver, then G is a *group of transformations* on $\mathbf{M}$ if there exists a continuous mapping

$$(16.1) \qquad G \times \mathbf{M} \ni (g,x) \longmapsto gx \in \mathbf{M}$$

with the following three properties:

$$(16.2) \qquad g(hx) = (gh)x, \quad g,h \in G, \quad x \in \mathbf{M}$$

$$(16.3) \qquad ex = x, \quad x \in \mathbf{M}$$

$$(16.4) \qquad g^{-1}(gx) = x, \quad g \in G, \quad x \in \mathbf{M}$$

where $G \times G \ni (g,h) \longmapsto gh \in G$ is the group operation on G, $e \in G$ is the neutral element of G, while $g^{-1} \in G$ is the inverse element of $g \in G$.

Here we should recall for a moment that important group transformations can happen to be defined *only locally*, thus they are more general than the situation in (16.1) - (16.4). Moreover, these *local* groups of transformations appear naturally and can give particularly relevant *symmetries* even in the case of global classical solutions of linear PDEs, see for instance the case of the *heat equation*, Olver [pp 122, 123]. However, in order to be able to present in a more clear and simple manner the basic ideas and methods related to the group invariance of global generalized solutions of nonlinear PDEs, we shall first deal with particular groups of transformations (16.1) - (16.4), namely, the so called *projectable* ones, see (16.52), (16.53). The way more general groups of transformations can be dealt with is indicated in Subsection 16.10.

Returning to the defining relations (16.1) - (16.4) in their unrestricted generality, it is clear that they do *not* need any differentiability conditions on the Lie group G or on the manifold M. However, the development of the theory of group transformations as it exists today, and ever since its inception due to Lie, requires a *high degree* of differentiability, that is, at least C^∞-smoothness, Hermann [1], Ibragimov, Olver, Bluman & Kumei. In this way, the mentioned narrow interpretation of Hilbert's fifth problem is indeed of a special interest even at present.

Nevertheless, we should not forget that the original - and still main - aim of Lie's theory of groups of transformations is *beyond* the mere study of the group transformations of finite dimensional manifolds, such as in (16.1).

Indeed, that main aim is actually in the study of the *group invariance of solutions of nonlinear* PDEs. As an illustration let us consider the nonlinear PDE, see (2.1), (2.2)

(16.5) $\quad T(x,D)U(x) = f(x), \quad x \in \Omega \subseteq \mathbb{R}^n$

this time with F in (2.2) being C^∞-smooth. If

(16.6) $\quad U : \Omega \longrightarrow \mathbb{R}$

is a C^∞-smooth solution of (16.5), then the problem of its group invariance can be formulated as follows. Let us take as manifold

(16.7) $\quad M = \Omega \times \mathbb{R} \subseteq \mathbb{R}^{n+1}$

which is the space of independent and dependent variables. Now the problem is to find *all Lie groups* G, as well as their *groups of transformations* (16.1) such that, whenever for a certain $g \in G$ the group transformation gU of U is again a C^∞-smooth function

(16.8) $\quad gU : \Omega \longrightarrow \mathbb{R}$

then it will follow that gU is also a solution of (16.5).

In fact, in the presentation of his fifth problem, Hilbert draws the attention to the following more *extended* formulation, relevant in the context of equations encountered in the calculus of variations, see Hilbert, or Browder [p 14]:

> 'In how far are the assertions which we can make in the case of differentiable functions true under proper modifications without this assumption?'

It is particularly *important* to note that group transformation and group invariance of solutions of linear and nonlinear PDEs are *fundamental* in physics. For instance, Einstein's General Relativity theory is intimately connected with *coordinate invariance* under C^2-smooth diffeomorphisms of the domain of independent variables. However, this issue of the general coordinate invariance of the equations of physics does not stop with the classical theories, and it is equally relevant for Quantum Mechanics. Nevertheless, here, the situation is far from satisfactory. Indeed, let us cite from Dyson's 1972 paper Missed Opportunities, [pp 645-647], which describes a situation that, in its main features, is still valid today:

'The most glaring incompatibility of concepts in contemporary physics is that between Einstein's principle of general coordinate invariance and all the modern schemes for a quantum mechanical description of nature. Einstein based his theory of general relativity on the principle that God did not attach any preferred labels to the points of space-time. This principle requires that the laws of physics should be invariant under the Einstein group E, which consists of all one-to-one and twice-differentiable transformations of the coordinates. By making full use of the invariance under E, Einstein was able to deduce the precise form of his law of gravitation from general requirements of mathematical simplicity without any arbitrariness. He was also able to reformulate the whole of classical physics (electromagnetism and hydrodynamics) in E-invariant fashion, and so determine unambiguously the mutual interactions of matter, radiation and gravitation within the classical domain. There is no part of physics more coherent mathematically and more satisfying aesthetically than this classical theory of Einstein based upon E-invariance.

On the other hand, all the currently viable formalisms for describing nature quantum-mechanically use a much smaller invariance group ... In practice all serious quantum-mechanical theories are based either on the Poincaré group P or the Galilei group G. This means that a class of preferred inertial coordinate-systems is postulated a priori, in flat contradiction to Einstein's principle. The contradiction is particularly uncomfortable, because Einstein's principle of general coordinate invariance has such an attractive quality of absoluteness. A physicist's intuition tells him that, if Einstein's principle is valid at all, it ought to be valid for the whole of physics, quantum-mechanical as well as classical. If the principle were not universally valid, it is difficult to understand why Einstein achieved such deeply coherent insights into nature by assuming it to be so ...'

Concerning the importance in physics of the group invariance of solutions of nonlinear PDEs, a whole range of examples and applications can be found in Hermann [1-5], Hermann & Hurt, Ibragimov, Olver, Bluman & Kumei, and in the literature cited there.

However, from the inception of Lie's theory there have been *two major difficulties* with the problem of the group invariance of solutions of *nonlinear* PDEs.

First, there was not available a theory for the existence of solutions for general nonlinear PDEs.

Secondly, many of those nonlinear PDEs whose solutions were known to exist, had solutions which were not C^∞-smooth on the whole of the domain of definition of the respective nonlinear PDEs, for instance, on the whole of Ω in the case of (16.5). Such a situation is rather typical and it is well illustrated in the case of the classical solutions given by the Cauchy-Kovalewskaia theorem for general analytic nonlinear PDEs. Here it should be recalled that the presence of this 'singularity' or 'blow up' phenomenon in the smooth solutions of nonlinear PDEs is a rather basic and deeply embedded *nonlinear effect*. Indeed, let us consider the simple case of the nonlinear scalar ODE

(16.9) $$D_t U(t) + U(t)^2 = 0, \quad t \in \mathbb{R}$$

with the initial value condition

$$(16.10) \quad U(t_0) = u_0$$

corresponding to any given $t_0 \in \mathbb{R}$ and $u_0 \in \mathbb{R}$, $u_0 \neq 0$. Then the solution of (16.9), (16.10) is

$$(16.11) \quad U(t) = \frac{u_0}{1 + u_0(t - t_0)}, \quad t \in \mathbb{R}\backslash\{t_0 - \frac{1}{u_0}\}$$

In other words, if we fix *any* $t_0 \in \mathbb{R}$, the solution (16.11) can 'blow up' *arbitrarily near to* t_0, provided that $|u_0|$ is sufficiently large.

In this way, smooth solutions of nonlinear PDEs can be expected to *fail* to exist outside of *arbitrarily small neighbourhoods* of points in the domain of definition of those equations.

In the context of Hilbert's fifth problem, this nonlinear phenomenon alone could, therefore, suffice in order to lead to the shift of interest from the main problem, which is the group invariance of solutions of nonlinear PDEs, to the narrow one, that is, merely the groups of transformations of the manifolds of independent and dependent variables. And as mentioned, until recently, such a shift of interest was to an extent justified due to the lack of theories which could give *global generalized solutions* for sufficiently large classes of *nonlinear* PDEs.

When, however, during the last decades *global generalized solutions* were obtained for several particular classes of nonlinear PDEs, see for instance Lions, the above two major difficulties were replaced with the following two new ones. Typically, these global generalized solutions are *no* longer usual functions, such as in (16.6), that is, defined on a *finite dimensional*, Euclidean space. Instead, they are *linear functionals*

$$(16.12) \quad U : \mathcal{F}(\Omega) \longrightarrow \mathbb{R}$$

where $\mathcal{F}(\Omega)$ is an *infinite dimensional* vector space. For instance, if we have distribution solutions $U \in \mathcal{D}'(\Omega)$ then $\mathcal{F}(\Omega) = \mathcal{D}(\Omega)$.

corresponding to arbitrary $\mathcal{C}^\infty$-smooth functions $\psi \in \mathcal{C}^\infty(\Omega)$. Then it is easy to see that

$$(16.23) \quad A = \mathcal{A}/\mathcal{I}^\infty_{nd}(\Omega)$$

is an associative and commutative algebra which has the unit element

$$1_A = u(1) + \mathcal{I}^\infty_{nd}(\Omega) \in A$$

Furthermore, we obtain the embedding of algebras

$$(16.24) \quad \mathcal{C}^\infty(\Omega) \ni \psi \longrightarrow u(\psi) + \mathcal{I}^\infty_{nd}(\Omega) \in A$$

since the following *neutrix* or *off diagonality* condition holds

$$(16.25) \quad \mathcal{I}^\infty_{nd}(\Omega) \cap \mathcal{U}^\infty(\Omega) = \{0\}$$

as an immediate consequence of (16.20). It should be noted that condition (16.22) can easily be satisfied by choosing, for instance, $\mathcal{A} = (\mathcal{C}^\infty(\Omega))^{\mathbb{N}}$.

An *important* and *nontrivial* property of the algebras A in (16.23), and which is a consequence of (16.25), Rosinger [3-6], is that they contain the L Schwartz distributions

$$(16.26) \quad \mathcal{D}'(\Omega) \subsetneq A$$

and this inclusion preserves both the multiplication and differentiation of $\mathcal{C}^\infty$-smooth functions.

However, here we shall rather be interested in the algebra embeddings (16.24) which will allow us suitable extensions of the mapping (16.17) in order to obtain global solutions for (16.14) - (16.16). For that purpose, let us take any subalgebra $\mathcal{B} \subsetneq (\mathcal{C}^\infty(\Omega))^{\mathbb{N}}$ such that

$$(16.27) \quad \mathcal{I}^\infty_{nd}(\Omega) \cup \mathcal{U}^\infty_{nd}(\Omega) \cup T(x,D)\mathcal{A} \subsetneq \mathcal{B}$$

a condition which is satisfied, for instance, by $\mathcal{B} = (\mathcal{C}^\infty(\Omega))^{\mathbb{N}}$. Then, similar to (16.23), we obtain an associative and commutative algebra

$$(16.28) \quad B = \mathcal{B}/\mathcal{I}^\infty_{nd}(\Omega)$$

with the unit element

$$1_B = u(1) + \mathcal{I}^\infty_{nd}(\Omega) \in B$$

and with the embedding of algebras

$$(16.29) \quad \mathcal{C}^\infty(\Omega) \ni \psi \longmapsto u(\psi) + \mathcal{I}^\infty_{nd}(\Omega) \in B$$

Now in view of (16.24), (16.29), we can extend the mapping (16.17) to the following commutative diagram

$$(16.30) \quad \begin{array}{ccc} \mathcal{C}^\infty(\Omega) & \xrightarrow{T(x,D)} & \mathcal{C}^\infty(\Omega) \\ \downarrow & & \downarrow \\ A & \xrightarrow[T]{} & B \end{array}$$

where T is defined by

$$(16.31) \quad s + \mathcal{I}^\infty_{nd}(\Omega) \longmapsto T(x,D)s + \mathcal{I}^\infty_{nd}(\Omega)$$

for $s = (\psi_\nu | \nu \in \mathbb{N}) \in \mathcal{A}$, where we defined $T(x,D)s = (T(x,D)\psi_\nu | \nu \in \mathbb{N})$. We note that (16.31) is a correct definition, since we have the property

$$(16.32) \quad \begin{array}{l} \forall\ t \in (\mathcal{C}^\infty(\Omega))^{\mathbb{N}},\ \ v \in \mathcal{I}^\infty_{nd}(\Omega)\ : \\ \quad T(x,D)(t + v) - T(x,D)t \in \mathcal{I}^\infty_{nd}(\Omega) \end{array}$$

which follows easily from (16.20) and (16.21).

The existence of global solutions for (16.14) - (16.16), see Rosinger [5, pp 265], [6, p 115], [7], is given in

Theorem 16.1

The analytic nonlinear PDE, see (16.14), (16.19)

$$(16.33) \quad T(x,D)U(x) = 0, \quad x = (t,x) \in \Omega$$

with the noncharacteristic analytic initial value problem

$$(16.34) \quad D_t^p U(t_0,y) = g_p(y), \quad 0 \leq p < m, \quad (t_0,y) \in S$$

where

$$(16.35) \quad S = \{x = (t,y) \in \Omega \mid t = t_0\} \neq \phi$$

can be associated with commutative diagrams (16.30) within which it has *global* generalized solutions

$$(16.36) \quad U \in A$$

defined on the whole of Ω. These global solutions U are *analytic* functions

$$(16.37) \quad \psi : \Omega \backslash \Gamma \longrightarrow \mathbb{C}$$

when restricted to suitable open dense sets $\Omega \backslash \Gamma$ in Ω. In other words, depending on U, we have

(16.38) $\Gamma \subset \Omega$ is closed, nowhere dense

Moreover, we can assume that

$$(16.39) \quad \text{mes } \Gamma = 0$$

where mes denotes the Lebesgue measure on $\mathbb{R}^n$. In particular $S \subset \Omega \backslash \Gamma$ and U satisfies the initial value problem (16.34) on S. ■

By looking more carefully at the way Theorem 16.1 is proved, see Rosinger [5, p 247-257], [6, pp 113-116], [7, pp 338-340], it is easy to note that it is a *particular* case of the following *existence* result.

Theorem 16.2

Suppose the nonlinear PDE, see (2.1), (2.2)

(16.40) $T(x,D)U(x) = 0, \quad x \in \Omega$

is $\mathcal{C}^\infty$-smooth, that is, F is $\mathcal{C}^\infty$-smooth. Further suppose that there exists

(16.41) $\psi \in \mathcal{C}^\infty(\Omega\backslash\Gamma), \quad$ with $\Gamma \subset \Omega$ closed, nowhere dense

which is a *classical* solution of (16.40) on the open dense set $\Omega\backslash\Gamma$ in Ω.

Then there exist commutative diagrams (16.30) and *global* generalized solutions

(16.42) $U \in A$

of (16.40), such that

(16.43) $U = \psi \quad$ on $\Omega\backslash\Gamma$ ■

The interest in this extension comes from the fact that a large variety of nonlinear PDEs, together with their global generalized solutions of applicative interest, are among those described in (16.40) - (16.43). One such case we shall deal with in the sequel is presented by the *delta waves*, Rauch & Reed, Oberguggenberger, of *semilinear hyperbolic* equations

$$U_t(t,x) + \lambda U_x(t,x) = F(U(t,x)), \quad t,x \in \mathbb{R}$$

where Γ will be the *characteristic curve* of this equation. Another case we shall also consider is the *nonlinear shock wave* equation

$$U_t(t,x) + U(t,x)U_x(t,x) = 0, \quad t > 0, \quad x \in \mathbb{R}$$

which is well known, Schaeffer, Lara Carrero, to possess a large class of *global weak solutions* of physical interest which are of the type (16.42), with Γ being the shock curves, where $\Omega = (0,\infty) \times \mathbb{R}$.

In general, the interest in results such as in Theorems 16.1 and 16.2 is due to the fact that the respective *closed nowhere dense singularity* sets $\Gamma \subset \Omega$ can have *arbitrary positive*, or in fact, *infinite* measure, restricted only by the condition

$$\text{mes}\ (\Gamma) < \text{mes}\ (\Omega)$$

in case $\text{mes}\ (\Omega) < \infty$, Oxtoby. It follows therefore that such *singularity* sets Γ can model not only *shocks* or *delta waves* but possibly Cantor set type *strange attractors* as well, which have lately been associated with *turbulence* or *chaos*, Temam.

16.3 Group Transformations for Functions

The group invariance of *global* generalized solutions of C^∞-*smooth nonlinear* PDEs in (16.40) can be implemented precisely in the same way as in the classical Lie theory, Olver, by considering groups of transformations on the manifold

$$\text{(16.44)} \qquad \mathbf{M} = \Omega \times \mathbb{R} \subseteq \mathbb{R}^{n+1}$$

of independent and dependent variables.

Given a Lie group G, a *group of transformations* on the manifold $\mathbf{M}$ is defined by a mapping

$$\text{(16.45)} \qquad \begin{array}{ccc} G \times \mathbf{M} & \longrightarrow & \mathbf{M} \\ (g,(x,u)) & \longmapsto & g(x,u) \end{array}$$

with $x \in \Omega$, $u \in \mathbb{R}$, which satisfies the following three conditions. If $g,h \in G$, $(x,u) \in \mathbf{M}$, then

$$\text{(16.46)} \qquad g(h(x,u)) = (gh)(x,u)$$

Further, if $e \in G$ is the neutral element in G, then

$$(16.47) \quad e(x,u) = (x,u), \quad (x,u) \in M$$

Finally, if $g \in G$, $(x,u) \in M$, then

$$(16.48) \quad g^{-1}(g(x,u)) = (x,u)$$

where $g^{-1} \in G$ is the inverse element of g.

Clearly, the mapping (16.45) is equivalent with the following two mappings

$$(16.49) \quad \begin{array}{l} G \times M \ni (g,(x,u)) \longmapsto g_1(x,u) \in \Omega \\ G \times M \ni (g,(x,u)) \longmapsto g_2(x,u) \in \mathbb{R} \end{array}$$

with

$$(16.50) \quad g(x,u) = (g_1(x,u),\ g_2(x,u)), \quad g \in G, \quad (x,u) \in M$$

In other words, the mapping (16.45) is in fact a *coordinate transform*

$$(16.51) \quad \begin{array}{l} x \\ u \end{array} \longmapsto \begin{array}{l} \bar{x} = g_1(x,u) \\ \bar{u} = g_2(x,u) \end{array}$$

involving *both* the independent variables x and the dependent variable u.

As in the classical Lie group theory, Olver, all the manifolds and mapping are considered to be $\mathcal{C}^\infty$-smooth.

For convenience, we shall first deal with *projectable* groups of transformations, that is, those for which (16.51) is of the particular form

$$(16.52) \quad \begin{array}{l} x \\ u \end{array} \longmapsto \begin{array}{l} \bar{x} = g_1(x) \\ \bar{u} = g_2(x,u) \end{array}$$

where

$$(16.53) \qquad \Omega \xrightarrow{\ g_1\ } \Omega$$

is a diffeomorphism. Clearly, the projectable groups of transformations contain as a particular case the groups of transformations of the form

$$(16.54) \qquad \begin{matrix} x \\ u \end{matrix} \longmapsto \begin{matrix} \bar{x} = g_1(x) \\ \bar{u} = u \end{matrix}$$

with g_1 as in (16.53), which, as is known, give the relevant *coordinate transforms* in General Relativity.

In order to implement the group invariance of the global solutions of (16.40), we have to *extend* the groups of transformations (16.45) acting on the manifold $\mathbf{M}$ to groups of transformations acting on the domain of definition of the nonlinear operators in (16.30), that is, to $\mathcal{C}^\infty(\Omega)$ and A, respectively.

The extension to $\mathcal{C}^\infty(\Omega)$ is rather straightforward. Indeed, we define a mapping

$$(16.55) \qquad \begin{matrix} G \times \mathcal{C}^\infty(\Omega) \longrightarrow \mathcal{C}^\infty(\Omega) \\ (g,U) \longmapsto gU \end{matrix}$$

by, see (16.52), (16.53)

$$(16.56) \qquad (gU)(x) = g_2(g_1^{-1}(x),\ U(g_1^{-1}(x))), \quad x \in \Omega$$

and can easily verify that the corresponding versions of (16.46) - (16.48) will hold for (16.55).

16.4 Group Transformations for Algebras of Generalized Functions

So far, we have dealt with constructions which are familiar in the classical Lie theory of group invariance. Now we shall extend them to the case of the *global generalized solutions* given by Theorem 16.2 above. In other words, as already mentioned, we have to further extend the groups of transformations (16.55) in order to obtain *groups of transformations* of *generalized functions*, see (16.30)

$$(16.57) \qquad \begin{array}{ccc} G \times A & \longrightarrow & A \\ (g,U) & \longmapsto & gU \end{array}$$

For that purpose, we recall the structure of A given in (16.23), namely

$$(16.58) \qquad A = \mathcal{A}/\mathcal{I}^{\infty}_{nd}(\Omega)$$

and note that, see (16.22)

$$(16.59) \qquad \mathcal{I}^{\infty}_{nd}(\Omega) \subset \mathcal{A} \subsetneq (\mathcal{C}^{\infty}(\Omega))^{\mathbb{N}}$$

Now (16.58) and (16.59) allow us to define the mapping (16.57) in *two steps*. First we simply extend the mapping (16.55) *termwise* to a mapping

$$(16.60) \qquad \begin{array}{ccc} G \times (\mathcal{C}^{\infty}(\Omega))^{\mathbb{N}} & \longrightarrow & (\mathcal{C}^{\infty}(\Omega))^{\mathbb{N}} \\ (g,s) & \longmapsto & gs \end{array}$$

where for $s = (\psi_\nu | \nu \in \mathbb{N}) \in (\mathcal{C}^{\infty}(\Omega))^{\mathbb{N}}$ we define

$$(16.61) \qquad gs = (g\psi_\nu | \nu \in \mathbb{N})$$

Then, using the quotient structure of A in (16.58) we define the mapping (16.57) as follows. Given

$$(16.62) \qquad U = s + \mathcal{I}^{\infty}_{nd}(\Omega) \in A = \mathcal{A}/\mathcal{I}^{\infty}_{nd}(\Omega)$$

with

$$(16.63) \qquad s = (\psi_\nu | \nu \in \mathbb{N}) \in \mathcal{A} \subsetneq (\mathcal{C}^{\infty}(\Omega))^{\mathbb{N}}$$

we set

$$(16.64) \quad gU = gs + \mathcal{I}^{\infty}_{nd}(\Omega) \in A = \mathcal{A}/\mathcal{I}^{\infty}_{nd}(\Omega)$$

where gs is defined in (16.61).

It is clear that the definition (16.62) - (16.64) of the mapping (16.57) is correct, if the following two conditions hold, see (16.60), (16.61)

$$(16.65) \quad g\mathcal{A} \subseteq \mathcal{A}, \quad g \in G$$

as well as

$$(16.66) \quad \begin{array}{l} \forall\ g \in G, \quad t \in \mathcal{A}, \quad v \in \mathcal{I}^{\infty}_{nd}(\Omega) : \\ \quad g(t + v) - gt \in \mathcal{I}^{\infty}_{nd}(\Omega) \end{array}$$

Now, condition (16.65) can easily be satisfied by taking, for instance, $\mathcal{A} = (\mathcal{C}^{\infty}(\Omega))^{\mathbb{N}}$. On the other hand, condition (16.66) comes about as one of the convenient properties of the nowhere dense ideal $\mathcal{I}^{\infty}_{nd}(\Omega)$. Indeed, according to (16.20), given $v = (\chi_{\nu} | \nu \in \mathbb{N}) \in \mathcal{I}^{\infty}_{nd}(\Omega)$ in (16.66), there exists $\Gamma \subset \Omega$ closed, nowhere dense, such that

$$(16.67) \quad \begin{array}{l} \forall\ x \in \Omega \backslash \Gamma : \\ \exists\ \mu \in \mathbb{N}, \quad V \subseteq \Omega\backslash\Gamma \text{ neighbourhood of } x : \\ \forall\ \nu \in \mathbb{N}, \quad \nu \geq \mu, \quad y \in V : \\ \quad \chi_{\nu}(y) = 0 \end{array}$$

Let us assume that $t = (\xi_{\nu} | \nu \in \mathbb{N}) \in (\mathcal{C}^{\infty}(\Omega))^{\mathbb{N}}$, then (16.61), (16.56) give

$$(16.68) \quad g(t + v) = (g(\xi_{\nu} + \chi_{\nu}) | \nu \in \mathbb{N})$$

with

$$(16.69) \quad (g(\xi_{\nu} + \chi_{\nu}))(x) = g_2(g_1^{-1}(x),\ \xi_{\nu}(g_1^{-1}(x)) + \chi_{\nu}(g_1^{-1}(x)))$$

for $\nu \in \mathbb{N}$ and $x \in \Omega$. However, in view of (16.53), it is clear that

$$\Sigma = g_1(\Gamma) \subset \Omega \quad \text{is closed, nowhere dense} \tag{16.70}$$

Furthermore

$$g_1\Big|_{\Omega \setminus \Gamma} : \Omega \setminus \Gamma \longrightarrow \Omega \setminus \Sigma \tag{16.71}$$

is again a diffeomorphism. Therefore (16.67) implies that

$$\begin{aligned} &\forall\ x \in \Omega\setminus\Sigma : \\ &\exists\ \mu \in \mathbb{N},\quad W = g_1(V) \subseteq \Omega\setminus\Sigma \ \text{ neighbourhood of } \ x : \\ &\forall\ \nu \in \mathbb{N},\quad \nu \geq \mu,\quad y \in W : \\ &\quad \chi_\nu(g_1^{-1}(y)) = 0 \end{aligned} \tag{16.72}$$

In this way (16.69), (16.72) and (16.56) will give

$$\begin{aligned} &\forall\ x \in \Omega\setminus\Sigma : \\ &\exists\ \mu \in \mathbb{N},\quad W \subseteq \Omega\setminus\Sigma \ \text{ neighbourhood of } \ x : \\ &\forall\ \nu \in \mathbb{N},\quad \nu \geq \mu,\quad y \in W : \\ &\quad (g(\xi_\nu + \chi_\nu))(y) = (g(\xi_\nu))(y) \end{aligned} \tag{16.73}$$

Therefore, applying (16.61) and (16.56) to t and taking into account (16.68), (16.69), (16.73) and the definition in (16.20) of the nowhere dense ideal $\mathcal{I}^\infty_{nd}(\Omega)$, the relation (16.66) will follow easily.

In this way, under the condition (16.65), the groups of transformations, see (16.57), (16.62) - (16.64)

$$\begin{aligned} G \times A &\longrightarrow A \\ (g,U) &\longmapsto gU \end{aligned} \tag{16.74}$$

are well defined on the *algebras* A of *generalized functions* which, according to Theorem 16.2, contain the *global* generalized solutions of the $\mathcal{C}^\infty$-*smooth nonlinear* PDEs in (16.40).

16.5 Group Invariance for Global Generalized Solutions

Let us recapitulate shortly the way *groups of transformations* can be constructed for algebras

$$(16.75) \quad A = \mathcal{A}/\mathcal{I}^{\infty}_{nd}(\Omega)$$

of *generalized functions*, algebras which, according to (16.26), contain the distributions, that is

$$(16.76) \quad \mathcal{D}'(\Omega) \subset A$$

and moreover, under the conditions in Theorem 16.2, contain as well the *global generalized solutions* of the $\mathcal{C}^{\infty}$-smooth nonlinear PDEs in (16.40).

First, we chose any subalgebra $\mathcal{A} \subseteq (\mathcal{C}^{\infty}(\Omega))^{\mathbb{N}}$ which satisfies the following two conditions, see (16.22), (16.65)

$$(16.77) \quad \mathcal{I}^{\infty}_{nd}(\Omega) \cup \mathcal{U}^{\infty}(\Omega) \subseteq \mathcal{A}$$

$$(16.78) \quad g\mathcal{A} \subseteq \mathcal{A}, \quad g \in G$$

conditions which are clearly satisfied for $\mathcal{A} = (\mathcal{C}^{\infty}(\Omega))^{\mathbb{N}}$, for instance.

Secondly, we define our algebra of *generalized functions* by the quotient

$$(16.79) \quad A = \mathcal{A}/\mathcal{I}^{\infty}_{nd}(\Omega)$$

Thirdly, given any *projectable* group of transformations, see (16.52), (16.53)

$$(16.80) \quad \begin{array}{ccc} G \times M & \longrightarrow & M \\ (g,(x,u)) & \longmapsto & g(x,u) \end{array}$$

where

$$(16.81) \quad M = \Omega \times \mathbb{R} \subseteq \mathbb{R}^{n+1}$$

is the manifold of the independent and dependent variables, we extend (16.80) to a group of transformations

$$\begin{array}{lccc} & G \times A & \longrightarrow & A \\ (16.82) & & & \\ & (g,U) & \longmapsto & gU \end{array}$$

as follows. If

$$(16.83) \qquad U = s + \mathcal{I}^{\infty}_{nd}(\Omega) \in A = \mathcal{A}/\mathcal{I}^{\infty}_{nd}(\Omega)$$

and

$$(16.84) \qquad s = (\psi_\nu | \nu \in \mathbb{N}) \in \mathcal{A} \subseteq (\mathcal{C}^{\infty}(\Omega))^{\mathbb{N}}$$

then

$$(16.85) \qquad gU = gs + \mathcal{I}^{\infty}_{nd}(\Omega) \in A = \mathcal{A}/\mathcal{I}^{\infty}_{nd}(\Omega)$$

where

$$(16.86) \qquad gs = (g\psi_\nu | \nu \in \mathbb{N}) \in \mathcal{A} \subseteq (\mathcal{C}^{\infty}(\Omega))^{\mathbb{N}}$$

while for $\nu \in \mathbb{N}$, we have, see (16.52), (16.53), (16.56)

$$(16.87) \qquad (g\psi_\nu)(x) = g_2(g_1^{-1}(x),\ \psi_\nu(g_1^{-1}(x))), \quad x \in \Omega$$

Now the *group invariance* of the *global generalized solutions* in A of $\mathcal{C}^{\infty}$*-smooth nonlinear* PDEs in (16.40) can be formulated as follows, see (16.5) - (16.8). Suppose that $U \in A$ is a global generalized solution of (16.40), or in other words, the following relation holds in B, see (16.30)

$$(16.88) \qquad TU = 0$$

Then U is called a *group invariant* solution of (16.40), if for every $g \in G$, the relation, see (16.82)

(16.89) $\quad T(gU) = 0$

will again hold in B.

Clearly, in view of the commutativity of the diagram (16.30), the above concept of group invariance contains the classical one, Olver, when U is a $\mathcal{C}^\infty$-smooth solution on Ω.

Next, in Subsection 16.6, we shall present the *extension* of the classical method of computing group invariant solutions, Olver, to the case of group invariant *global generalized solutions* of (16.40).

The above construction in (16.75) - (16.87), culminating with the groups of transformations (16.82) on algebras of generalized functions and the group invariance (16.89) of generalized solutions, is a *first in the literature* concerning *nonlinear* groups of transformations of spaces of generalized functions, spaces which are large enough in order to contain the distributions. Moreover, the *simplicity* of the above construction *contrasts* quite sharply even with those earlier constructions which only aim to define distributions on manifolds, and do so in a coordinate independent manner, see for instance De Rham, and for more recent attempts, Biagioni, Aragona & Biagioni, Damsma [1,2], De Roever & Damsma, Damyanov, Colombeau & Meril.

16.6 Computing Symmetry Groups for Global Generalized Solutions

In this Subsection we show that those *projectable groups of transformations* which are *classical symmetry groups* of the $\mathcal{C}^\infty$-*smooth nonlinear* PDEs in (16.40) - that is, which yield the *group invariance* of their *classical solutions* - can be *extended* to symmetry groups of the *global generalized solutions* of these equations.

The basic idea of this *extension* of symmetry groups is rather simple and straightforward and it is the same with the one used in Subsections 16.4 and 16.5 where we extended the group transformations to algebras of generalized functions. Namely, we use the fact that generalized solutions are classes of sequences of $\mathcal{C}^\infty$-smooth functions determined modulo the *nowhere dense ideal*.

Indeed, let us suppose given a $\mathcal{C}^\infty$-smooth nonlinear PDE, see (16.40)

$$(16.90) \quad T(x,D)U(x) = 0, \quad x \in \Omega$$

and a *projectable* group of transformations, see (16.44) - (16.48), (16.52), (16.53)

$$(16.91) \quad \begin{array}{ccc} G \times M & \longrightarrow & M \\ (g,(x,u)) & \longmapsto & g(x,u) \end{array}$$

Then, in view of (16.55), (16.56), we can obtain the extension of (16.91) to the classical, that is, $\mathcal{C}^\infty$-smooth functions

$$(16.92) \quad \begin{array}{ccc} G \times \mathcal{C}^\infty(\Omega) & \longrightarrow & \mathcal{C}^\infty(\Omega) \\ (g,U) & \longmapsto & gU \end{array}$$

Let us now suppose that the Lie group G in (16.91) is a *symmetry group* of the nonlinear PDE in (16.90) in the *classical* sense. This means, Olver, that for every classical solution $U \in \mathcal{C}^\infty(\Omega')$ of the nonlinear PDE in (16.90), on any nonvoid open subset $\Omega' \subseteq \Omega$, that is

$$(16.93) \quad T(x,D)U(x) = 0, \quad x \in \Omega'$$

this classical solution will be invariant with respect to the Lie group G, see (16.5) - (16.8)

$$(16.94) \quad T(x,D)(gU)(x) = 0, \quad x \in \Omega'', \quad g \in G$$

In this case we can show that the Lie group G will also be a *symmetry group* for the *global generalized solutions* of the PDE in (16.90).

Indeed, let us consider associated with the PDE in (16.90) any given commutative diagram (16.30) under the conditions in Theorem 16.2. Further, let $U \in A = \mathcal{A}/\mathcal{I}^\infty_{nd}(\Omega)$ be a global generalized solution of (16.90), that is

$$(16.95) \quad TU = 0 \quad \text{in} \quad B$$

In view of (16.23), we can assume that

$$(16.96)\qquad U = s + \mathcal{I}^{\infty}_{\mathrm{nd}}(\Omega) \in A = \mathcal{A}/\mathcal{I}^{\infty}_{\mathrm{nd}}(\Omega)$$

where

$$(16.97)\qquad s = (\psi_\nu \mid \nu \in \mathbb{N}) \in \mathcal{A} \subseteq (\mathcal{C}^{\infty}(\Omega))^{\mathbb{N}}$$

Then, according to (16.31), the relation (16.95) is equivalent with

$$(16.98)\qquad (T(x,D)\psi_\nu \mid \nu \in \mathbb{N}) \in \mathcal{I}^{\infty}_{\mathrm{nd}}(\Omega)$$

It follows that, see (16.20), there exists $\Gamma \subset \Omega$ closed, nowhere dense, for which

$$(16.99)\qquad \begin{array}{l} \forall\ x \in \Omega\setminus\Gamma\ : \\ \exists\ \mu \in \mathbb{N},\ \ V \subseteq \Omega\setminus\Gamma \ \text{ neighbourhood of } x: \\ \forall\ \nu \in \mathbb{N},\ \ \nu \geq \mu,\ \ y \in V\ : \\ \quad T(y,D)\psi_\nu(y) = 0 \end{array}$$

In other words each ψ_ν, with $\nu \in \mathbb{N}$, is a $\mathcal{C}^{\infty}$-smooth, therefore *classical solution* of (16.90) on Ω, outside of a suitably small neighbourhood of Γ. Let us indicate that situation more clearly by defining for $\nu \in \mathbb{N}$, the *open* subset of Ω, see (16.99)

$$(16.100)\qquad \Omega_\nu = \left\{ x \in \Omega\setminus\Gamma \;\middle|\; \begin{array}{l} \exists\ V \subseteq \Omega\setminus\Gamma \ \text{ neighbourhood of } x : \\ T(y,D)\psi_\nu(y) = 0,\ \ y \in V \end{array} \right\}$$

Then, in view of the characterization of $\mathcal{I}^{\infty}_{\mathrm{nd}}(\Omega)$ in Rosinger [5], it follows that (16.99) is equivalent with the relation

$$(16.101)\qquad \bigcup_{\nu\in\mathbb{N}} \bigcap_{\substack{\mu\in\mathbb{N}\\ \mu\geq\nu}} \Omega_\mu = \Omega\setminus\Gamma$$

and in particular, for $\nu \in \mathbb{N}$, we obtain

$$(16.102) \quad T(x,D)\psi_\nu = 0 \quad \text{on} \quad \Omega_\nu$$

Let us now take any given $g \in G$. Then the group transformation gU of U through the action of g is given by (16.85) - (16.87).

We recall that, in view of (16.53), we obtain

$$(16.103) \quad \Sigma = g_1(\Gamma) \subset \Omega \quad \text{closed, nowhere dense}$$

and also

$$(16.104) \quad \Omega \backslash \Gamma \xrightarrow{\ g_1\ } \Omega \backslash \Sigma$$

is a diffeomorphism of open sets.

In order to prove (16.89), it suffices to show that, see (16.31), (16.85) - (16.87)

$$(16.105) \quad (T(x,D)(g\psi_\nu) \mid \nu \in \mathbb{N}) \in \mathcal{I}^\infty_{nd}(\Omega)$$

However (16.101), (16.104) give

$$(16.106) \quad \bigcup_{\nu \in \mathbb{N}} \bigcap_{\substack{\mu \in \mathbb{N} \\ \mu \geq \nu}} g_1(\Omega_\nu) = \Omega \backslash \Sigma$$

while (16.87), (16.102), (16.104) and the fact that (16.94) was assumed to hold for the given group of transformations (16.91) and for $\mathcal{C}^\infty$-smooth solutions of (16.93) on any nonvoid open subset $\Omega' \subseteq \Omega$, will yield for $\nu \in \mathbb{N}$ the relation

$$(16.107) \quad T(x,D)(g\psi_\nu) = 0 \quad \text{on} \quad g_1(\Omega_\nu)$$

But then, in view of (16.104), the relations (16.106) and (16.107) will imply (16.105). Therefore, the proof of (16.89) is completed.

In this way, we have indeed proved that the *projectable* Lie group G is a *symmetry group* of the *global generalized solutions* of the $\mathcal{C}^\infty$-*smooth nonlinear* PDEs in (16.90).

Finally, it is *important* to note the following. The results in this Subsection only show that *each* projectable Lie group of symmetry of *classical* solutions of nonlinear PDEs *remains* a symmetry group of *generalized solutions*, when the respective PDEs are considered within the *algebras of generalized functions*. This, however, need *not* necessarily mean that within these algebras the global generalized solutions of the nonlinear PDEs may not have *other* symmetry groups as well. The study of these possible *additional* symmetry groups will be undertaken elsewhere, with certain first details being presented in Subsection 16.10.

16.7 Application to Delta Waves of Semilinear Hyperbolic PDEs

A first example of *nonlinear group invariance of global weak solutions of nonlinear* PDEs implemented within *finite dimensional* manifolds is presented next.

Let us consider the *semilinear hyperbolic* equation

$$(16.108) \quad U_t(t,x) + \lambda U_x(t,x) = F(U(t,x)), \quad t > 0, \quad x \in \mathbb{R}$$

with $\lambda \in \mathbb{R}$ given and $U : (0,\infty) \times \mathbb{R} \longrightarrow \mathbb{R}$ the unknown function. About $F : \mathbb{R} \longrightarrow \mathbb{R}$ we shall assume that it is $\mathcal{C}^\infty$-smooth, with both F and F' *bounded* on $\mathbb{R}$. Then, as is known, given any *initial value* condition

$$(16.109) \quad U(0,x) = u(x), \quad x \in \mathbb{R}$$

with u a $\mathcal{C}^\infty$-smooth function, there exists a unique $\mathcal{C}^\infty$-smooth classical solution U of (16.108), (16.109), defined on the whole of the domain $t,x \in \mathbb{R}$.

In the case of semilinear hyperbolic equations (16.108), especially *systems*, that is, with the unknown function $U : \Omega \longrightarrow \mathbb{R}^n$, there exists an interest in *rough initial value* conditions (16.109), when u is *no longer* a classical function but rather a certain type of *distribution* in $\mathcal{D}'(\mathbb{R})$, more precisely, a distribution with *support* given by a *finite number* of points in $\mathbb{R}$. The typical result in this regard, see Oberguggenberger [2, p 137] for the general formulation, is the following.

Theorem 16.3

If $u \in \mathcal{D}'(\mathbb{R})$ is a distribution with the *support a finite set of points* then

$$(16.110) \quad \begin{array}{l} U_t(t,x) + \lambda U_x(t,x) = F(U(t,x)), \quad t,x \in \mathbb{R} \\ U(0,x) = u(x), \quad x \in \mathbb{R} \end{array}$$

has a solution

$$(16.111) \quad U = V + W \quad \text{on} \quad \mathbb{R}^2$$

where $W \in \mathcal{C}^\infty(\mathbb{R}^2)$ is the unique *classical* solution of the *semilinear* problem

$$(16.112) \quad \begin{array}{l} W_t(t,x) + \lambda W_x(t,x) = F(W(t,x)), \quad t,x \in \mathbb{R} \\ W(0,x) = 0, \quad x \in \mathbb{R} \end{array}$$

while $V \in \mathcal{D}'(\mathbb{R}^2)$ is the *distributional* solution of the *linear* problem

$$(16.113) \quad \begin{array}{l} V_t(t,x) + \lambda V_x(t,x) = 0, \quad t,x \in \mathbb{R} \\ V(0,x) = u(x), \quad x \in \mathbb{R} \end{array}$$

■

We recall that solutions of the type (16.111) - (16.113), corresponding to the above kind of *rough initial values* $u \in \mathcal{D}'(\mathbb{R})$, are called *delta waves*, Rauch & Reed. These solutions can be interpreted in a number of ways. Three definitions in this regard are those in (16.30) and Oberguggenberger [2, p 137, p 270]. In the rest of this Subsection, however, we shall only need the definition in (16.30).

Now, it is easy to see that the *classical* symmetry groups of the semilinear hyperbolic PDE in (16.108) are generated by the following two *nonlinear one dimensional* Lie groups

$$(16.114) \quad (t,x,u) \longmapsto (t, \lambda t + a^{-1}(a(x - \lambda t) + \epsilon), u)$$

$$(16.115) \quad (t,x,u) \longmapsto (\frac{1}{\lambda}(x - b^{-1}(b(x - \lambda t) + \epsilon)), x, u)$$

where $(t,x,u) \in M = \mathbb{R}^3$, $g = \epsilon \in G = (\mathbb{R},+)$, while $a,b : \mathbb{R} \longrightarrow \mathbb{R}$ are arbitrary $\mathcal{C}^\infty$-smooth diffeomorphisms, with a^{-1}, b^{-1} denoting the inverse functions of a,b respectively.

Clearly, both of the group transformations (16.114) and (16.115) are *projectable*, see (16.52), (16.53). Therefore, we can apply the results in Subsections 16.3 - 16.6 and obtain the *nonlinear group invariance* of the *delta wave* solutions (16.111) with respect to the groups of transformations (16.114), (16.115).

Let us start with the group transformations (16.114). With the notation in (16.52), (16.53) we have for $g = \epsilon \in G = (\mathbb{R},+)$

$$\begin{array}{ccc} t & & \bar{t} = t \\ & \longmapsto & \\ x & & \bar{x} = \lambda t + a^{-1}(a(x - \lambda t) + \epsilon) \end{array}$$

or in other words

$$(16.116) \quad \begin{array}{ccc} \Omega & \xrightarrow{\quad g_1 \quad} & \Omega \\ (t,x) & \longmapsto & (t, \lambda t + a^{-1}(a(x - \lambda t) + \epsilon)) \end{array}$$

where $\Omega = \mathbb{R}^2$. It follows that

$$(16.117) \quad \begin{array}{ccc} \Omega & \xrightarrow{\quad g_1^{-1} \quad} & \Omega \\ (t,x) & \longmapsto & (t, \lambda t + a^{-1}(a(x - \lambda t) - \epsilon)) \end{array}$$

For simplicity, let us first consider the situation when in (16.110) we have

$$(16.118) \quad u = c\delta \in \mathcal{D}'(\mathbb{R}), \quad c \in \mathbb{R}, \quad c \neq 0$$

that is, our rough initial value is a multiple of the Direc delta distribution with support at $0 \in \mathbb{R}$. Then clearly, the solution of (16.113) is given by

$$(16.119) \quad V(t,x) = c\delta(x - \lambda t), \quad t,x \in \mathbb{R}$$

It follows that the solution (16.111) is of the type described in Theorem 16.2, with

$$\Gamma = \{(t,x) \in \mathbb{R}^2 \mid x = \lambda t\}$$

being a *characteristic curve* of (16.108).

Let us now take any $\varphi \in \mathcal{D}(\mathbb{R})$, such that $\text{supp}\, \varphi \subsetneq [-1,1]$ and

$$\int \varphi(x)dx = 1$$

and define $\varphi_\nu \in \mathcal{D}(\mathbb{R}^2)$, with $\nu \in \mathbb{N}$, by

$$\varphi_\nu(t,x) = \nu\varphi(\nu(x - \lambda t)), \quad t,x \in \mathbb{R}$$

Further, let us define the sequence of functions

$$(16.120) \quad s = (\psi_\nu | \nu \in \mathbb{N}) \in (\mathcal{C}^\infty(\mathbb{R}^2))^{\mathbb{N}}$$

by, see (16.111), (16.112)

$$(16.121) \quad \psi_\nu(t,x) = W(t,x) + c\varphi_\nu(t,x), \quad \nu \in \mathbb{N}, \quad t,x \in \mathbb{R}$$

Then clearly

$$(16.122) \quad s \text{ converges weakly in } \mathcal{D}'(\mathbb{R}^2) \text{ to } U = V + W$$

and moreover

$$(16.123) \quad U = s + \mathcal{I}^\infty_{\text{nd}}(\mathbb{R}^2) \in A = \mathcal{A}/\mathcal{I}^\infty_{\text{nd}}(\mathbb{R}^2)$$

is a solution of (16.108) in the sense of Theorem 16.2, provided that the algebra A is chosen in a suitable manner.

In view of Subsection 16.6, we know that (16.114) is a *symmetry group* of the *global generalized solution* (16.123). We shall now check that this is indeed true. For that purpose we recall (16.56), (16.61), (16.64) and obtain for $\epsilon \in \mathbb{R}$

$$(16.124) \quad \epsilon U = \epsilon s + \mathcal{I}^{\infty}_{nd}(\mathbb{R}^2) \in A = \mathcal{A}/\mathcal{I}^{\infty}_{nd}(\mathbb{R}^2)$$

where

$$(16.125) \quad \epsilon s = (\epsilon \psi_\nu \mid \nu \in \mathbb{N}) \in \mathcal{A}$$

while for $\nu \in \mathbb{N}$, we have, see (16.117)

$$(16.126) \quad (\epsilon \psi_\nu)(t,x) = \psi_\nu(t, \lambda t + a^{-1}(a(x - \lambda t) - \epsilon)), \quad t,x \in \mathbb{R}$$

Therefore, it only remains to show that, see (16.89)

$$(16.127) \quad T(D)(\epsilon U) = 0, \quad \epsilon \in \mathbb{R}$$

where $T(D)$ is the partial differential operator associated with (16.108), that is

$$T(D)U = U_t + \lambda U_x - F(U), \quad U \in \mathcal{C}^{\infty}(\mathbb{R}^2)$$

However, in view of (16.31) and (16.124), it follows that, for $\epsilon \in \mathbb{R}$, we have

$$(16.128) \quad T(D)(\epsilon U) = (T(D)(\epsilon \psi_\nu) \mid \nu \in \mathbb{N}) + \mathcal{I}^{\infty}_{nd}(\mathbb{R}^2)$$

On the other hand (16.121), (16.126) yield

$$(16.129) \quad (\epsilon \psi_\nu)(t,x) = W(t, \lambda t + a^{-1}(a(x - \lambda t) - \epsilon)) + \\ + \nu c \varphi(\nu a^{-1}(a(x - \lambda t) - \epsilon))$$

for $\epsilon \in \mathbb{R}$, $\nu \in \mathbb{N}$ and $t, x \in \mathbb{R}$.

Let us assume that

(16.130) a is strictly increasing on $\mathbb{R}$

Then clearly, the condition

$$-1 \leq \nu a^{-1}(a(x - \lambda t) - \epsilon) \leq 1$$

is equivalent with

(16.131) $a^{-1}(a(-1/\nu) + \epsilon) \leq x - \lambda t \leq a^{-1}(a(1/\nu) + \epsilon)$

hence in view of (16.129), it follows that

(16.132) $(\epsilon \psi_\nu)(t,x) = W(t, \lambda t + a^{-1}(a(x - \lambda t) - \epsilon))$ if $x - \lambda t$ does not satisfy (16.131)

Further we note that

(16.133) $\lim_{\nu \to \infty} a^{-1}(a(-1/\nu) + \epsilon) = \lim_{\nu \to \infty} a^{-1}(a(1/\nu) + \epsilon) = a^{-1}(a(0) + \epsilon)$

Finally, through a direct computation we obtain from (16.132) the relation

(16.134) $T(D)(\epsilon \psi_\nu)(t,x) = T(D)W(\bar{t}, \bar{x})$ if $x - \lambda t$ does not satisfy (16.131)

where $\bar{t} = t$, $\bar{x} = \lambda t + a^{-1}(a(x - \lambda t) - \epsilon)$. The relation (16.134) is actually a direct consequence of the fact that (16.114) is a symmetry group of the PDE in (16.108), and therefore of (16.112) as well.

In this way (16.129), (16.133) and (16.134) will indeed give (16.127).

We note that, in case (16.130) does not hold, then instead, we shall have

$$(16.135) \quad a \text{ is strictly decreasing on } \mathbb{R}$$

since the existence of the inverse function a^{-1} is assumed in (16.114). Furthermore, it is easy to see that the above group invariance property (16.127) of the delta wave solutions U will again hold in the case of (16.135).

The above group invariance results can be easily extended to *arbitrary* rough initial values

$$(16.136) \quad u \in \mathcal{D}'(\mathbb{R}), \quad \text{supp } u \text{ is a finite set of points in } \mathbb{R}$$

Finally, all the above group invariance results will hold for the nonlinear one dimensional Lie groups (16.115) as well.

16.8 Application to Shock Waves

Let us consider the scalar *nonlinear shock wave* equation

$$(16.137) \quad U_t(t,x) + U(t,x)U_x(t,x) = 0, \quad t > 0, \quad x \in \mathbb{R}$$

with $U : (0,\infty) \times \mathbb{R} \longrightarrow \mathbb{R}$. As is well known, Smoller, typical shock wave solutions of (16.137) are given by weak solutions

$$(16.138) \quad U \in \mathcal{D}'((0,\infty) \times \mathbb{R})$$

for which

$$(16.139) \quad U\big|_{\Omega/\Gamma} \in \mathcal{C}^\infty(\Omega\setminus\Gamma)$$

where $\Omega = (0,\infty) \times \mathbb{R} \subset \mathbb{R}^2$, while $\Gamma \subset \Omega$ is the closed, nowhere dense set of shock curves. In this way, such shock wave solutions enter within the *global generalized solutions* dealt with in Theorem 16.2.

Now, as is well known, the *classical* symmetry groups of the shock wave equation (16.137) are generated by the following four *nonlinear one dimensional* Lie groups:

$$(16.140)\quad (t,x,u) \longmapsto (te^{a(u)\epsilon},\ xe^{a(u)\epsilon},\ u)$$

$$(16.141)\quad (t,x,u) \longmapsto (t + b(u)\epsilon,\ x,\ u)$$

$$(16.142)\quad (t,x,u) \longmapsto (t,\ x + c(u)\epsilon,\ u)$$

$$(16.143)\quad (t,x,u) \longmapsto (t,\ x + t(d^{-1}(d(u) + \epsilon) - u),\ d^{-1}(d(u) + \epsilon))$$

where $(t,x,u) \in M = (0,\infty) \times \mathbb{R}^2 \subset \mathbb{R}^3$, $g = \epsilon \in G = (\mathbb{R},+)$ is the one dimension group parameter, while $a,b,c,d \in \mathcal{C}^\infty(\mathbb{R})$ are arbitrary functions.

It is clear that *none* of these four nonlinear one dimensional Lie groups is projectable, except in the trivial case when a, b, c or d are constant functions. In this way, with respect to these four nonlinear one dimensional Lie groups we cannot apply the theory developed earlier in Subsections 16.3 - 16.6. Nevertheless, in order to attempt going beyond the mentioned theory, we shall consider the *particular* case of shock wave solutions (16.138), (16.139) given by the *Riemann solvers*, and then enquire into their *group invariance* with respect to some of the *classical* groups (16.140) - (16.143) of the nonlinear shock wave equation.

As is well known, the *Riemann solvers* for the *nonlinear shock wave equation*

$$(16.144)\quad U_t(t,x) + U(t,x)U_x(t,x) = 0, \quad t > 0, \quad x \in \mathbb{R}$$

correspond to the *discontinuous initial value* conditions

$$(16.145)\quad U(0,x) = u_\ell + (u_r - u_\ell)H(x - x_0), \quad x \in \mathbb{R}$$

where $x_0, u_\ell, u_r \in \mathbb{R}$, $u_\ell > u_r$, while H denotes the Heaviside function

$$H(x) = \begin{cases} 0 & \text{if } x \leq 0 \\ 1 & \text{if } x > 0 \end{cases}$$

Since the initial value in (16.145) is *decreasing*, the *global weak solution* of (16.144), (16.145) will contain *shocks*. It is easy to see that this weak solution which satisfies both the Rankine-Hugoniot jump condition and the Entropy condition is given by the Riemann solver

$$(16.146) \quad U(t,x) = u_\ell + (u_r - u_\ell)H(x - x_0 - (u_\ell + u_r)t/2), \quad t > 0, \quad x \in \mathbb{R}$$

therefore, the set of *shock* points is

$$(16.147) \quad \Gamma = \{(t,x) \in \Omega = (0,\infty) \times \mathbb{R} \mid x - x_0 = (u_\ell + u_r)t/2\}$$

which is obviously closed and nowhere dense in Ω. Moreover, it is clear that

$$(16.148) \quad U \in \mathcal{D}'(\Omega), \quad U|_{\Omega\setminus\Gamma} \in \mathcal{C}^\infty(\Omega\setminus\Gamma)$$

It follows that (16.146) - (16.148) is indeed a particular case of (16.138), (16.139), and therefore, of Theorem 16.2.

Now, we shall give a suitable explicit construction of the *generalized solution* (16.41) - (16.43) which corresponds to (16.146) - (16.148).

Let $\omega \in \mathcal{D}(\mathbb{R})$ be such that

$$\omega \geq 0, \quad \text{supp}\ \omega \subseteq [-1,1], \quad \int_{\mathbb{R}} \omega(x)dx = 1$$

and define $\omega_\nu \in \mathcal{D}(\mathbb{R})$, for $\nu \in \mathbb{N}$, by

$$\omega_\nu(x) = \nu\omega(\nu x), \quad x \in \mathbb{R}$$

Further, let us define

$$(16.149) \quad s = (\psi_\nu \mid \nu \in \mathbb{N}) \in (\mathcal{C}^\infty(\Omega))^{\mathbb{N}}$$

where

$$(16.150) \quad \psi_\nu(t,x) = \chi_\nu(x - x_0 - (u_\ell + u_r)t/2), \quad \nu \in \mathbb{N}, \quad t > 0, \quad x \in \mathbb{R}$$

while χ_ν, for $\nu \in \mathbb{N}$, is defined by the convolution

$$(16.151) \quad \chi_\nu = (u_\ell + (u_r - u_\ell)H_{x_0}) * \omega_\nu \in \mathcal{C}^\infty(\mathbb{R})$$

where H_{x_0} is the shift or translate of the Heaviside function H, i.e.

$$H_{x_0}(x) = H(x - x_0), \quad x \in \mathbb{R}$$

We note that, for $\nu \in \mathbb{N}$, we obtain

$$(16.152) \quad \begin{aligned} &\chi_\nu \text{ nonincreasing on } \mathbb{R} \\ &\chi_\nu(y) = \left| \begin{array}{ll} u_\ell & \text{if } y \leq x_0 - 1/\nu \\ u_r & \text{if } y \geq x_0 + 1/\nu \end{array} \right. \end{aligned}$$

Now, in view of (16.146) - (16.152), it follows easily that one can construct commutative diagrams (16.30) such that the Riemann solver (16.146)

$$(16.153) \quad U = s + \mathcal{I}^\infty_{nd}(\Omega) \in A = \mathcal{A}/\mathcal{I}^\infty_{nd}(\Omega)$$

is a *global generalized solution* of (16.144) in the sense of Theorem 16.2.

Let us consider the group invariance of the generalized solution U in (16.153) with respect, for instance, the Lie group (16.141), that is

$$(16.154) \quad \begin{array}{ccc} G \times M & \longrightarrow & M \\ (\epsilon,(t,x,u)) & \longmapsto & \epsilon(t,x,u) \end{array}$$

with

$$(16.155) \quad \epsilon(t,x,u) = (t + b(u)\epsilon,x,u)$$

Then we can obtain the group transform gU in the following manner. Let $g = \epsilon \in G = (\mathbb{R},+)$ be given and fixed. Let us compute, see (16.149), (16.61)

$$(16.156) \quad gs = (g\psi_\nu \mid \nu \in \mathbb{N})$$

by extending the procedure in (16.56), which is based on (16.52), (16.53), to the general case of (16.49) - (16.51). For that purpose, and in view of (16.154) - (16.156), we have first to study whether the mappings

$$(16.157) \quad \Omega \ni (t,x) \longmapsto (t + b(\psi_\nu(t,x))\epsilon, x) \in \Omega, \quad \nu \in \mathbb{N}$$

are invertible.

Let us, for instance, assume that we have

$$(16.158) \quad b \text{ nondecreasing on } \mathbb{R} \text{ and } u_\ell + u_r \geq 0$$

In this case, for the Lie group (16.154), (16.155), the relations (16.151) will become, see (16.157)

$$(16.159) \quad \begin{aligned} \bar{t} &= t + b(\psi_\nu(t,x))\epsilon \\ \bar{x} &= x \\ \bar{u} &= \psi_\nu(t,x) \end{aligned}$$

and we have to obtain t as a function of $\bar{t}$ and $\bar{x} = x$ from the first relation in (16.159). However, (16.159), (16.150) will give

$$\frac{\partial \bar{t}}{\partial t} = 1 + \epsilon b'(\psi_\nu(t,x)) \frac{\partial \psi_\nu(t,x)}{\partial t} =$$

$$= 1 - \epsilon b'(\chi_\nu(x - x_0 - (u_\ell+u_r)t/2))\chi_\nu'(x - x_0 - (u_\ell+u_r)t/2)\frac{u_\ell + u_r}{2}$$

therefore, in view of (16.152), (16.158), we obtain uniformly in $\nu \in \mathbb{N}$

$$(16.160) \quad \frac{\partial \bar{t}}{\partial t} \geq 1 > 0, \quad (t,x) \in \Omega, \quad \epsilon \geq 0$$

Now in view of (16.160), we can apply the implicit function theorem to the first relation in (16.159), and obtain

$$(16.161) \quad t = \tau_\nu(\bar{t},\bar{x},\epsilon), \quad \nu \in \mathbb{N}, \quad (\bar{t},\bar{x}) \in \Omega, \quad \epsilon \geq 0$$

where τ_ν are $\mathcal{C}^\infty$-smooth.

In this way it is clear that the mappings, see (16.157)

$$(16.162) \quad \Omega \ni (t,x) \longmapsto (t + b(\psi_\nu(t,x))\epsilon,x) \in \Omega, \quad \nu \in \mathbb{N}, \quad \epsilon \geq 0$$

are *diffeomorphism*.

It is also clear that in the case of the nonlinear one dimensional Lie group (16.142), we can in a similar manner obtain the *diffeomorphisms*

$$(16.163) \quad \Omega \ni (t,x) \longmapsto (t,x + c(\psi_\nu(t,x))\epsilon) \in \Omega, \quad \nu \in \mathbb{N}, \quad \epsilon \geq 0$$

provided that, see (16.158)

$$(16.164) \quad c \;\; \text{nonincreasing on} \;\; \mathbb{R}$$

Let us now see under which possible further conditions will the nonlinear one dimensional Lie groups (16.141), (16.142) lead through (16.162), (16.163) respectively, to *symmetry solutions* of the nonlinear shock wave equations, obtained from the global generalized solutions given by the Riemann solvers (16.153). Here we should first note that, not surprisingly, the *group transformations*, when implemented within *generalized solutions*, can bring with them a picture which is more *sophisticated* than the usual, classical one. Moreover, it can happen that *reinterpretations* within the classical framework of the picture within generalized solutions can *miss* important points. This phenomenon has been noted earlier, Rosinger [3-8], Colombeau [1,2], Biagioni, Oberguggenberger

[2], in connection with various rather basic manifestations of the *conflict* between *discontinuity, multiplication and differentiation*, see Rosinger [6, pp 1-20, 62-64, 79-93], and also Rosinger [6, pp 361-366] which is relevant to the particular case of Colombeau's generalized functions.

Typically, for instance, the Heaviside function satisfies in *classical* terms the relation

$$H^m = H, \quad m \in \mathbb{N}, \quad m \geq 2$$

while this relation need *not* hold in algebras of generalized functions. As one of the consequences of this *difference*, it is known in classical shock wave theory, Smoller, that the weak solutions of the nonlinear conservation laws

$$U_t + UU_x = 0, \quad t > 0, \quad x \in \mathbb{R}$$

and

$$(U^m)_t + U(U^m)_x = 0, \quad t > 0, \quad x \in \mathbb{R}, \quad m \in \mathbb{N}, \quad m \geq 2$$

are *not* the same, although their classical solutions coincide. Therefore, it may seem quite natural that *new aspects* of the mentioned *conflict* can emerge when in addition to discontinuity, multiplication and differentiation one will bring into the picture *group invariance* as well.

Let us return to the Riemann solvers (16.153) and see, for instance, the way they are transformed by the Lie groups (16.142), which under the condition (16.164), will lead to the diffeomorphisms (16.163). Given $\nu \in \mathbb{N}$ and $\epsilon \geq 0$, we can return to the general setting in (16.49) - (16.51), and obtain

$$\begin{aligned} &\bar{t} = t \\ (16.165)\quad &\bar{x} = x + c(\psi_\nu(t,x))\epsilon \\ &\bar{u} = \psi_\nu(t,x) \end{aligned}$$

On the other hand (16.150) - (16.152) result in

$$(16.166)\quad \psi_\nu(t,x) = \begin{vmatrix} u_\ell & \text{if} \quad x - x_0 < (u_\ell + u_r)t/2 - 1/(\nu + 1) \\ u_r & \text{if} \quad x - x_0 > (u_\ell + u_r)t/2 + 1/(\nu + 1) \end{vmatrix}$$

Now (16.165), (16.166) will give the following two transformations, each valid in a part of $\Omega = (0,\infty) \times \mathbb{R}$. First, we have

$$(16.167)\quad \begin{aligned} \bar{t} &= t \\ \bar{x} &= x + c(u_\ell)\epsilon \\ \bar{u} &= u_\ell \end{aligned}$$

valid in the left domain

$$(16.168)\quad \Omega_{\ell,\nu} = \{(t,x) \in \Omega \mid x - x_0 < (u_\ell + u_r)t/2 - 1/(\nu + 1)\}$$

then secondly, we have

$$(16.169)\quad \begin{aligned} \bar{t} &= t \\ \bar{x} &= x + c(u_r)\epsilon \\ \bar{u} &= u_r \end{aligned}$$

which will hold in the right domain

$$(16.170)\quad \Omega_{r,\nu} = \{(t,x) \in \Omega \mid x - x_0 > (u_\ell + u_r)t/2 + 1/(\nu + 1)\}$$

It follows that in terms of the coordinates (t,x,u), the transformation (16.167) is a *nonlinear* shift or translation along the x-axis, with magnitude $c(u_\ell)\epsilon$. Similarly, (16.169) is a *nonlinear* shift or translation along the x-axis, with magnitude $c(u_r)\epsilon$. Moreover, both of them *preserve* the slope $(u_\ell + u_r)/2$ of the *shock*.

We note however that, in view of (16.166) and the assumption that $u_\ell > u_r$, we have

$$(16.171) \quad c(u_\ell) \leq c(u_r)$$

therefore, unless

$$(16.172) \quad c(u_\ell) = c(u_r)$$

the shift in the left domain $\Omega_{\ell,\nu}$ will be *slower* than that in the right domain $\Omega_{r,\nu}$. This is clearly going to cause a *gap* between the two domains and this gap will remain when $\nu \longrightarrow \infty$, unless $\epsilon \downarrow 0$. Therefore, the resulting picture does *not* have any classical interpretation.

Let us assume that the arbitrary $\mathcal{C}^\infty$-smooth function c in the nonlinear one dimensional Lie group (16.142) satisfies (16.164). Also, for the given values $u_\ell > u_r$ in the Riemann solver (16.146), (16.153), we shall assume that (16.172) holds, that is

$$(16.173) \quad c(u_\ell) = c(u_r) = c_o$$

without however c necessarily being a constant function. Then based on (16.163), we can compute for the Riemann solver U in (16.153), the group transformation ϵU, with $\epsilon \geq 0$, according to the following extension of (16.85), (16.86)

$$(16.174) \quad \epsilon U = (\epsilon\psi_\nu | \nu \in \mathbb{N}) + \mathcal{I}^\infty_{nd}(\Omega) \in A = \mathcal{A}/\mathcal{I}^\infty_{nd}(\Omega)$$

where, for $\nu \in \mathbb{N}$ and $(t,x) \in \Omega$, we have

$$(16.175) \quad (\epsilon\psi_\nu)(t,x) = \psi_\nu(t,\sigma_\nu(t,x,\epsilon))$$

while σ_ν results form the diffeomorphism

$$(16.176) \quad \Omega \ni (t,x) \longmapsto (t,\sigma_\nu(t,x,\epsilon)) \in \Omega$$

which is the inverse of the mapping in (16.163).

We note that the definition in (16.175), (16.176) is in line with the way group transformations (16.51), which are more general than the projectable ones in (16.53), may act on functions, see Olver [p 94]. However, this issue is taken up in more detail in the next Subsection.

Now in view of (16.166) - (16.170) and (16.173) it is clear that, for $\nu \in \mathbb{N}$, $\epsilon \geq 0$, we have

$$(16.177) \quad x + c(\psi_\nu(t,x))\epsilon = x + c_0\epsilon, \quad (t,x) \in \Omega_{\ell,\nu} \cup \Omega_{r,\nu}$$

hence

$$(16.178) \quad \sigma_\nu(t,x,\epsilon) = x - c_0\epsilon, \quad (t,x - c_0\epsilon) \in \Omega_{\ell,\nu} \cup \Omega_{r,\nu}$$

therefore (16.174) - (16.178) and (16.31) will give, see (16.30)

$$(16.179) \quad T(D)(\epsilon U) = 0 \quad \text{in} \quad B, \quad \epsilon \geq 0$$

This means that, if we denote by $T(D)$ the nonlinear partial differential operator associated with the shock wave equation (16.137), then the *semigroup transformation*, see (16.174)

$$(16.180) \quad [0,\infty) \ni \epsilon \longmapsto \epsilon U \in A = \mathcal{A}/\mathcal{I}^\infty_{nd}(\Omega)$$

as generated by the nonlinear one dimensional Lie group (16.142), is a *symmetry semigroup* for the global generalized solution given by the Riemann solver (16.146), (16.153).

A similar result can be obtained for the nonlinear one dimensional Lie group (16.141).

16.9 The Case of Local Groups of Transformations

Motivated by the example in the previous Subsection, we shall now present a theoretical support for extensions of group invariance beyond those of the projectable groups studied in Subsections 16.3 - 16.6.

For the sake of a more comprehensive approach, we shall start with arbitrary *local* groups of transformations. Later, however, we shall concentrate on *local* groups of transformations applied to *global* generalized solutions of $\mathcal{C}^\infty$-smooth nonlinear PDEs covered by Theorem 16.2. This case obviously contains the group invariance of the Riemann solvers of the nonlinear shock wave equation, studied in the previous Subsection.

Let us, for convenience, recall, Olver [p 19], that a *local* Lie group is a finite dimensional manifold with a distinguished element $e \in G$, two open subsets $G_0, G_1 \subseteq G$, with

$$(16.181) \quad e \in G_0 \subseteq G_1$$

as well as two mappings

$$(16.182) \quad \begin{array}{l} G_1 \times G_1 \ni (g,h) \longmapsto gh \in G \\ G_0 \ni g \longmapsto g^{-1} \in G_1 \end{array}$$

for which the following three conditions hold. If $g,h,k,gh,hk \in G_1$ then

$$(16.183) \quad (gh)k = g(hk)$$

further

$$(16.184) \quad ge = eg = g, \quad g \in G_1$$

and finally, if $g \in G_0$ then

$$(16.185) \quad gg^{-1} = g^{-1}g = e$$

Given now a local Lie group G, then a *local group of transformations* on a finite dimensional manifold M is defined by an *open* subset $\mathcal{U}$, with

$$(16.186) \quad \{e\} \times M \subseteq \mathcal{U} \subseteq G \times M$$

and $\mathcal{C}^\infty$-smooth mapping

$$(16.187)\quad \mathcal{U} \ni (g,x) \longmapsto gx \in \mathbf{M}$$

which satisfies the following three conditions, Olver [p 21]: If (h,x), $(g,hx) \in \mathcal{U}$, gh is defined, that is, $g,h \in G_1$, see (16.181), and also $(gh,x) \in \mathcal{U}$, then

$$(16.188)\quad g(hx) = (gh)x$$

Further

$$(16.189)\quad ex = x, \quad x \in \mathbf{M}$$

And finally, if $(g,x) \in \mathcal{U}$ and g^{-1} is defined, that is, $g \in G_0$, see (16.182) then $(g^{-1},gx) \in \mathcal{U}$ and

$$(16.190)\quad g^{-1}(gx) = x$$

In the context of Theorem 16.2, the manifolds $\mathbf{M}$ of interest are

$$(16.191)\quad \mathbf{M} = \Omega \times \mathbb{R} \subset \mathbb{R}^{n+1}$$

in which case the mappings (16.187) defining the *local* groups of transformations will have the form

$$(16.192)\quad \begin{aligned} \mathcal{U} &\longrightarrow \mathbf{M} \\ (g,(x,u)) &\longmapsto (g_1(x,u),\ g_2(x,u)) \end{aligned}$$

where $g \in G$, $x \in \Omega$, $u \in \mathbb{R}$, $g_1(x,u) \in \Omega$ and $g_2(x,u) \in \mathbb{R}$. It is useful to consider the following open sets. Given $g \in G$ and $(x,u) \in \mathbf{M}$, let

$$(16.193)\quad \mathbf{M}_g = \{(x',u') \in \mathbf{M} \mid (g,(x',u')) \in \mathcal{U}\}$$

$$(16.194)\quad G_{(x,u)} = \{g' \in G \mid (g',(x,u)) \in \mathcal{U}\}$$

Then clearly

$$(16.195) \quad (x,u) \in M_g \Longleftrightarrow g \in G_{(x,u)}$$

Now the meaning of the mapping in (16.192) becomes obvious. If $g \in G$ is fixed, then the mapping

$$(16.196) \quad M_g \ni (x,u) \longmapsto g(x,u) = (g_1(x,u), g_2(x,u)) \in M$$

is in fact a *coordinate transform*, see (16.51), involving both the independent variable x and the dependent variable u, namely

$$(16.197) \quad \begin{matrix} x \\ u \end{matrix} \longmapsto \begin{matrix} \bar{x} = g_1(x,u) \\ \bar{u} = g_2(x,u) \end{matrix}$$

Moreover, if g^{-1} exists in G then, in view of (16.190), it is clear that (16.196), thus (16.197) as well, define a homeomorphism

$$(16.198) \quad M_g \ni (x,u) \longmapsto g(x,u) \in M_{g^{-1}}$$

Just as earlier with (16.1) - (16.4), the above definition and construction in (16.186) - (16.198) do *not* require more than $\mathcal{C}^0$-smoothness. However, in line with the usual Lie group theory, we shall assume that all manifolds and mappings under consideration are $\mathcal{C}^\infty$-smooth. Nevertheless, this will not prevent us from being able to define groups of transformations and group invariance for the *global* generalized solutions of *arbitrary* $\mathcal{C}^\infty$-*smooth* nonlinear PDEs in (16.40).

As a consequence of the above blanket $\mathcal{C}^\infty$-smoothness assumption, it follows that in case g^{-1} exists in G, the mapping

$$(16.199) \quad M_g \ni (x,u) \longmapsto g(x,u) \in M_{g^{-1}}$$

is in fact a $\mathcal{C}^\infty$-smooth diffeomorphism.

16.10 Local Group Invariance

Given a local group of transformations (16.192) associated with the nonlinear PDE in (16.40), in order to deal with the group invariance of the solutions of this equation, first we have to *extend* the action of the local Lie group G from the manifold M in (16.191) to functions

$$U : \Omega \longrightarrow \mathbb{R}$$

Here, however, we have to note that in the case of non projectable groups of transformations, rather basic difficulties may arise. Indeed, quite independently of any possible PDE under consideration, the action of an element of a general local Lie group on a function will not always result in a function. Indeed, let us take $\Omega = \mathbb{R}$ and $U : \Omega \longrightarrow \mathbb{R}$ given by $U(x) = ax$, for $x \in \mathbb{R}$, where $a > 0$ is fixed. In this case $M = \mathbb{R}^2$ and let us take G as the Lie group of the rotations of the plane. Then, in order to still remain a function from Ω to $\mathbb{R}$, the function U can *only* be rotated with an angle θ for which

$$-\frac{\pi}{2} - \alpha < \theta < \pi/2 - \alpha$$

where $0 < \alpha < \pi/2$ and $\operatorname{tg} \alpha = a$. In this way, the larger $a > 0$, the less can U be rotated for $\theta > 0$. It follows, in particular, that the function $U : \Omega \longrightarrow \mathbb{R}$ defined by $U(x) = x^2$, for $x \in \mathbb{R}$, *cannot* be rotated at all, lest it ceases to be a function from Ω to $\mathbb{R}$, since its positive slope is unbounded for $x > 0$. Here we should note that in case we consider a *local* version of the function $U(x) = x^2$ defined on a *bounded* open subset of $\mathbb{R}$, then certain rotations of U will again turn it into a function. Needless to say, there are other groups of transformations, such as for instance translations, which turn the whole function $\mathbb{R} \ni x \xmapsto{U} x^2 \in \mathbb{R}$ into another function.

It follows that, the *local* nature of the groups of transformation may have to be *further restricted*, this time depending on the particular functions upon which they happen to act.

With *locality* therefore so deeply embedded in the *general* situation, let us allow it for functions and generalized functions as well, and in order to accommodate not only global generalized solutions, let us consider functions which are of the form

(16.200) $U : \Delta \longrightarrow \mathbb{R}$

where $\Delta \subseteq \Omega$ is nonvoid, *open*.

Let us now note that each function U in (16.200) can be uniquely associated with the subset of $\mathbf{M}$ given by

(16.201) $\gamma_U = \{(x,U(x)) | x \in \Delta\} \subset \mathbf{M}$

a subset called the *graph* of U. It follows that, if $g \in G$ and, see (16.193)

(16.202) $\gamma_U \subseteq \mathbf{M}_g$

then we can define, see (16.192), (16.196)

(16.203) $g\gamma_U = \{g(x,U(x)) | x \in \Delta\} \subseteq \mathbf{M}$

Moreover, if g^{-1} exists then, see (16.198)

(16.204) $g\gamma_U \subseteq \mathbf{M}_{g^{-1}}$

Now the question is whether or not the subset $g\gamma_U$ of $\mathbf{M}$ is the graph of a function

(16.205) $U' : \Delta' \longrightarrow \mathbb{R}$

where Δ' is a nonvoid *open* subset of Ω, in other words, whether there exists a function U' in (16.205) such that

(16.206) $g\gamma_U = \gamma_{U'}$

in which case we can define the group transformation gU of U by

(16.207) $gU = U'$

For our purposes it will be enough to find *sufficient conditions* for (16.205) - (16.207).

In this regard first we note that we at least have a trivial solution, as for $g = e$ we obtain $U' = U$, since (16.205) - (16.207) will clearly hold.

In general, let us take any $g \in G_0$ satisfying (16.202), see (16.181), then (16.200), (16.197), (16.198) will give the *injective mappings*

(16.208)
$$\begin{array}{ccccc} \Delta & \xrightarrow{\text{inj}} & M_g & \xrightarrow{\text{inj}} & M_{g^{-1}} \\ x & \longmapsto & (x,U(x)) & \longmapsto & (g_1(x,U(x)),\ g_2(x,U(x))) \end{array}$$

In particular, we obtain the mappings

(16.209) $\Delta \ni x \longmapsto g_1(x,U(x)) \in \Omega$

(16.210) $\Delta \ni x \longmapsto g_2(x,U(x)) \in \mathbb{R}$

On the other hand, the existence of U' in (16.205) is the same with the existence of the *injective* mapping

(16.211) $\Delta' \ni x' \xmapsto{\text{inj}} (x',U'(x)) \in M$

Therefore, a *sufficient* condition for (16.205) - (16.207) is the existence of the *inverse* mapping in (16.209). Indeed this will allow us to obtain U' in (16.205) through the commutative diagram, see (16.210), (16.211)

$$
(16.212)\qquad
\begin{array}{ccc}
\Delta' \ni x' & \longmapsto & U'(x') = g_2(x,U(x)) \in \mathbb{R} \\
\downarrow & & \uparrow \\
\Delta \ni x = (g_1 \circ (\mathrm{id}_\Delta \times U))^{-1}(x') & \longmapsto & (x,U(x)) \in M_g
\end{array}
$$

In other words, a sufficient condition for (16.205) - (16.207) is the possibility of *solving the system*

$$
(16.213)\qquad
\begin{aligned}
x' &= g_1(x,U(x)) \\
U'(x') &= g_2(x,U(x))
\end{aligned}
$$

by *eliminating* $x \in \Omega$ and thus obtaining a mapping

$$
\Delta' \ni x' \longmapsto U'(x') \in \mathbb{R}
$$

that is, the function in (16.205).

It follows that (16.205) - (16.207) is connected with the *invertibility* of the mapping in (16.209). Let us therefore define the *open* subset of Ω given by

$$
(16.214)\qquad \Omega_{g,U} = \left\{ y \in \Omega \;\middle|\; \begin{array}{l} \exists\, W \subseteq \Omega \text{ neighbourhood of } y : \\ (g_1 \circ (\mathrm{id}_\Delta \times U))^{-1} \text{ exists on } W \end{array} \right\}
$$

Obviously, for $g = e$ we obtain

$$
(16.215)\qquad g_1 \circ (\mathrm{id}_\Delta \times U) = (g_1 \circ (\mathrm{id}_\Delta \times U))^{-1} = \mathrm{id}_\Delta, \quad \Omega_{g,U} = \Delta
$$

Further, for *projectable* groups of transformations, see (16.52), (16.53), we have

$$
(16.216)\qquad
\begin{aligned}
& g_1 \circ (\mathrm{id}_\Delta \times U) = g_1\big|_\Delta, \quad (g_1 \circ (\mathrm{id}_\Delta \times U))^{-1} = g_1^{-1}\big|_{\Delta'} \\
& \Delta' = g_1(\Delta) = \Omega_{g,U}
\end{aligned}
$$

Now in view of (16.215), we can in general assume the existence of a subset $G_U \subseteq G_o$, with $e \in G_U$, such that

$$(16.217) \quad \Omega_{g,U} \neq \phi, \quad g \in G_U$$

It follows that (16.217) is a *sufficient* condition for (16.205) - (16.207). Indeed, if $g \in G_U$ then clearly we can take

$$(16.218) \quad \begin{aligned} \Delta' &= \Omega_{g,U} \\ U' &= (g_2 \circ (id_\Delta \times U)) \circ (g_1 \circ (id_\Delta \times U))^{-1} \end{aligned}$$

Therefore, if we denote, see (16.200)

$$(16.219) \quad \mathcal{C}^\infty_{par}(\Omega) = \left\{ U : \Delta \longrightarrow \mathbb{R} \;\middle|\; \begin{array}{l} *) \ \Delta \subseteq \Omega \ \text{ nonvoid, open} \\ **) \ U \in \mathcal{C}^\infty(\Delta) \end{array} \right\}$$

then we can assume the existence of a nonvoid set $\mathcal{V}$ such that, see (16.186)

$$(16.220) \quad \{e\} \times \mathcal{C}^\infty_{par}(\Omega) \subseteq \mathcal{V} \subseteq G \times \mathcal{C}^\infty_{par}(\Omega)$$

together with a mapping, see (16.187), (16.207)

$$(16.221) \quad \begin{array}{ccc} \mathcal{V} & \longrightarrow & \mathcal{C}^\infty_{par}(\Omega) \\ (g,U) & \longmapsto & gU \end{array}$$

It is easy to verify that the mapping in (16.221) will satisfy the corresponding versions of the conditions (16.188) - (16.190).

For clarity, let us summarize the way the mapping (16.221), that is, the action of the *local* Lie group G on *partial* functions on Ω, is defined, namely, through the commutative diagram

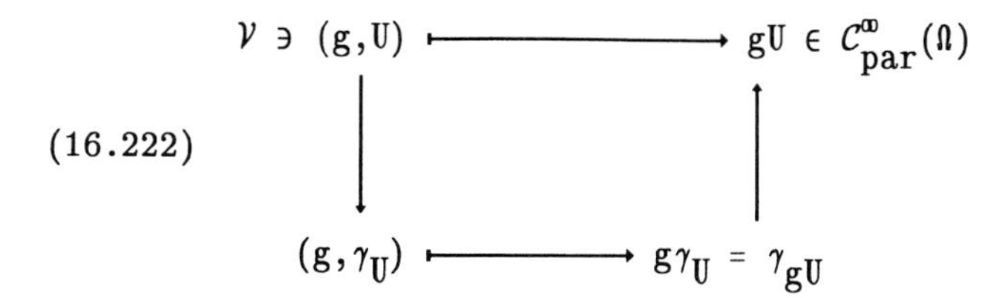

for the validity of which we only need the conditions in (16.202) and (16.217). As we noted in (16.215) and (16.216), the previous two conditions are trivially satisfied for $g = e$, or for any *projectable* group of transformations.

Finally, we note that, similar to the usual situation in Lie group theory, Olver [pp 92-95], when the local group action (16.186), (16.187) is *extended* from the manifold **M** to functions as in (16.220), (16.221), we do *not* assume any kind of topology on the respective space of functions $\mathcal{C}^{\infty}_{par}(\Omega)$. This is due to the fact that the functions U in (16.200) are identified with their graphs γ_u in (16.201), and these graphs are subsets in the manifold **M**. In this way, the topology of **M**, and of course of G, will prove to suffice in our considerations, since we only perform *algebraic* operations on the functions U. Therefore, their properties which are of interest to us, such as for instance smoothness, can be dealt with by considering each such function *individually*.

In the application of group invariance to Riemann solver type global weak solutions of nonlinear shock wave equations, we encountered a situation which can be described by the following particular case of the construction in (16.181) - (16.222).

Let us consider an arbitrary *nonlocal* group transformation, see (16.44) - (16.51)

$$(16.223)\qquad \begin{array}{ccc} G \times \mathbf{M} & \longrightarrow & \mathbf{M} \\ (g,(x,u)) & \longmapsto & g(x,u) = (g_1(x,u),\ g_2(x,u)) \end{array}$$

which, when extended to $\mathcal{C}^{\infty}$-smooth functions, will only act *locally*, that is, admits a nonvoid set $\mathcal{V}$ such that, see (16.220), (16.221)

$$(16.224) \quad \{e\} \times \mathcal{C}^{\infty}(\Omega) \subsetneqq \mathcal{V} \subsetneqq G \times \mathcal{C}^{\infty}(\Omega)$$

and for which we have the mapping

$$(16.225) \quad \begin{array}{ccc} \mathcal{V} & \longrightarrow & \mathcal{C}^{\infty}(\Omega) \\ (g,U) & \longmapsto & gU \end{array}$$

satisfying the condition, see (16.209), (16.214)

$$(16.226) \quad \Omega \xrightarrow{(g_1 \circ (id_{\Omega} \times U))^{-1}} \Omega \quad \text{exists for} \quad (g,U) \in \mathcal{V}$$

and therefore giving, see (16.218)

$$(16.227) \quad gU = (g_2 \circ (id_{\Omega} \times U)) \circ (g_1 \circ (id_{\Omega} \times U))^{-1} \quad \text{on} \quad \Omega, \text{ for} \quad (g,U) \in \mathcal{V}$$

and in particular, see (16.214), (16.217)

$$(16.228) \quad \Omega_{g,U} = \Omega, \quad (g,U) \in \mathcal{V}$$

We note that the *projectable* groups of transformations, see (16.52), (16.53), satisfy (16.223) - (16.228) with, see (16.55), (16.56)

$$(16.229) \quad \mathcal{V} = G \times \mathcal{C}^{\infty}(\Omega)$$

therefore, they are *nonlocal* both with respect to the group transformation (16.223) of the manifold M of independent and dependent variables, as well as their extension (16.224), (16.225) to $\mathcal{C}^{\infty}$-smooth functions.

Now it is easy to see that under the general assumptions (16.223) - (16.228), the construction in Subsections 16.3 - 16.5 can be extended as follows. First we start with the *localized* version of (16.60). Let us for that purpose define

$$(16.230) \quad \mathcal{V}_{\infty} \subsetneqq G \times (\mathcal{C}^{\infty}(\Omega))^{\mathbb{N}}$$

as being the set of all pairs (g,s), with $g \in G$ and $s = (\psi_\nu | \nu \in \mathbb{N}) \in (\mathcal{C}^\infty(\Omega))^{\mathbb{N}}$, such that, see (16.224), (16.225)

$$(16.231) \quad (g,\psi_\nu) \in \mathcal{V}, \quad \nu \in \mathbb{N}$$

Then it is clear that

$$(16.232) \quad \{e\} \times (\mathcal{C}^\infty(\Omega))^{\mathbb{N}} \subseteq \mathcal{V}_\infty \subseteq G \times (\mathcal{C}^\infty(\Omega))^{\mathbb{N}}$$

moreover (16.225) yields the desired *localized* version of (16.60), namely

$$(16.233) \quad \begin{array}{ccc} \mathcal{V}_\infty & \longrightarrow & (\mathcal{C}^\infty(\Omega))^{\mathbb{N}} \\ (g,s) & \longmapsto & gs = (g\psi_\nu | \nu \in \mathbb{N}) \end{array}$$

Now we can proceed with the *localized* version of (16.57). For that purpose we define

$$(16.234) \quad \mathcal{V}_A \subseteq G \times A$$

as the set of all pairs (g,U), with $g \in G$ and $U \in A$, for which there exists $s = (\psi_\nu | \nu \in \mathbb{N}) \in \mathcal{A}$, such that

$$(16.235) \quad \begin{array}{l} U = s + \mathcal{I}^\infty_{nd}(\Omega) \in A = \mathcal{A}/I^\infty_{nd}(\Omega) \\ (g,s) \in \mathcal{V}_\infty \end{array}$$

Then in view of (16.233), we can finally define the *localized* version of (16.57) by

$$(16.236) \quad \begin{array}{ccc} \mathcal{V}_A & \longrightarrow & A \\ (g,U) & \longmapsto & gU \end{array}$$

where

$$(16.237) \quad gU = gs + \mathcal{I}^\infty_{nd}(\Omega) \in A = \mathcal{A}/\mathcal{I}^\infty_{nd}(\Omega)$$

Obviously, the above definition (16.236), (16.237) is correct, provided that we have, see (16.65)

$$(16.238) \quad \{gs \mid s \in \mathcal{A},\ (g,s) \in \mathcal{V}_\infty\} \subseteq \mathcal{A}, \quad g \in G$$

as well as, see (16.66)

$$(16.239) \quad \begin{array}{l} \forall\ g \in G,\ t \in \mathcal{A},\ v \in \mathcal{I}^\infty_{nd}(\Omega) : \\ \quad (g,t+v),\ (g,t) \in \mathcal{V}_\infty \Rightarrow g(t+v) - gt \in \mathcal{I}^\infty_{nd}(\Omega) \end{array}$$

However, condition (16.238) is clearly satisfied, for instance, for $\mathcal{A} = (\mathcal{C}^\infty(\Omega))^{\mathbb{N}}$. Concerning condition (16.239), this time the situation is different from that in (16.66), where the convenient properties of the nowhere dense ideal $\mathcal{I}^\infty_{nd}(\Omega)$ were sufficient in order to secure that latter condition. Moreover, in the case of (16.66), we obtain its equivalent form (16.73) owing to the simplicity of the respective group transformations of functions in (16.56), see also (16.69).

This time however, we have (16.227) instead of (16.56), as the way group transformations act on functions. And then, in general, the equivalent of (16.67) cannot be obtained. Therefore, we can secure (16.239) only by assuming that, if needed, we have already replaced $\mathcal{V}_\infty$ given in (16.230), (16.231) by an appropriate *subset* of it, a subset which satisfies (16.239). Here it should be noted that, just as in the case of *general local* groups of transformations (16.186) - (16.190), see also Olver [p 21], such *nonvoid* subsets of $\mathcal{V}_\infty$ will always exist, even if only trivially, that is, given by

$$\{e\} \times (\mathcal{C}^\infty(\Omega))^{\mathbb{N}}$$

However, as seen in Subsection 16.8, one can as well have *nontrivial* choices of $\mathcal{V}_\infty$. Indeed, in the case of each of the nonlinear one dimensional Lie groups (16.141) and (16.142), we shall have

$$(16.240) \quad [0,\infty) \times \mathcal{S} \subseteq \mathcal{V}_\infty \subseteq G \times (\mathcal{C}^\infty(\Omega))^{\mathbb{N}}$$

where $\mathcal{S}$ is the set of sequences $s \in (\mathcal{C}^{\infty}(\Omega))^{\mathbb{N}}$ in (16.149) - (16.152) corresponding to the Riemann solvers (16.146) of the nonlinear shock wave equation.

Finally, in view of (16.179), it is clear that under the conditions in (16.223) - (16.239) one can *extend* certain *nonprojectable* nonlinear classical symmetry groups of nonlinear PDEs to *local* symmetry groups of *global generalized solutions* of such equations.

The results in this Section were obtained partly in collaboration with the doctoral student Yvonne E Walus.

17. GROUP INVARIANCE OF GENERALIZED SOLUTIONS OBTAINED THROUGH THE ALGEBRAIC METHOD : AN ALTERNATIVE APPROACH

17.0 Preliminaries

In Section 16 we presented the group invariance of generalized solutions obtained through the algebraic method based on the so called *nowhere dense ideals* of sequences of smooth functions. This algebraic method is a *particular* case of the general algebraic theory initiated and developed in Rosinger [1-8]. One of the *advantages* of this particular case based on nowhere dense ideals is that it can deliver the *global existence* result in Theorem 16.1. This global existence result concerning generalized solutions of arbitrary *analytic nonlinear* PDEs is, so far, the most general *type independent* global existence result known in the literature, except for the global existence results in Sections 5 and 13, which are still more general, as they are valid for arbitrary *continuous nonlinear* PDEs.

In this way, one of the main interests in the results on group invariance presented in Section 16 is that, owing to Theorem 16.1, they apply to *global generalized solutions* of large classes of nonlinear PDEs, solutions about which now we *do know* that they exist.

Another *particular* case of the general algebraic theory in Rosinger [1-8] was developed independently in Colombeau [1,2,3]. As mentioned for instance in Rosinger [5,6], Biagioni, Oberguggenberger [2], this particular theory has a number of important advantages and in a certain way it occupies a natural central position with respect to all the possible algebraic theories presented and characterized in Rosinger [1-8]. The particular algebraic method in Colombeau [1,2,3] delivers global generalized solutions to various classes of linear and nonlinear PDEs, notably, nonlinear wave equations as well as more general hyperbolic and parabolic equations.

Therefore, there is an interest in implementing group invariance in the case of those nonlinear PDEs for which the particular algebraic method in Colombeau [1,2,3] can provide global generalized solutions. Indeed, such an interest is justified by the mentioned advantages of the algebraic method in Colombeau [1,2,3], as well as by the fact that it can given an alternative to the approach in Section 16.

A general type independent existence result comparable with Theorems 16.1 and 16.2, or with those in Sections 5 and 13, has so far not been proved in the setting of Colombeau, when the differential operators involved are considered in their natural differential algebraic sense, see Section 4 and (16.31). On the other hand, when the derivatives in the partial differential operators are replaced by mollified, that is, regularized versions, a general existence and uniqueness result for arbitrary linear and nonlinear PDEs of evolution type holds in the setting of Colombeau, see Colombeau & Heibig, also Colombeau, Heibig & Oberguggenberger. However, the mollifying or regularizing of the derivatives is a linear operation which, therefore, can hardly at all be subjected to nonlinear group invariance, a group invariance typical in the case of nonlinear PDEs. In view of that, we shall not consider here Colombeau's mentioned type independent global existence results. Rather, we shall be concerned with the method in Colombeau [1,2,3], which can give global existence results for particular classes of linear or nonlinear PDEs, and does so by obtaining solutions in the natural differential algebraic sense.

It should be mentioned that one of the main *limitations* of Colombeau's method is in the *growth* conditions it imposes on the basic asymptotical processes it involves. And as we shall see in this Section, the mentioned growth conditions will *limit* both the nonlinear PDEs and the group transformations acting on their solutions. On the other hand, *none* of these limitations was needed in Section 16. Similarly, they will *not* be needed when in Section 18 we shall study group invariance based on the order completion method.

17.1 A Brief Review of the Structure of Colombeau's Generalized Functions

For convenience, we shall only deal with generalized functions and solutions defined on the whole $\mathbb{R}^n$. However, with minor technical adaptations, the generalized solutions defined on open subsets $\Omega \subseteq \mathbb{R}^n$ can be dealt with in a similar manner, see Biagioni, Oberguggenberger [2] for the way the Colombeau algebra of generalized functions is treated on such open subsets $\Omega \subseteq \mathbb{R}^n$.

In order to define Colombeau's algebra of generalized functions on $\mathbb{R}^n$, we proceed as follows.

For $m \in \mathbb{N}_+ = \mathbb{N}\setminus\{0\}$, let

$$(17.1) \qquad \Phi_m = \left\{ \phi \in \mathcal{D}(\mathbb{R}^n) \;\middle|\; \begin{array}{l} *) \int_{\mathbb{R}^n} \phi(x)dx = 1 \\ **) \ \forall \ p \in \mathbb{N}^n, \ \ 1 \leq |p| \leq m : \\ \qquad \int_{\mathbb{R}^n} x^p \phi(x)dx = 0 \end{array} \right\}$$

and let us denote

$$(17.2) \qquad \Phi = \Phi_1$$

Now we define the *subalgebra* $\mathcal{A}$ in $(\mathcal{C}^\infty(\mathbb{R}^n))^\Phi$ as the set of functions

$$(17.3) \qquad \Phi \times \mathbb{R}^n \ni (\phi,x) \longmapsto f(\phi,x) \in \mathbb{R}, \text{ with } f(\phi,\cdot) \in \mathcal{C}^\infty(\mathbb{R}^n) \text{ for } \phi \in \Phi$$

which satisfy the condition

$$(17.4) \qquad \begin{array}{l} \forall \ K \subset \mathbb{R}^n \text{ compact}, \ \ p \in \mathbb{N}^n : \\ \exists \ m \in \mathbb{N}_+ : \\ \forall \ \phi \in \Phi_m : \\ \exists \ \eta, c > 0 : \\ \forall \ x \in K, \ \ \epsilon \in (0,\eta) : \\ \quad |D_x^p \ f(\phi_\epsilon,x)| \leq c/\epsilon^m \end{array}$$

where $\phi_\epsilon \in \Phi_m$ is defined by

$$(17.5) \qquad \phi_\epsilon(x) = \phi(x/\epsilon)/\epsilon^n, \quad \epsilon > 0, \quad x \in \mathbb{R}^n$$

Further, we define the *ideal* $\mathcal{I}$ in $\mathcal{A}$ as the set of $f \in \mathcal{A}$ for which

$$(17.6)\quad \begin{array}{l} \forall\ K \subset \mathbb{R}^n \ \text{compact},\ \ p \in \mathbb{N}^n : \\ \exists\ \ell \in \mathbb{N}_+,\ \ \beta \in B : \\ \forall\ m \in \mathbb{N}_+,\ \ m \geq \ell,\ \ \phi \in \Phi_m : \\ \exists\ \eta, c > 0 : \\ \forall\ x \in K,\ \ \epsilon \in (0,\eta) : \\ \quad |D_x^p\ f(\phi_\epsilon, x)| \leq c\epsilon^{\beta(m)-\ell} \end{array}$$

where we denoted

$$(17.7)\qquad B = \left\{\beta : \mathbb{N}_+ \longrightarrow (0,\infty) \ \middle|\ \begin{array}{l} *)\ \beta \ \text{increasing} \\ **)\ \lim_{m\to\infty} \beta(m) = \infty \end{array} \right\}$$

Now Colombeau's algebra of generalized functions on $\mathbb{R}^n$ is given by

$$(17.8)\qquad \mathcal{G}(\mathbb{R}^n) = \mathcal{A}/\mathcal{I}$$

The class of *nonlinear* PDEs which can be defined on $\mathcal{G}(\mathbb{R}^n)$ can be obtained as follows. If $p \in \mathbb{N}^n$ then the partial derivative operator

$$(17.9)\qquad D^p : \mathcal{G}(\mathbb{R}^n) \longrightarrow \mathcal{G}(\mathbb{R}^n)$$

is defined by

$$(17.8)\qquad D^p(f + \mathcal{I}) = (D_x^p f) + \mathcal{I},\ \ f \in \mathcal{A}$$

We note that in this way we obtain the embedding of *differential algebras*

$$(17.9)\qquad \mathcal{C}^\infty(\mathbb{R}^n) \ni f \longmapsto \bar{f} + \mathcal{I} \in \mathcal{G}(\mathbb{R}^n)$$

where, see (17.3)

$$(17.10)\qquad \bar{f}(\phi,x) = f(x),\ \ \phi \in \Phi,\ \ x \in \mathbb{R}^n$$

The L Schwartz distributions $\mathcal{D}'(\mathbb{R}^n)$ can be embedded as a vector subspace in $\mathcal{G}(\mathbb{R}^n)$ according to the mapping

(17.11) $\mathcal{D}'(\mathbb{R}^n) \ni T \longmapsto f_T + \mathcal{I} \in \mathcal{G}(\mathbb{R}^n)$

where, see (17.3)

(17.12) $f_T(\phi,x) = T_y(\phi(y - x)), \quad \phi \in \Phi, \quad x \in \mathbb{R}^n$

An important property of the *linear* embedding (17.11) is that is preserves arbitrary differentiation of the distributions in $\mathcal{D}'(\mathbb{R}^n)$.

Given now $\ell \in \mathbb{N}_+$ and $P \in \mathcal{C}^\infty(\mathbb{R}^\ell)$, we call P *slowly increasing* if for each of its partial derivatives D^qP, with $q \in \mathbb{N}^\ell$, we have an estimate

(17.13) $|D^qP(\xi)| \le c(1 + |\xi|)^d, \quad \xi \in \mathbb{R}^\ell$

for certain $c,d > 0$. It follows that in such a case, one can define the mapping

(17.14) $P : (\mathcal{G}(\mathbb{R}^n))^\ell \longrightarrow \mathcal{G}(\mathbb{R}^n)$

by

(17.15) $P(f_1 + \mathcal{I},\dots,f_\ell + \mathcal{I}) = P(f_1,\dots f_\ell) + \mathcal{I}, \quad f_1,\dots,f_\ell \in \mathcal{A}$

Indeed, (17.15) is a valid definition owing to the following two properties

(17.16) $\forall \; f_1,\dots,f_\ell \in \mathcal{A} :$
$\quad P(f_1,\dots,f_\ell) \in \mathcal{A}$

and

(17.17) $\forall \; f_1,\dots,f_\ell \in \mathcal{A}, \; h_1,\dots,h_\ell \in \mathcal{I} :$
$\quad P(f_1 + h_1,\dots,f_\ell + h_\ell) - P(f_1,\dots,f_\ell) \in \mathcal{I}$

see for details Colombeau [2, pp 27-31].

If follows that in Colombeau's algebra $\mathcal{G}(\mathbb{R}^n)$ of generalized functions one can define nonlinear PDEs in (2.1), (2.2), as long as F is C^∞-*smooth* and *slowly increasing*. Therefore, in this Section, we shall always assume that the nonlinear PDEs in (2.1), (2.2) which we consider are defined by *slowly increasing* functions F as mentioned above.

17.2 Group Transformations on $\mathcal{G}(\mathbb{R}^n)$

In the sequel, whenever dealing with a nonlinear PDE considered within $\mathcal{G}(\mathbb{R}^n)$, we shall assume that it possesses generalized solutions in $\mathcal{G}(\mathbb{R}^n)$. In other words, in this Section we shall limit ourselves to those nonlinear PDEs in $\mathcal{G}(\mathbb{R}^n)$ which are known to have generalized solutions in $\mathcal{G}(\mathbb{R}^n)$. However, as mentioned in Subsection 17.1, the class of such PDEs contains a variety of cases of particular interest.

Here we shall only consider *projectable* groups of transformations. Therefore, let be given the nonlinear PDE, see (2.1), (2.2)

$$(17.18) \qquad T(x,D)U(x) = 0, \quad x \in \mathbb{R}^n$$

where F is *slowly increasing*, as specified in Subsection 17.1, while $U : \mathbb{R}^n \longrightarrow \mathbb{R}$ is the unknown function.

In view of (17.18), the manifold of the independent and dependent variables is

$$(17.19) \qquad M = \mathbb{R}^n \times \mathbb{R} = \mathbb{R}^{n+1}$$

therefore the projectable groups of transformations of interest will be of the form, see (16.52), (16.53)

$$(17.20) \qquad \begin{array}{ccc} G \times M & \longrightarrow & M \\ (g,(x,u)) & \longmapsto & (g_1(x),\ g_2(x,u)) \end{array}$$

where

$$(17.21) \quad \mathbb{R}^n \ni x \longmapsto g_1(x) \in \mathbb{R}^n$$

is a diffeomorphism.

Now, we *extend* the groups of transformations (17.20), (17.21) to the algebra of generalized functions $\mathcal{G}(\mathbb{R}^n)$. This extension follows the same basic idea as in Subsection 16.4. Here we recall that the action of G on functions in $\mathcal{C}^\infty(\mathbb{R}^n)$ was defined in (16.55), (16.56). Therefore, the action of G on sequences of functions in $(\mathcal{C}^\infty(\mathbb{R}^n))^\Phi$ follows, similar to (16.60), (16.61), in a natural termwise manner, i.e.

$$(17.22) \quad (gf)(\phi,x) = (gf(\phi,\cdot))(x), \quad \phi \in \Phi, \quad x \in \mathbb{R}^n$$

for $g \in G$ and f in (17.3). To be more precise, let us denote by $k : \Phi \times \mathbb{R}^n \longrightarrow \mathbb{R}$ the function defined in (17.22), in other words, given by

$$k(\phi,x) = (gf(\phi,\cdot))(x), \quad \phi \in \Phi, \quad x \in \mathbb{R}^n$$

then (17.20), (17.21), (16.56) result in

$$(17.23) \quad k(\phi,x) = g_2(g_1^{-1}(x),f(\phi,g_1^{-1}(x))), \quad \phi \in \Phi, \quad x \in \mathbb{R}^n$$

However, because of reasons similar to those in (17.13) - (17.14), we shall have to *restrict* the projectable groups of transformations (17.20), (17.21) by requiring that, see (17.13), (17.23)

$$(17.24) \quad g_2 \text{ is slowly increasing}$$

at least in the second variable u, and uniformly on compact sets with respect to the first variable x.

Clearly, these *slowly increasing projectable* groups of transformations (17.20), (17.21), (17.24) contain as a particular case the groups of transformations (16.54) relevant in General Relativity.

Now in view of (17.23), (17.24), we note that similar to (17.16), we have, see (17.4), (16.65)

$$(17.25) \quad g\mathcal{A} \subseteq \mathcal{A}, \quad g \in G$$

Also, similar to (17.17), the following property holds, see (17.6), (16.66)

$$(17.26) \quad \begin{array}{l} \forall\ g \in G, \quad f \in \mathcal{A}, \quad h \in \mathcal{I} : \\ \quad g(f + h) - gf \in \mathcal{I} \end{array}$$

Therefore, in view of (17.25), (17.26), we can define *group transformations* for the algebra $\mathcal{G}(\mathbb{R}^n)$ of generalized functions according to, see (16.82) - (16.87)

$$(17.27) \quad \begin{array}{ccc} G \times \mathcal{G}(\mathbb{R}^n) & \longrightarrow & \mathcal{G}(\mathbb{R}^n) \\ (g,U) & \longmapsto & gU \end{array}$$

where for

$$(17.28) \quad U = f + \mathcal{I} \in \mathcal{G}(\mathbb{R}^n) = \mathcal{A}/\mathcal{I}$$

we have

$$(17.29) \quad gU = gf + \mathcal{I} \in \mathcal{G}(\mathbb{R}^n) = \mathcal{A}/\mathcal{I}$$

17.3 Group Invariance of Global Generalized Solutions

Similar to Subsection 16.5, and in accordance with the restrictions needed in the case of Colombeau's algebra of generalized functions, we can now define the concept of *group invariant* generalized solutions.

Let us therefore consider given a $\mathcal{C}^\infty$-*smooth* and *slowly increasing* nonlinear PDE, see (17.18)

$$(17.30) \quad T(x,D)U(x) = 0, \quad x \in \mathbb{R}^n$$

and let us assume that it has a *generalized solution*

$$(17.31) \quad U = f + \mathcal{I} \in \mathcal{G}(\mathbb{R}^n) = \mathcal{A}/\mathcal{I}$$

where $f \in \mathcal{A}$. We note that in view of (17.14) - (17.17), if we assume that the nonlinear partial differential operator $T(x,D)$ in (17.30) is defined by, see (2.1), (2.2)

$$(17.32) \qquad T(x,D)U(x) = F(x,U(x),\ldots,D_x^pU(x),\ldots), \quad x \in \mathbb{R}^n$$

for $U \in \mathcal{C}^\infty(\mathbb{R}^n)$, then we obtain the commutative diagram

$$(17.33) \qquad \begin{array}{ccc} \mathcal{C}^\infty(\mathbb{R}^n) & \xrightarrow{\quad T(x,D) \quad} & \mathcal{C}^\infty(\mathbb{R}^n) \\ \downarrow & & \downarrow \\ \mathcal{G}(\mathbb{R}^n) & \xrightarrow{\quad T \quad} & \mathcal{G}(\mathbb{R}^n) \end{array}$$

where the embedding $\mathcal{C}^\infty(\mathbb{R}^n) \longrightarrow \mathcal{G}(\mathbb{R}^n)$ is defined in (17.9), while the mapping T is given by (17.14), (17.15), with P corresponding in an obvious manner to F in (17.32)

Let now be given a *slowly increasing projectable* group of transformations, see (17.19) - (17.21), (17.24)

$$(17.34) \qquad G \times M \longrightarrow M$$

where $M = \mathbb{R}^n \times \mathbb{R}$ is the manifold of the independent and dependent variables.

Then U is called a *group invariant* solution of (17.30), if for every $g \in G$, the relation, see (17.33), (17.27)

$$(17.35) \qquad T(gU) = 0$$

will hold.

In view of the commutative diagram (17.33), it is obvious that the above definition contains as a particular case the classical one valid for $\mathcal{C}^\infty$-smooth solutions U on $\mathbb{R}^n$.

17.4 Symmetry Groups for Generalized Solutions

In the case of Colombeau's algebra of generalized functions $\mathcal{G}(\mathbb{R}^n) = \mathcal{A}/\mathcal{I}$, the structure of the ideal $\mathcal{I}$ *prevents* us from the possibility of a *direct* extension of classical symmetry groups to symmetry groups of generalized functions, an extension which, as seen in Subsection 16.6, does quite easily take place within the algebras of generalized functions defined through the nowhere dense ideal. Indeed, the *difficulty* with the structure of Colombeau's ideal $\mathcal{I}$ is that it contains elements $h \in \mathcal{I}$ which do *not* vanish anywhere, namely, which are such that

$$(17.36) \quad h(\phi,x) \neq 0, \quad \phi \in \Phi, \quad x \in \mathbb{R}^n$$

In this way, the line of argument in (16.99) - (16.107) valid for *all* projectable groups of transformations, $\mathcal{C}^\infty$-smooth nonlinear PDEs and their global generalized solutions within Theorem 16.2, *cannot* be used in the case of Colombeau's algebra $\mathcal{G}(\mathbb{R}^n)$.

However, in the context of generalized solutions in $\mathcal{G}(\mathbb{R}^n)$, it may be of interest to note the following positive fact with respect to the class of linear partial differential operators which admit a continuous *right inverse*, see Meise, Taylor & Vogt for the definition of this class. Let $P = P(x,D)$ be a linear partial differential operator with $\mathcal{C}^\infty$-smooth coefficients which has a right inverse Q, assumed to be continuous in the $\mathcal{C}^\infty$-topology. Further, let G be a classical symmetry group for P. Then we can show that G will be a symmetry group also for the generalized solutions $U \in \mathcal{G}(\mathbb{R}^n)$, that is, for which

$$(17.37) \quad P(x,D)U = 0 \text{ in } \mathcal{G}(\mathbb{R}^n)$$

Indeed, assume U has the representation in (17.28). Then (17.37) is equivalent with

$$Pf = w \in \mathcal{I}$$

Now the right inverse property of Q and the linearity of P will give

$$P(f - Qw) = 0 \in \mathcal{I}$$

In particular

$$(17.38) \quad P(f(\phi,x) - Qw(\phi,x)) = 0, \quad \phi \in \Phi, \quad x \in \mathbb{R}^n$$

Given any $g \in G$, the classical solutions

$$f(\phi,\cdot) - Qw(\phi,\cdot), \quad \phi \in \Phi$$

in (17.38) will be invariant with respect to g. In other words

$$(17.39) \quad P(g(f(\phi,\cdot) - Qw(\phi,\cdot))) = 0, \quad g \in G, \quad \phi \in \Phi$$

However, the $\mathcal{C}^\infty$-continuity of Q results in $Qw \in \mathcal{T}$, thus $f - Qw \in U$ and hence $g(f - Qw) \in gU$, see (17.29).

In this way (17.39) yields

$$P(x,D)(gU) = 0 \quad \text{in} \quad \mathcal{G}(\mathbb{R}^n)$$

for all $g \in G$, which means that, indeed, G is as well a symmetry group of the generalized solutions in $\mathcal{G}(\mathbb{R}^n)$.

18. GROUP INVARIANCE OF GLOBAL GENERALIZED SOLUTIONS OBTAINED THROUGH THE ORDER COMPLETION METHOD

18.1 Introduction

Following the previous two Sections, here we present a *third* different and independent method for the group invariance of global generalized solutions of large classes of nonlinear PDEs. One of the interests in this third method comes from the fact that it is based on the order completion method in solving arbitrary continuous nonlinear PDEs, a method which gives *existence* results, see Sections 5 and 13, which so far are the *most general* in the literature. In particular, they are more general than those in Subsection 16.2, or in Colombeau [1,2,3] which were assumed in Section 17.

The relevant setup is given by commutative diagrams

(18.1)

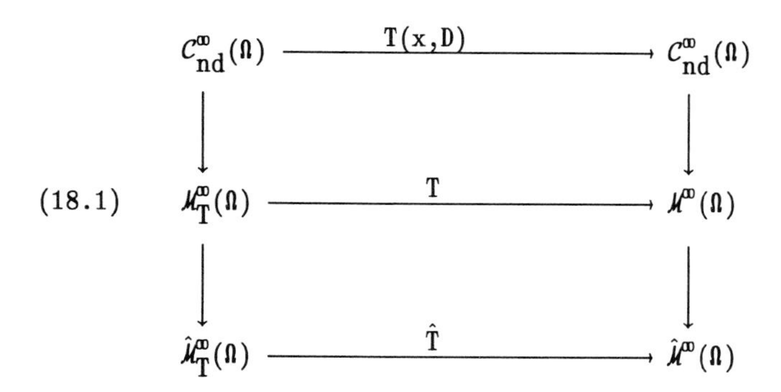

which correspond to the $\mathcal{C}^\infty$-*smooth* version of the nonlinear PDEs, see (2.1), (2.2)

(18.2) $\quad T(x,D)U(x) = f(x), \quad x \in \Omega \subseteq \mathbb{R}^n$

where both F in (2.2) and f are $\mathcal{C}^\infty$-smooth. Indeed, (18.1) follows directly from (4.6), (4.15) or (7.80) as a particularization corresponding to the $\mathcal{C}^\infty$-smooth nonlinear PDEs in (18.2).

As in Section 16, we shall start with *projectable* groups of transformations of the *finite* dimensional manifolds of independent and dependent variables

(18.3) $\quad M = \Omega \times \mathbb{R} \subseteq \mathbb{R}^{n+1}$

In other words, we shall first consider Lie groups G which act on M according to

$$(18.4)\qquad \begin{array}{ccc} G \times M & \longrightarrow & M \\ (g,(x,u)) & \longmapsto & g(x,u) \end{array}$$

where for $g \in G$, $(x,u) \in M$ we have

$$(18.5)\qquad g(x,u) = (g_1(x),\ g_2(x,u))$$

and

$$(18.6)\qquad \Omega \xrightarrow{\ g_1\ } \Omega$$

is a diffeomorphism, while $g_2(x,u) \in \mathbb{R}$.

18.2 Group Transformations of Spaces of Generalized Functions

The extension of the groups of transformations (18.3) - (18.6) to the diagrams (18.1) requires the extension of the group transformations to the *domains* $\mathcal{C}^\infty(\Omega)$, $\mathcal{C}^\infty_{nd}(\Omega)$, $\mathcal{M}^\infty_T(\Omega)$ and $\hat{\mathcal{M}}^\infty_T(\Omega)$ of the respective operators $T(x,D)$, T and $\hat{T}$. This extension can be obtained in a few steps as follows. First we recall that the extension, see (16.55), (16.56)

$$(18.7)\qquad \begin{array}{ccc} G \times \mathcal{C}^\infty(\Omega) & \longrightarrow & \mathcal{C}^\infty(\Omega) \\ (g,U) & \longmapsto & gU \end{array}$$

is immediate, and it is given by

$$(18.8)\qquad (gU)(x) = g_2(g_1^{-1}(x), U(g_1^{-1}(x))),\quad x \in \Omega$$

Further, the extension

$$(18.9)\qquad \begin{array}{ccc} G \times \mathcal{C}^\infty_{nd}(\Omega) & \longrightarrow & \mathcal{C}^\infty_{nd}(\Omega) \\ (g,U) & \longmapsto & gU \end{array}$$

is an obvious generalization of (18.7), (18.8). Indeed, assume that, see (3.1), $U \in \mathcal{C}^\infty(\Omega \backslash \Gamma)$ where $\Gamma \subset \Omega$ is closed, nowhere dense. Then in view of (18.6), it clearly results that $\Sigma = g_1(\Gamma) \subset \Omega$ is also closed and nowhere dense. In this way, we can extend (18.8) by

$$(18.10) \qquad (gU)(x) = g_2(g_1^{-1}(x), U(g_1^{-1}(x))), \quad x \in \Omega \backslash \Sigma$$

and obtain $gU \in \mathcal{C}^\infty(\Omega \backslash \Sigma)$.

Before we go further, we note that (18.9), (18.10) allow us the immediate further extension leading to the groups of transformations

$$(18.11) \qquad \begin{array}{ccc} G \times \mathcal{M}^\infty(\Omega) & \longrightarrow & \mathcal{M}^\infty(\Omega) \\ (g,U) & \longmapsto & gU \end{array}$$

where gU is the $\sim$ equivalence class, see (3.3), of gu, where $u \in U$. This definition is correct, in view of (18.10).

Now the next step would be to define the groups of transformations

$$\begin{array}{ccc} G \times \mathcal{M}_T^\infty(\Omega) & \longrightarrow & \mathcal{M}_T^\infty(\Omega) \\ (g,U) & \longmapsto & gU \end{array}$$

However, here we encounter the following difficulty: if $U,V \in \mathcal{M}^\infty(\Omega)$ and $T(x,D)U \sim T(x,D)V$, that is, $U \sim_T V$, see (4.3), then it will *not* necessarily result that we have as well

$$T(x,D)(gU) \sim T(x,D)(gV), \quad g \in G$$

or in other words, we need *not* have, see Example 18.1 below

$$gU \sim_T gV, \quad g \in G$$

Example 18.1

Let $\Omega = \mathbb{R}$, $T(x,D) = D_x$, $G = (\mathbb{R},+)$, while (18.4), (18.5) is given for $g = \epsilon \in G = (\mathbb{R},+)$ by

$$g_1(x) = x, \quad g_2(x,u) = ((x + u)^3 + \epsilon)^{1/3} - x$$

Further, let us take $U,V \in \mathcal{C}^\infty(\Omega)$ given by

$$U(x) = a, \quad V(x) = b, \quad x \in \Omega$$

where $a,b \in \mathbb{R}$, $a \neq b$. Then clearly

$$T(x,D)U = T(x,D)V = 0 \quad \text{on} \quad \Omega$$

However, if $g = \epsilon \in G = (\mathbb{R},+)$, $\epsilon \neq 0$, then for $x \in \Omega$ we obtain

$$(gU)(x) = ((x + a)^3 + \epsilon)^{1/3} - x, \quad (gV)(x) = ((x + b)^3 + \epsilon)^{1/3} - x$$

hence

$$T(x,D)(gU)(x) \neq T(x,D)(gV)(x), \quad x \in \Omega \qquad \blacksquare$$

In view of this *failure* of the equivalence relationship $\sim_T$ to be compatible with arbitrary group transformations, we have to go back to the diagram (18.1) and implement certain *modifications*.

The general idea for the modifications has been presented in Section 13, through the construction in (13.8) - (13.14). Here, that construction will be particularized and adapted to the case of groups of transformations applied to the order completion method used in solving continuous nonlinear PDEs.

Given a group transformation (18.3) - (18.6), we define an equivalence relation $\sim_{T,G}$ on $\mathcal{C}^\infty_{nd}(\Omega)$ as follows: if $u,v \in \mathcal{C}^\infty_{nd}(\Omega)$ then, see (18.9),

$$(18.12) \qquad u \sim_{T,G} v \iff \left[\begin{array}{l} \forall \ g \in G : \\ \quad T(x,D)(gu) \sim T(x,D)(gv) \end{array} \right]$$

It is obvious that, see (13.8)

$$(18.13) \qquad \begin{array}{l} \forall \ u,v \in \mathcal{C}^{\infty}_{nd}(\Omega) : \\ \quad u \sim_{T,G} v \Longrightarrow u \sim_T v \end{array}$$

Therefore, if we define, see (13.9)

$$(18.14) \qquad \mathcal{M}^{\infty}_{T,G}(\Omega) = \mathcal{C}^{\infty}_{nd}(\Omega)/\sim_{T,G}$$

we obtain the *surjective* mapping

$$(18.15) \qquad \begin{array}{ccc} \mathcal{M}^{\infty}_{T,G}(\Omega) & \xrightarrow{\text{sur}} & \mathcal{M}^{\infty}_{T}(\Omega) \\ U & \longmapsto & U' \end{array}$$

where U' is the $\sim_T$ equivalence class of any $u \in U$. In this way the commutative diagram results, see (13.10)

(18.16)

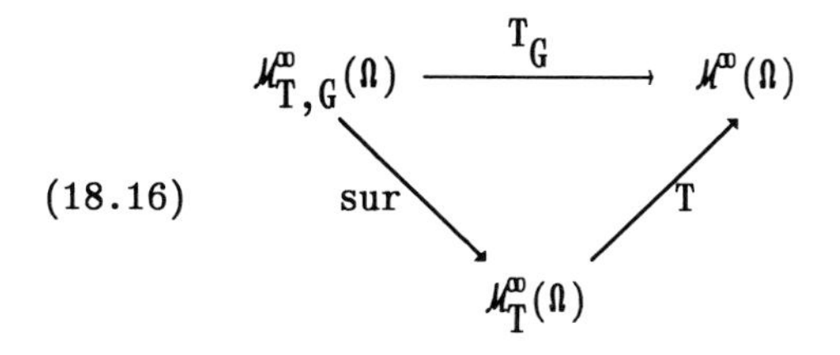

where similar to (4.5), (4.6) and (13.10), we have

$$(18.17) \qquad \begin{array}{ccc} \mathcal{M}^{\infty}_{T,G}(\Omega) & \xrightarrow{T_G} & \mathcal{M}^{\infty}(\Omega) \\ U & \longmapsto & T_G U \end{array}$$

with $T_G U$ being the $\sim$ equivalence class of any $T(x,D)u$, where $u \in U$.

The advantage of the spaces of *generalized functions* $\mathcal{M}^{\infty}_{T,G}(\Omega)$ is that they admit groups of transformations

$$(18.18)\qquad \begin{array}{ccc} G \times \mathcal{M}^{\infty}_{T,G}(\Omega) & \longrightarrow & \mathcal{M}^{\infty}_{T,G}(\Omega) \\ (g,U) & \longmapsto & gU \end{array}$$

defined as follows: if $u \in U$ then gU is the $\sim_{T,G}$ equivalence class of gu, see (18.7). And it is easy to see that in view of (18.12), the definition in (18.18) is correct.

At this stage, similar to (13.11) and partly recalling (4.7), we can define the partial order $\leq_{T,G}$ on $\mathcal{M}^{\infty}_{T,G}(\Omega)$ by

$$(18.19)\qquad U \leq_{T,G} V \Longleftrightarrow \left[\begin{array}{l} *)\ U = V \\ \text{or} \\ **)\ \forall\ g \in G : \\ \qquad T_G(gU) \underset{\neq}{\leq} T_G(gV) \quad \text{in} \quad \mathcal{M}^{\infty}(\Omega) \end{array}\right]$$

The effect of this definition is that the commutative diagram (18.16) becomes a commutative diagram of *increasing* mappings

$$(18.20)\qquad \begin{array}{ccc} (\mathcal{M}^{\infty}_{T,G}(\Omega), \leq_{T,G}) & \xrightarrow{\ T_G\ } & (\mathcal{M}^{\infty}(\Omega), \leq) \\ & \text{sur}\searrow \qquad \nearrow T & \\ & (\mathcal{M}^{\infty}_{T}(\Omega), \leq_T) & \end{array}$$

More importantly, we can replace the commutative diagrams (18.1) with the following more general ones, see the argument leading to (13.14)

(18.21) 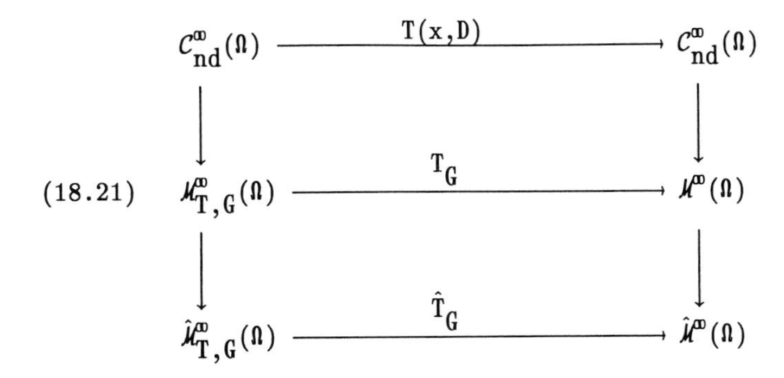

where T_G and $\hat{T}_G$ are *increasing* mappings.

And in view of (18.18), it only remains to extend the group transformations (18.3) - (18.6) to the space of generalized functions $\hat{\mathcal{M}}^\infty_{T,G}(\Omega)$. This extension will be presented in the next Subsection.

Remark 18.1

1) It is easy to see that, if with the notation in Section 13, we restrict ourselves from $\mathcal{C}^m_{nd}(\Omega)$ to $\mathcal{C}^\infty_{nd}(\Omega)$ and take

$$\sim_{T,*} = \sim_{T,G}$$

then the partial order $\leq_{T,G}$ defined in (18.19) need *not* coincide with the partial order $\leq_{T,*}$ in (13.11), when the latter is restricted to $\mathcal{M}^\infty_{T,*}(\Omega)$. However $\leq_{T,G}$ will always be *smaller* than $\leq_{T,*}$, see also Remark 18.3.

2) As in the case of $\leq_{T,*}$, see Remark 13.1, the partial order $\leq_{T,G}$ is in general *not* the 'pull-back' through T_G of the partial order $\leq$ on $\mathcal{M}^\infty(\Omega)$. In fact, this follows clearly from pct. 1) above.

Example 18.2

In order to illustrate the construction in the previous part of this Subsection, we shall apply it in a simple case of (18.2) given by the *semilinear* PDE

$$(18.22) \quad U_t(t,x) + \lambda U_x(t,x) = F(U(t,x)), \quad t > 0, \quad x \in \mathbb{R}$$

where $\lambda \in \mathbb{R}$, $\lambda \neq 0$ and $F : \mathbb{R} \longrightarrow \mathbb{R}$ is $\mathcal{C}^\infty$-smooth.

As seen in (16.114), among the symmetry groups of (18.22) are the following

$$(18.23) \quad \begin{array}{ccc} G \times M & \longrightarrow & M \\ (\epsilon,(t,x,u)) & \longmapsto & (t,x(\epsilon),u) \end{array}$$

where $M = \Omega \times \mathbb{R}$, $\Omega = (0,\infty) \times \mathbb{R}$, $\epsilon = g \in G = (\mathbb{R},+)$ while

$$(18.24) \quad \frac{dx(\epsilon)}{d\epsilon} = a(x(\epsilon) - \lambda t), \quad \epsilon \in \mathbb{R}$$

with $a : \mathbb{R} \longrightarrow \mathbb{R}$ an arbitrary $\mathcal{C}^\infty$-smooth function. It follows that (18.23) is a *projectable* group of transformations, therefore, we can apply to it the construction in the previous part of this Subsection. Moreover, in view of (18.24), it is clear that (18.23) contains *nonlinear* groups of transformations. For instance, if $a(\xi) = \xi^2$, for $\xi \in \mathbb{R}$, then

$$(18.25) \quad x(\epsilon) = \lambda t + (x - \lambda t)/(1 - (x - \lambda t)\epsilon), \quad \epsilon \in \mathbb{R}$$

In fact (18.25) gives an instance of the group of transformations (18.23) which is not only nonlinear, but also *local*, since it requires that

$$(18.26) \quad \epsilon \neq 1/(x - \lambda t), \quad \text{if} \quad x - \lambda t \neq 0$$

However, it is easy to see that (18.23) contains as well *global* group transformations. For instance, by taking $a(\xi) = \xi$, for $\xi \in \mathbb{R}$, we obtain

$$(18.27) \quad x(\epsilon) = \lambda t + (x - \lambda t)e^\epsilon, \quad \epsilon \in \mathbb{R}$$

Let us now see the specific form the diagrams (18.20) and (18.21) will take in the particular case of (18.23), (18.27). First we note that (18.8) becomes

$$(18.28) \qquad (\epsilon U)(t,x) = U(t,\lambda t + (x - \lambda t)e^{-\epsilon}), \quad t > 0, \quad x \in \mathbb{R}$$

for $U \in \mathcal{C}^\infty(\Omega)$ and $\epsilon \in \mathbb{R}$. It follows that

$$(18.29) \qquad T(D)(\epsilon U)\Big|_{(t,x)} = T(D)U\Big|_{(t,\lambda t + (x - \lambda t)e^{-\epsilon})}, \quad t > 0, \quad x \in \mathbb{R}$$

for $U \in \mathcal{C}^\infty(\Omega)$, $\epsilon \in \mathbb{R}$, where

$$T(D)U = U_t + \lambda U_x - F(U)$$

is the partial differential operator associated with (18.22).

Now, in view of (18.29) it is clear that, see (18.12), (18.13)

$$(18.30) \qquad u \sim_{T,G} v \Longleftrightarrow u \sim_T v$$

for $u,v \in \mathcal{C}^\infty_{nd}(\Omega)$. Therefore, see (18.4), (18.16), (18.17)

$$(18.31) \qquad \mathcal{M}^\infty_{T,G}(\Omega) = \mathcal{M}^\infty_T(\Omega), \quad T_G = T$$

And then, in view of (18.19), (18.30), (18.31), we obtain

$$(18.32) \qquad U \leq_{T,G} V \Longleftrightarrow U \leq_T V$$

for $U,V \in \mathcal{M}^\infty_T(\Omega) = \mathcal{M}^\infty_{T,G}(\Omega)$.

In this way, in the particular case of (18.22), (18.23) and (18.27) the diagram (18.21) reduces to that in (18.1). ∎

Remark 18.2

The *invariance* relation (18.29) is the reason why in the particular case of the Example 18.1, the diagram (18.21) reduces to (18.1). However, it is important to note that, in general, an invariance relation such as in (18.29) will *not* hold. Indeed, given a nonlinear PDE in (18.2) and a projectable group of transformations (18.4) - (18.6), this group is a *symmetry* group of the PDE in (18.2), if and only if for every $U \in \mathcal{C}^\infty(\Omega)$ and $g \in G$ we have, see (18.7)

$$(18.33) \quad T(x,D)U = f \text{ on } \Omega \Longrightarrow T(x,D)(gU) = f \text{ on } \Omega$$

Nevertheless, this need *not* in general mean that, similar with (18.29), we would also have for $U \in \mathcal{C}^\infty(\Omega)$ and $g \in G$ the relation, see (18.6)

$$(18.34) \quad T(x,D)(gU)(x) = T(g_1^{-1}(x),D)U(g_1^{-1}(x)), \quad x \in \Omega$$

Indeed, the heat equation, for instance

$$(18.35) \quad U_t(t,x) = U_{xx}(t,x), \quad t > 0, \quad x \in \mathbb{R}$$

has as one of its symmetry groups, Olver [p 122], given by

$$(18.36) \quad \begin{matrix} t \\ x \\ u \end{matrix} \longmapsto \begin{matrix} \bar{t} = te^{2\epsilon} \\ \bar{x} = xe^{\epsilon}, \quad \epsilon \in \mathbb{R} \\ \bar{u} = u \end{matrix}$$

Therefore, denoting by $L(D)$ the partial differential operator associated with (18.35), namely

$$L(D)U = U_t - U_{xx}, \quad U \in \mathcal{C}^\infty(\Omega)$$

it follows that, *unlike* in (18.34), we have

$$(18.37) \quad L(D)(\epsilon U)(t,x) = \epsilon^{-2\epsilon}L(D)U(te^{-2\epsilon},xe^{-\epsilon})$$

for $U \in \mathcal{C}^\infty(\Omega)$, $t > 0$, $x \in \mathbb{R}$ and $\epsilon \in \mathbb{R}$. However, (18.37) still leads to (18.30), and therefore, to (18.31), (18.32) and the fact that (18.21) will reduce to (18.1).

Finally, we note that in general, the diagram (18.21) does *not* reduce to (18.1). Indeed, as seen for instance in Example 18.1 above, the relation (18.30) need not always hold. Here we note that this failure of (18.30) has been obtained for a most simple linear differential operator, namely, $T(x,D) = D_x$ and a nonlinear group transformation which is not necessarily related to the respective differential operator. However, even in case the linear partial differential operator and the group transformation are related through symmetry, the relation (18.30) can still fail. Indeed, let us return to the linear heat equation (18.35), and let us recall that another symmetry group of it is given by

$$(18.38) \qquad \begin{matrix} t \\ x \\ u \end{matrix} \longmapsto \begin{matrix} t \\ x + 2t\epsilon \\ ue^{-x\epsilon - t\epsilon^2} \end{matrix}, \quad \epsilon \in \mathbb{R}$$

Now it is easy to see that with the symmetry group (18.38) the following implication will in general *not* hold

$$(18.39) \qquad L(D)U = L(D)V \Longrightarrow L(D)(\epsilon U) = L(D)(\epsilon V), \quad \epsilon \in \mathbb{R}$$

for $U, V \in \mathcal{C}^\infty(\Omega)$. ∎

18.3 Group Transformations of Dedekind Order Completions

Our aim now is to define the group transformations

$$(18.40) \qquad \begin{matrix} G \times \hat{\mathcal{M}}^\infty_{T,G}(\Omega) & \longrightarrow & \hat{\mathcal{M}}^\infty_{T,G}(\Omega) \\ (g,F) & \longmapsto & gF \end{matrix}$$

Let us therefore take any $g \in G$ and $F \in \hat{\mathcal{M}}^\infty_{T,G}(\Omega)$. Since $F \subseteq \mathcal{M}^\infty_{T,G}(\Omega)$, we can use (18.18) and define

$$(18.41)\qquad gF = \{gU \mid U \in F\} \subseteq \mathcal{M}^{\infty}_{T,G}(\Omega)$$

Now we shall show that, see (A.11)

$$(18.42)\qquad gF \text{ is a cut in } (\mathcal{M}^{\infty}_{T,G}(\Omega), \leq_{T,G})$$

For that purpose, let us first prove for $U, V \in \mathcal{M}^{\infty}_{T,G}(\Omega)$ the relation

$$(18.43)\qquad U \leq_{T,G} V \Longleftrightarrow \left[\begin{array}{l} \forall\ g \in G : \\ \quad gU \leq_{T,G} gV \end{array}\right]$$

The implication '$\Longleftarrow$' is clear, if we choose $g = e$. For the implication '$\Longrightarrow$' we recall the two cases in (18.19). First, if $U = V$, then clearly $gU = gV$, with $g \in G$, and the implication '$\Longrightarrow$' holds. Otherwise, we have in $\mathcal{M}^{\infty}(\Omega)$ the relation

$$T_G(hU) \underset{\neq}{\leq} T_G(hV),\ h \in G$$

But then, as G is a group, it clearly follows that, for any given $g \in G$, we have in $\mathcal{M}^{\infty}(\Omega)$ the relation

$$T_G(hgU) \underset{\neq}{\leq} T_G(hgV),\quad h \in G$$

and then again, the implication '$\Longrightarrow$' will hold.

Based on (18.43), we can prove in $(\mathcal{M}^{\infty}_{T,G}(\Omega), \leq_{T,G})$ the relation

$$(18.44)\qquad g(A^{\ell}) = (gA)^{\ell},\quad g \in G,\quad A \subseteq \mathcal{M}^{\infty}_{T,G}(\Omega)$$

For the inclusion '$\subseteq$' let $U \in g(A^{\ell})$. Then $U = gV$, for a certain $V \in A^{\ell}$. This means that

$$V \leq_{T,G} W,\quad W \in A$$

But then (18.43) yields

$$gV \leq_{T,G} gW, \quad W \in A$$

or since (18.18) is a group transformation, it follows that

$$gV \leq_{T,G} Z, \quad Z \in gA$$

In other words $U = gV \in (gA)^{\ell}$. For the converse inclusion '$\supseteq$', let $U \in (gA)^{\ell}$. Then

$$U \leq_{T,G} V, \quad V \in gA$$

which in view of (18.43) gives

$$g^{-1}U \leq_{T,G} g^{-1}V, \quad V \in gA$$

or

$$g^{-1}U \leq_{T,G} W, \quad W \in A$$

In this way $g^{-1}U \in A^{\ell}$, which means that $U \in g(A^{\ell})$, and the proof of (18.44) is completed. Similarly, one can prove that the relation

$$(18.45) \qquad g(B^{u}) = (gB)^{u}, \quad g \in G, \quad B \subseteq \mathcal{M}^{\infty}_{T,G}(\Omega)$$

holds in $(\mathcal{M}^{\infty}_{T,G}(\Omega), \leq_{T,G})$.

In view of (18.44), (18.45), the relation (18.42) follows easily, namely

$$\begin{aligned} gF &= g(F^{u\ell}) = g((F^{u})^{\ell}) = (g(F^{u}))^{\ell} = \\ &= ((gF)^{u})^{\ell} = (gF)^{u\ell} \end{aligned}$$

We can in view of (18.42) conclude that (18.41) is a correct definition of the group transformation (18.40).

We end this Subsection by presenting a result on the nature of group transformations of the spaces of generalized functions in the basic diagram (18.21).

Proposition 18.1

Given the $\mathcal{C}^{\infty}$-smooth nonlinear PDE in (18.2) and the projectable group of transformations (18.3) - (18.6). Then for every $g \in G$, we obtain the *two order isomorphisms*, see (18.18)

$$(18.46) \qquad \mathcal{M}^{\infty}_{T,G}(\Omega) \ni U \stackrel{g}{\longmapsto} gU \in \mathcal{M}^{\infty}_{T,G}(\Omega)$$

and, see (18.40)

$$(18.47) \qquad \hat{\mathcal{M}}^{\infty}_{T,G}(\Omega) \ni F \stackrel{g}{\longmapsto} gF \in \hat{\mathcal{M}}^{\infty}_{T,G}(\Omega)$$

Proof

First we note that both mappings in (18.46) and (18.47) are bijections, as it follows easily from the fact that (18.18) and (18.40) are group transformations.

Now (18.46) follows directly from (18.43).

In order to prove (18.47), we shall show that

$$(18.48) \qquad \begin{array}{l} \forall \; F, H \in \hat{\mathcal{M}}^{\infty}_{T,G}(\Omega) : \\ \quad F \leq_{T,G} H \Longleftrightarrow \left[\begin{array}{l} \forall \; g \in G : \\ \quad gF \leq_{T,G} gH \end{array} \right] \end{array}$$

Indeed, as above in (18.43), the implication '$\Longleftarrow$' is immediate and we only deal with the converse implication '$\Longrightarrow$'. For that we note the equivalence, see (A.22)

$$(18.49) \qquad F \leq_{T,G} H \Longleftrightarrow F \subseteq H$$

where we recall that $F, H \subseteq \mathcal{M}^{\infty}_{T,G}(\Omega)$. However, in view of (18.41) we obtain

$$F \subseteq H \Longleftrightarrow \left[\begin{array}{c} \forall \ g \in G : \\ gF \subseteq gH \end{array} \right]$$

and then applying (18.49) to gF and gH, the proof of (18.48), as well as of (18.47), is completed. ∎

18.4 Group Invariance of Global Generalized Solutions

Let us consider the $\mathcal{C}^{\infty}$-*smooth* nonlinear PDE, see (18.2)

$$(18.50) \quad T(x,D)U(x) = f(x), \quad x \in \Omega$$

and let us assume given any *projectable* group of transformations (18.3) - (18.6). Then we can construct the commutative diagram (18.21).

Further, let us assume given a *global generalized solution* of (18.50) within (18.21), that is, we are given

$$(18.51) \quad F \in \hat{\mathcal{M}}^{\infty}_{T,G}(\Omega)$$

such that

$$(18.52) \quad \hat{T}_G F = f \ \text{ in } \ \hat{\mathcal{M}}^{\infty}(\Omega)$$

Then F is called a *group invariant* global generalized solution of (18.2), if for every $g \in G$ we have, see (18.40)

$$(18.53) \quad \hat{T}_G(gF) = f \ \text{ in } \ \hat{\mathcal{M}}^{\infty}(\Omega)$$

18.5 Computing Symmetry Groups for Global Generalized Solutions

We shall show that, similar to the situation in Subsection 16.6, those *projectable* groups of transformations which are *classical symmetry* groups of the $\mathcal{C}^{\infty}$-smooth nonlinear PDE in (18.50) will *extend* to symmetry groups of the class of *global generalized solutions* of (18.50) specified next.

Let us recall that according to Theorem 16.1, in case the nonlinear PDE in (18.50) is *analytic* then it will admit a large class of solutions

$$(18.54) \quad T(x,D)u(x) = f(x), \quad x \in \Omega\setminus\Gamma$$

where $\Gamma \subset \Omega$ is closed nowhere dense while

$$(18.55) \quad u : \Omega\setminus\Gamma \longrightarrow \mathbb{C} \text{ is analytic}$$

It follows that $u \in \mathcal{C}^{\infty}_{nd}(\Omega)$, hence by taking $U \in \mathcal{M}^{\infty}_{T,G}(\Omega)$ as the $\sim_{T,G}$ equivalence class of u, we obtain

$$T_G\, U = f \quad \text{in} \quad \mathcal{M}^{\infty}(\Omega)$$

Therefore, in a way which recalls (5.1), (13.15) and (13.37), we have for *arbitrary analytic nonlinear* PDEs in (18.50), when considered within the diagrams (18.21), the following inclusions

$$(18.56) \quad T_G(\mathcal{M}^{\infty}_{T,G}(\Omega)) \supset \mathcal{A}n(\Omega)$$

where $\mathcal{A}n(\Omega)$ denotes the set of *analytic* functions on Ω.

Here we recall that, along with Theorem 16.1, the existence result in (18.56) is at present by far the most general, that is, *type independent* result on *global solutions* of *nonlinear* PDEs.

However, in view of (18.54) - (18.56), and since the theory of Lie group transformations does *not* require analyticity but only $\mathcal{C}^{\infty}$-smoothness, we shall assume about our $\mathcal{C}^{\infty}$-smooth nonlinear PDE in (18.50) that it can be associated with an inclusion

$$(18.57) \quad T_G(\mathcal{M}^{\infty}_{T,G}(\Omega)) \supset \mathcal{F}(\Omega)$$

for a *relevantly large* subset of functions

$$(18.58) \quad \mathcal{F}(\Omega) \subseteq \mathcal{C}^{\infty}(\Omega)$$

(18.76) $U \in C^{\infty}_{nd}(\Omega)$

and

(18.77) $T(D)U = 0$ on $\Omega\backslash\Sigma$

where, see (18.72)

$$T(D)U = U_{tt} - U_{xx} - F(U)$$

for $U \in C^{\infty}(\Omega)$.

It follows that the solutions (18.74), (18.75) of the Klein-Gordon equation (18.72) are also included in the framework of (18.57), (18.58). Moreover, the classical symmetry group of the Klein-Gordon equation (18.72) is given by the *projectable* group transformation

(18.78) $(t,x,u) \longmapsto (t(\epsilon),x(\epsilon),u(\epsilon))$

where $(t,x,u) \in M = \Omega \times \mathbb{R} \subset \mathbb{R}^3$, $g = \epsilon \in G = (\mathbb{R},+)$, while

$$t(\epsilon) = \frac{at + x + b}{2a} e^{a\epsilon} + \frac{at - x + b}{2a} e^{-a\epsilon} - \frac{b}{a}$$

(18.79) $$x(\epsilon) = \frac{ax + t + c}{2a} e^{a\epsilon} + \frac{ax - t + c}{2a} e^{-a\epsilon} - \frac{c}{a}$$

$$u(\epsilon) = u$$

where $a,b,c \in \mathbb{R}$, $a \neq 0$ are arbitrary. In this way, according to Subsection 18.4, the *global generalized solutions* (18.74) - (18.77) of the Klein-Gordon equations will be invariant with respect to the group transformation (18.79).

Let us turn now to the Riemann solvers (16.146) of the nonlinear shock wave equation

(18.80) $U_t(t,x) + U(t,x)U_x(t,x) = 0, \quad t > 0, \quad x \in \mathbb{R}$

which, as mentioned, are encompassed by the framework in (18.57), (18.58). This time, however, the classical symmetry groups (16.140) - (16.143) of (18.80) are in general *no longer* projectable. Therefore we cannot apply the theory in Subsection 18.5.

However, as in Subsection 16.8, in the particular case of the Riemann solvers (16.146), we shall try to go beyond the method in Subsection 18.5.

For that purpose, let us for instance consider the group of transformations (16.142), given by the one dimensional nonlinear Lie group

$$(18.81) \qquad \Omega \times \mathbb{R} \ni (t,x,u) \longmapsto (t,x + c(u)\epsilon,u) \in \Omega \times \mathbb{R}$$

where $\Omega = (0,\infty) \times \mathbb{R}$, $\epsilon \in \mathbb{R}$ is the group parameter, while $c \in \mathcal{C}^\infty(\mathbb{R})$ is an arbitrarily given function. We shall now clarify the way this group of transformations can be applied to a given Riemann solver, see (16.146)

$$(18.82) \qquad U(t,x) = u_\ell + (u_r - u_\ell)H(x - x_0 - (u_\ell + u_r)t/2), \quad t > 0, \quad x \in \mathbb{R}$$

where $x_0, u_\ell, u_r \in \mathbb{R}$, with $u_\ell > u_r$, are fixed.

We note that the *closed nowhere dense* subset of Ω given by

$$(18.83) \qquad \Gamma = \{(t,x) \in \Omega \mid x - x_0 = (u_\ell + u_r)t/2\}$$

is the set of *shock* points of the Riemann solver U in (18.82). In particular, we have

$$(18.84) \qquad U \in \mathcal{C}^\infty(\Omega \setminus \Gamma)$$

Let us now define the open subsets of Ω

$$(18.85) \qquad \begin{aligned} \Omega_\ell &= \{(t,x) \in \Omega \mid x - x_0 < (u_\ell + u_r)t/2\} \\ \Omega_r &= \{(t,x) \in \Omega \mid x - x_0 > (u_\ell + u_r)t/2\} \end{aligned}$$

which form a partition of $\Omega \setminus \Gamma$. Further, let us assume that, see (16.173)

(18.86) $c(u_\ell) = c(u_r) = c_0$

Now we apply the group transformations (18.81) to the Riemann solver U in (18.82) on the open set $\Omega\backslash\Gamma = \Omega_\ell \cup \Omega_r$ and obtain for $\epsilon \in \mathbb{R}$ the relation, Olver [p 94]

(18.87) $(\epsilon U)(t,x) = U(t,x - c_0\epsilon), \quad (t,x) \in \Omega\backslash\Gamma_\epsilon$

noting that

(18.88) $\Omega\backslash\Gamma \ni (t,x) \longmapsto (t,x + c_0\epsilon) \in \Omega\backslash\Gamma_\epsilon$

is a diffeomorphism, where

(18.89) $\Gamma_\epsilon = \{(t,x + c_0\epsilon) \mid (t,x) \in \Gamma\}$

is a closed nowhere dense subset of Ω.

Now in view of (18.87) - (18.89), it is obvious that the group of transformations (18.81) will under the condition (18.86) transform the Riemann solver (18.82) into the Riemann solvers (18.87) which have the shocks Γ_ϵ with the same slope as that of Γ.

Here we should note that the above construction of group invariance of Riemann solvers is *simpler* than the corresponding one in Subsection 16.8. Moreover, unlike there, we obtain in (18.87) a *symmetry group* and not only a symmetry semigroup. Furthermore, the above construction is *more general* as well, since it does not require conditions such as (16.164).

A similar group invariance result for Riemann solvers can be obtained for the group of transformations (16.141).

The results in this and the previous Section were obtained partly in collaboration with the doctoral student Mynhard Rudolph.

'The time has come', the Walrus said,
'To talk of many things:
Of shoes - and ships - and sealing-wax
Of cabbages - and kings -'

Lewis Carroll, Alice
in Wonderland

Appendix. MacNeille Order Completion Through Dedekind Cuts and Related Results

The abstract, general result on order completion which was used essentially in this work is presented now. It was established in 1937, in MacNeille. For convenience, we recall it here in those of its features which are relevant to this work. A detailed proof of it can be found in Luxemburg & Zaanen.

A given partially ordered set, or in short *poset* $(X,\leq)$ is called *order complete*, if and only if every subset $A \subseteq X$ has an infimum and a supremum in X. In case every subset $A \subseteq X$ which is bounded from above will have a supremum in X, and at the same time, every subset $B \subseteq X$ which is bounded from below will have an infimum in X, we shall call the poset $(X,\leq)$ *Dedekind order complete.* For instance, $\mathbb{R}$ and $\mathcal{M}es\ (\Omega)$, with $\Omega \subseteq \mathbb{R}^n$ open, are Dedekind order complete, but not order complete, when considered with their respective usual order. On the other hand $\bar{\mathbb{R}} = [-\infty,+\infty]$ is order complete as well. Obviously, an order complete poset will also be Dedekind order complete. However, as the case of $\mathbb{R}$ shows it, for instance, the converse is not true.

In order to avoid possible confusion, we mention that the method used in MacNeille is based on Dedekind cuts. However, this methods delivers a poset which is order complete and not only Dedekind order complete.

Given two nonvoid posets $(X,\leq)$ and $(Y,\leq)$, they are called *order isomorphic*, if and only if there exists a bijection $\varphi : X \longrightarrow Y$ such that, for $x,x' \in X$, we have

$$(A.1) \qquad x \leq x' \iff \varphi(x) \leq \varphi(x')$$

in which case the mapping $\varphi : X \longrightarrow Y$ is called an *order isomorphism.*

We shall often be interested in the more general case of an injective mapping $\varphi : X \longrightarrow Y$ which is an order isomorphism between X and $\varphi(X)$, without however having $\varphi(X) = Y$, that is, without φ being surjective. Such a mapping we shall call *order isomorphical embedding.*

A mapping $\varphi : X \longrightarrow Y$ is said *to preserve infima and suprema*, if and only if, for $x \in X$ and $A \subset X$, we have

(A.2) $\quad x = \sup A \Rightarrow \varphi(x) = \sup \varphi(A)$

(A.3) $\quad x = \inf A \Rightarrow \varphi(x) = \inf \varphi(A)$.

It is important to note that the properties of a mapping $\varphi : X \longrightarrow Y$ of being an order isomorphical embedding, or of preserving infima and suprema are in general independent, as rather trivial examples can show it, see Luxemburg & Zaanen and the examples at the end of this Appendix.

However, for any $\varphi : X \longrightarrow Y$ injective, we have the following two results.

In case the poset X is a lattice, if φ preserves infima and suprema, then φ is an order isomorphical embedding.

Suppose now that both posets X and Y are lattices, and in addition, for $y \in Y$, we have

(A.4) $\quad y = \sup \{\varphi(x) \mid \varphi(x) \leq y\} = \inf \{\varphi(x) \mid y \leq \varphi(x)\}$

Then φ preserves infima and suprema, if and only if φ is an order isomorphical embedding.

Finally we note that an order isomorphism $\varphi : X \longrightarrow Y$ will always preserve infima and suprema.

With the above preparation, we can now turn to the construction needed in the MacNeille theorem.

Suppose $(X,\leq)$ is a nonvoid poset. For convenience and without loss of generality, we further suppose that X does not have a smallest or a largest element.

Let us introduce the following compact notation for half unbounded intervals in the poset X. Given $a \in X$, we denote

$$(A.5) \qquad \begin{aligned} <a] &= \{x \in X \mid x \leq a\} \\ [a> &= \{x \in X \mid a \leq x\} \end{aligned}$$

Two dual operators on $\mathcal{P}(X)$ will play a crucial role. They are defined as follows

$$(A.6) \qquad X \supset A \longmapsto A^u = \bigcap_{a \in A} [a> \subset X$$

$$(A.7) \qquad X \supset A \longmapsto A^\ell = \bigcap_{a \in A} <a] \subset X$$

It is now obvious that for $A \subset X$, we have

$$(A.8) \qquad A^u = X \Longleftrightarrow A^\ell = X \Longleftrightarrow A = \phi.$$

while

$$(A.9) \qquad A^u = \phi \Longleftrightarrow A \text{ unbounded from above}$$

$$(A.10) \qquad A^\ell = \phi \Longleftrightarrow A \text{ unbounded from below.}$$

Our interest is in the subsets $A \subset X$ which satisfy the condition

$$(A.11) \qquad A^{u\ell} = A$$

and which subsets of X are called *cuts* in the poset X. In particular, we shall be interested in the subset $\hat{X} \subset \mathcal{P}(X)$ defined by

$$(A.12) \qquad \hat{X} = \{A \subset X \mid A \text{ is a cut}\}$$

that is, the set of all cuts, which will in fact supply the order completion of the poset X. In view of (A.8) - (A.10), we obviously have

$$(A.13) \qquad \phi,\ X \in \hat{X}$$

therefore $\hat{X} \neq \phi$.

Before going further, we establish a few results which can give a better insight into the structure of the set $\hat{X}$ of cuts of the poset X. For that, it is useful to specify several properties of the dual pair of operators in (A.6) and (A.7). Given $A, B \subset X$, we obviously have

$$(A.14) \qquad A \subset B \Longrightarrow A^u \supset B^u, \quad A^\ell \supset B^\ell.$$

Further, for $A \subset X$, we have

$$(A.15) \qquad A \subset A^{u\ell}, \quad A \subset A^{\ell u}$$

Indeed, if $x \in X$ then

$$x \in A^{u\ell} \Longleftrightarrow (\forall\ \bar{a} \in A^u : x \leq \bar{a}).$$

But $a \leq \bar{a}$, for $a \in A$, $\bar{a} \in A^u$. Hence, if $x \in A$, then $x \leq \bar{a}$, for $\bar{a} \in A^u$. In this way $x \in A \Longrightarrow x \in A^{u\ell}$. By a dual argument we obtain $A \subset A^{\ell u}$ as well.

An important property valid for $A \subset X$ is the following

$$(A.16) \qquad A^{u\ell u} = A^u, \quad A^{\ell u\ell} = A^\ell$$

In order to prove it, we note that the inclusion $A^u \subset (A^u)^{\ell u}$ follows from (A.15). For the converse inclusion, we note that $A \subset A^{u\ell}$ in (A.15) and the implication in (A.14) yield $A^u \supset (A^{u\ell})^u$. The second equality in (A.16) is obtained by duality.

The following two properties of the set $\hat{X}$ of cuts result now easily

(A.17)

$\forall\ A \subset X$:

*) $\quad A^{u\ell} \in \hat{X}$

**) $\quad \forall\ B \in \hat{X}$:

$$A \subset B \Longrightarrow A^{u\ell} \subset B$$

$$B \subset A \Longrightarrow B \subset A^{u\ell}$$

Indeed, in view of (A.16), we have

$$(A^{u\ell})^{u\ell} = (A^{u\ell u})^{\ell} = (A^u)^{\ell} = A^{u\ell}$$

Further, $A \subset B$ and (A.14) yield $A^u \supset B^u$, hence $A^{u\ell} \subset B^{u\ell} = B$. The other implication follows in a similar way.

We note that for $x \in X$, we have

$$\text{(A.18)} \qquad \{x\}^u = [x\rangle,\ \{x\}^{\ell} = \langle x],\quad [x\rangle^{\ell} = \langle x],\ \langle x]^u = [x\rangle$$

hence

$$\text{(A.19)} \qquad \{x\}^{u\ell} = \langle x],\ \{x\}^{\ell u} = [x\rangle.$$

For simplicity, we shall denote $\{x\}^u = x^u$, $\{x\}^{\ell} = x^{\ell}$, $\{x\}^{u\ell} = x^{u\ell}$, $\{x\}^{\ell u} = x^{\ell u}$, etc.

An important property of cuts is the following one. Given $A \in \hat{X}$, then

$$\text{(A.20)} \qquad \phi \neq A \underset{\neq}{\subset} X \Longleftrightarrow \begin{bmatrix} \exists\ a,b \in X : \\ \langle a] \subseteq A \subseteq \langle b] \end{bmatrix}$$

where the implication $\Longleftarrow$ is obvious. For the converse implication $\Longrightarrow$ we note that

$$A^{u\ell} = A \underset{\neq}{\subset} X$$

hence (A.8) gives

$$A^u \neq \phi$$

and then (A.9) implies that A is indeed bounded from above. On the other hand $A \neq \phi$ yields $a \in A$. Then obviously $\langle a] \subseteq A$.

Finally, X will be embedded into $\hat{X}$ according to the mapping

$$\text{(A.21)} \qquad X \ni x \stackrel{\varphi}{\longmapsto} x^{u\ell} = x^{\ell} = \langle x] \in \hat{X}$$

which is injective, as seen easily.

At this stage it only remains to define the appropriate partial order $\leq$ on $\hat{X}$, which will be done in the following natural way : given $A,B \in \hat{X}$, then

(A.22) $\quad A \leq B \text{ in } \hat{X} \iff A \subset B \text{ in } X$

With the above preparations, we can now formulate the basic result on Dedekind order completion.

Theorem (H M MacNeille, 1937)

The poset $(\hat{X},\leq)$ is order complete, that is, each nonvoid subset in $\hat{X}$ has an infimum and a supremum.

The embedding $X \xrightarrow{\varphi} \hat{X}$ preserves infima and suprema, moreover, it is an order isomorphism between X and $\varphi(X) \subset \hat{X}$, in other words, $X \xrightarrow{\varphi} \hat{X}$ is an order isomorphical embedding.

For $A \in \hat{X}$ we have with the order in $\hat{X}$ the relations

$$\text{(A.23)} \quad \begin{aligned} A &= \sup \{x^{\ell} \mid x \in X, \ x^{\ell} \subset A\} \\ &= \inf \{x^{\ell} \mid x \in X, \ A \subset x^{\ell}\} \end{aligned} \qquad \blacksquare$$

Remark A.1

Suppose the poset X is already order complete. Then the mapping, see (A.21)

(A.24) $\quad X \ni x \xrightarrow{\text{bij}} \langle x] \in \hat{X}$

preserves infima and suprema, and it is an order isomorphism between X and $\hat{X}$. Indeed, in view of MacNeille's theorem, we only have to show that the above mapping is surjective. Therefore, let $A \in \hat{X}$, then (A.23) gives

$$A = \sup_{\hat{X}} \{\langle x] \mid x \in A\}.$$

But

$$\sup_{\hat{X}} \{<x] \mid x \in A\} = <\sup_X A]$$

since the mapping (A.24) preserves infima and suprema. It follows that

$$A = <x], \quad \text{where} \quad x = \sup_X A \in X .$$ ■

A useful consequence is presented now. For $A \subseteq X$ we have the relation

(A.25) $$A^{u\ell} = \sup_{\hat{X}} \{x^{\ell} \mid x \in A\}.$$

Indeed, in view of (A.14) and (A.19), we obtain

$$x \in A \Longrightarrow x^{\ell} \subseteq A^{u\ell} .$$

Hence the inequality '$\geq$' in (A.25). For the converse inequality, take $B \in \hat{X}$, such that

$$x \in A \Longrightarrow x^{\ell} \subseteq B .$$

Then obviously $A \subseteq B$, therefore **) in (A.17) gives $A^{u\ell} \subseteq B$.

The relevance of the order completion procedure in MacNeille's theorem can be better understood as follows. Given $A, B \subset X$, we call (A,B) a *Dedekind pair*, if and only if

(A.26) $$A^u = B, \quad A = B^{\ell}$$

in other words

$$A = \{x \in X \mid \forall\, y \in B: x \leq y\}, \quad B = \{y \in X \mid \forall\, x \in A: x \leq y\}$$

Then (A.26) implies that for a Dedekind pair (A,B) we have

(A.27) $\quad A^{u\ell} = B^{\ell} = A, \quad B^{\ell u} = A^{u} = B$

therefore A is a cut, that is $A \in \hat{X}$.

In general, given $A \subset X$, in view of (A.16), it follows that

(A.28) $\quad (A^{u\ell},\ A^{u})$ is a Dedekind pair

(A.29) $\quad (A^{\ell},\ A^{\ell u})$ is a Dedekind pair.

Moreover, in view of (A.26), it is clear that *every* Dedekind pair can be written in *both* forms (A.28) and (A.29).

In view of (A.24), (A.26) and (A.29), it is now obvious that the above construction of order completion, given in MacNeille's theorem, has the following dual variant, in which

(A.30) $\quad \hat{X} = \{A \subset X \mid A^{\ell u} = A\}$

(A.31) $\quad X \in x \xmapsto{\varphi} x^{\ell u} = x^{u} = [x\rangle \in \hat{X}$

while the relations (A.23) will take the form: for $A \in \hat{X}$ we have with the order in $\hat{X}$ that

(A.32)
$$\begin{aligned} A &= \sup \{x^{u} \mid x \in X,\ A \subset x^{u}\} = \\ &= \inf \{x^{u} \mid x \in X,\ x^{u} \subset A\}. \end{aligned}$$

Coming back to MacNeille's theorem, an important consequence is the following property, which is a direct result of the definition (A.22) of the order on the set $\hat{X}$ of cuts, as well as of **) in (A.17). Given a family of cuts $A_i \in \hat{X}$, with $i \in I$, then, with the order in $\hat{X}$, we have

(A.33) $\quad \sup_{i\in I} A_i = \inf \{A \in \hat{X} \mid \bigcup_{i\in I} A_i \subset A\} = (\bigcup_{i\in I} A_i)^{u\ell}$

similarly

$$(A.34) \qquad \inf_{i \in I} A_i = \sup \{A \in \hat{X} \mid A \subset \bigcap_{i \in I} A_i\} = (\bigcap_{i \in I} A_i)^{u\ell} = \bigcap_{i \in I} A_i \ .$$

We shall present now a number of results related to the MacNeille order completion. These results, which we could not find to be available in the literature, have been developed here in view of their important use within the Parts I, II and III of this work.

The following general result on the *extension* of order isomorphical embeddings to order completions will be useful.

Suppose given an arbitrary mapping

$$(A.35) \qquad \varphi : X \longrightarrow Y$$

between two posets X and Y, neither of which has a minimum or maximum element. Let us denote by $\hat{X}$ and $\hat{Y}$ the order completions of X and Y respectively, constructed according to MacNeille's theorem. Further, let us define the mapping

$$(A.36) \qquad \hat{\varphi} : \mathcal{P}(X) \longrightarrow \hat{Y}$$

as follows: for $A \subseteq X$, we have

$$(A.37) \qquad \hat{\varphi}(A) = (\varphi(A))^{u\ell} = \sup_{\hat{Y}} \{<\varphi(x)] \mid x \in A\}$$

which is well defined, owing to (A.12), *) in (A.17), as well as (A.25). We note that in view of (A.19) we have obtained the following commutative diagram

$$\begin{array}{ccc} X \ni x & \longmapsto & \varphi(x) \in Y \\ \downarrow & & \downarrow \\ \mathcal{P}(X) \ni \{x\} & \longmapsto & \hat{\varphi}(\{x\}) = <\varphi(x)] \in \hat{Y} \end{array}$$

without any assumption on the mapping φ in (A.35).

We recall that φ in (A.35) is *increasing*, if and only if

$$\text{(A.38)} \qquad x,y \in X, \quad x \leq y \Rightarrow \varphi(x) \leq \varphi(y).$$

It should be noted that an increasing and injective mapping need not be an order isomorphical embedding, see the example at the end of this Appendix.

The general extension result can be stated now:

Proposition A.1

The mapping

$$\hat{\varphi} : \mathcal{P}(X) \longrightarrow \hat{Y}$$

is increasing, if on $\mathcal{P}(X)$ we take the partial order defined by the inclusion $\subseteq$.

If φ is increasing, then $\hat{\varphi}$ is an extension of φ, that is, we have the commutative diagram

$$\text{(A.39)} \qquad \begin{array}{ccc} X \ni x & \xrightarrow{\quad \varphi \quad} & \varphi(x) \in Y \\ \big\downarrow & & \big\downarrow \\ \hat{X} \ni <x] & \xrightarrow[\quad \hat{\varphi} \quad]{} & \hat{\varphi}(<x]) = <\varphi(x)] \in \hat{Y} \end{array}$$

If φ is an order isomorphical embedding, then $\hat{\varphi}$ restricted to $\hat{X}$, that is

$$\text{(A.40)} \qquad \hat{\varphi} : \hat{X} \longrightarrow \hat{Y}$$

is also an order isomorphical embedding.

Proof

Let us first show that $\hat{\varphi}$ is increasing. Let $A,B \in \mathcal{P}(X)$ and $A \subseteq B \subseteq X$, hence

$$\varphi(A) \subseteq \varphi(B) \subseteq Y$$

which together with (A.14) and (A.37) give

$$\hat{\varphi}(A) = (\varphi(A))^{u\ell} \subseteq (\varphi(B))^{u\ell} = \hat{\varphi}(B).$$

In this way, according to (A.22), we obtain in $\hat{Y}$ the inequality

$$\hat{\varphi}(A) \leq \hat{\varphi}(B).$$

Assume now that φ is increasing.

Take $x \in X$, then (A.37) gives

$$\text{(A.41)} \qquad \hat{\varphi}(<x]) = (\varphi(<x]))^{u\ell}.$$

But (A.38) implies obviously

$$\varphi(<x]) \subseteq <\varphi(x)]$$

hence (A.14) and (A.18) yield

$$\text{(A.42)} \qquad (\varphi(<x]))^{u\ell} \subseteq <\varphi(x)].$$

Now, in view of (A.15), we have

$$\varphi(x) \in \varphi(<x]) \subseteq (\varphi(<x]))^{u\ell}$$

therefore, owing to (A.17) and (A.19), we obtain

$$<\varphi(x)] \subset (\varphi(<x]))^{u\ell}$$

which together with (A.41) and (A.42) clearly yield (A.39).

Assume now that φ is an order isomorphical embedding.

We show that $\hat{\varphi}$ is injective. For that, let $A,B \in \hat{X}$ and $A \neq B$. Then one of the following relations must hold

$$A \setminus B \neq \phi \quad \text{or} \quad B \setminus A \neq \phi.$$

Let us assume that $A \setminus B \neq \phi$ and take $x \in A \setminus B$. Then in view of (A.12), we have $x \notin B^{u\ell}$, hence

$$\exists\, z \in B^u : x \nleq z.$$

But according to (A.6), we have

$$\forall\, y \in B : y \leq z.$$

Since φ is an order isomorphical embedding, we clearly have

$$\varphi(x) \nleq \varphi(z) \quad \text{and} \quad \forall\, y \in B : \varphi(y) \leq \varphi(z).$$

It follows that $\varphi(z) \in (\varphi(B))^u$, thus

$$\text{(A.43)} \qquad \varphi(x) \notin (\varphi(B))^{u\ell} = \hat{\varphi}(B).$$

However, in view of (A.15), we have

$$\text{(A.44)} \qquad \varphi(x) \in \varphi(A) \subseteq (\varphi(A))^{u\ell} = \hat{\varphi}(A).$$

Therefore, we can conclude that indeed

$$\hat{\varphi}(A) \neq \hat{\varphi}(B).$$

Finally we show that for $A,B \in \hat{X}$, we have the implication

$$\hat{\varphi}(A) \leq \hat{\varphi}(B) \Longrightarrow A \leq B.$$

Assume indeed that $A \nleq B$ in $\hat{X}$. Then in view of (A.22), it follows that $A \setminus B \neq \phi$. If we take now $x \in A \setminus B$, then the above argument leading to (A.43) and (A.44) will give

$$\varphi(x) \in \hat{\varphi}(A) \setminus \hat{\varphi}(B) \neq \phi.$$

Thus in view of (A.22), we have in $\hat{Y}$

$$\hat{\varphi}(A) \not\leq \hat{\varphi}(B).$$

■

We present two useful consequences of the Proposition A.1 above.

Suppose again given an arbitrary mapping

$$\text{(A.45)} \qquad \varphi : X \longrightarrow Y$$

between two posets X and Y, and suppose that X does not have a minimum or maximum, while Y is order complete, in particular, we have the order isomorphism, see (A.24)

$$\text{(A.46)} \qquad Y \ni y \xrightarrow[\text{bij}]{\beta} \langle y] \in \hat{Y}$$

which therefore, will also preserve infima and suprema.

Corollary A.1

The following diagram is commutative

$$\text{(A.47)} \qquad \begin{array}{ccc} X & \xrightarrow{\varphi} & Y \\ & \nearrow_{i} & \downarrow \beta \\ \hat{X} & \xrightarrow[\hat{\varphi}]{} & \hat{Y} \end{array}$$

where for $A \in \hat{X}$ we define

$$\text{(A.48)} \qquad i(A) = y \quad \text{if} \quad \hat{\varphi}(A) = \beta(y) = \langle y] .$$

In this case we also have

$$\text{(A.49)} \qquad i(A) = \sup_{Y} \varphi(A), \quad \hat{\varphi}(A) = \langle i(A)], \quad A \in \hat{X}$$

and

(A.50) $\quad i : \hat{X} \longrightarrow Y$ is increasing.

If φ is increasing then the following diagram is also commutative

$$\text{(A.51)} \qquad \begin{array}{ccc} X & \xrightarrow{\ \varphi\ } & Y \\ \alpha \big\downarrow & \overset{i}{\nearrow} & \big\downarrow \beta \\ \hat{X} & \xrightarrow[\ \hat{\varphi}\ ]{} & \hat{Y} \end{array}$$

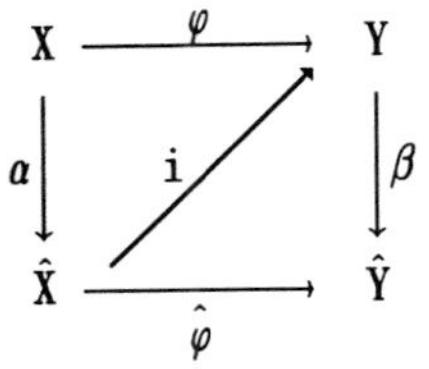

where α is defined in (A.21).

Finally, if φ is an order isomorphical embedding then

(A.52) $\quad i : \hat{X} \longrightarrow Y$ is an order isomorphical embedding.

Proof

We note that in view of (A.46), the definition of i in (A.48) is correct, hence the commutativity of (A.47).

The second relation in (A.49) follows directly from (A.48). Let us now prove the first relation in (A.49).

For that we use (A.37), which together with (A.48) gives

$$i(A) = y \quad \text{if} \quad \sup_{\hat{Y}} \{<\varphi(x)] \mid x \in A\} = <y] \ .$$

Therefore (A.46) will imply

$$i(A) = \sup_{Y} \{\varphi(x) \mid x \in A\} \ .$$

Let us now prove (A.50). Take $A, B \in \hat{X}$, with $A \subseteq B$, then $\varphi(A) \subseteq \varphi(B)$, hence (A.14), (A.37) give

$$\hat{\varphi}(A) = (\varphi(A))^{u\ell} \subseteq (\varphi(B))^{u\ell} = \hat{\varphi}(B) \ .$$

Therefore, if $i(A) = y$ and $i(B) = z$, then the second relation in (A.49) implies

$$\langle y] = \hat{\varphi}(A) \subseteq \hat{\varphi}(B) = \langle z]$$

which means that

$$i(A) = y \leq z = i(B) \ .$$

Assuming that φ is increasing, we prove now (A.51). But in view of (A.47), we only have to show that

$$\varphi(x) = i(\langle x]), \quad x \in X \ .$$

Now, according to (A.49), we have

$$i(\langle x]) = \sup_Y \varphi(\langle x]), \quad x \in X$$

hence we only have to prove the relation

$$\text{(A.53)} \qquad \sup_Y \varphi(\langle x]) = \varphi(x), \quad x \in X \ .$$

However, since φ is increasing, we obtain

$$\varphi(\langle x]) \subseteq \langle \varphi(x)], \quad x \in X$$

thus the inequality '$\leq$' in (A.53). But obviously $\varphi(x) \in \varphi(\langle x])$, hence the proof of (A.53), and therefore of (A.51) are completed.

We turn finally to (A.52), which follows easily from (A.40), (A.47) and (A.49).

■

A still more involved general result on the extension of order isomorphical embeddings will be useful as well.

Suppose given arbitrary mappings

(A.54) $\quad X \xrightarrow{\ \varphi\ } Y \xrightarrow{\ \psi\ } Z$

where X, Y, Z are posets. Suppose that X and Y do not have minimum or maximum, while Z is order complete, namely, the mapping, see (A.24)

(A.55) $\quad Z \ni z \underset{\text{bij}}{\overset{\gamma}{\longmapsto}} \langle z] \in \hat{Z}$

is an order isomorphism which therefore preserves infima and suprema. Let us then define the mapping

(A.56) $\quad \chi = \psi \cdot \varphi : X \longrightarrow Z$

which leads to the commutative diagram

(A.57)
$$\begin{array}{ccccc} & & \chi & & \\ X & \xrightarrow[\varphi]{} & Y & \xrightarrow[\psi]{} & Z \end{array}$$

Corollary A.2

If φ and ψ are increasing then the diagram is commutative

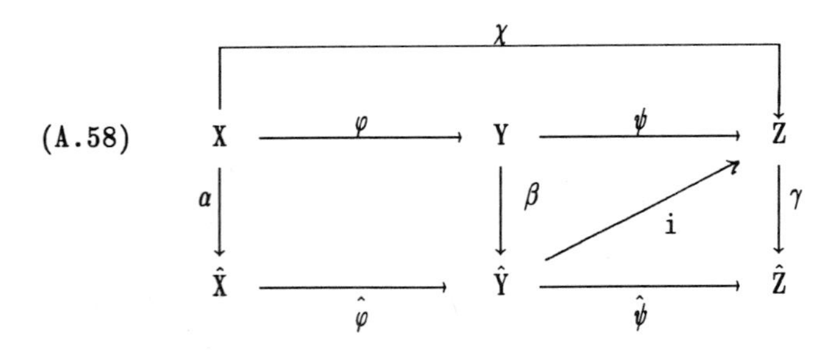

If φ and ψ are order isomorphical embeddings and $\hat{\chi}$ is surjective then the diagram is commutative

(A.59)

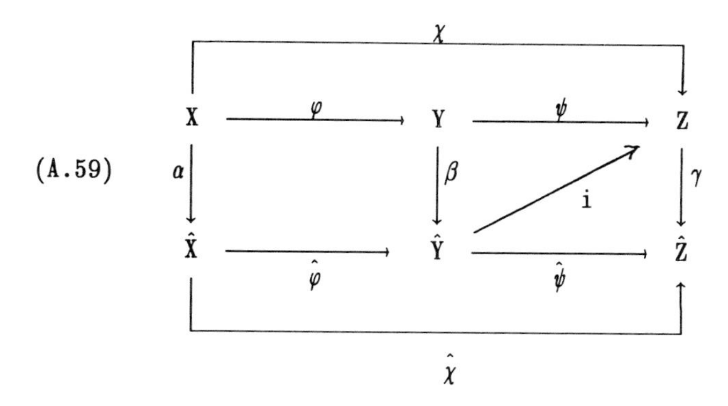

in particular

(A.60) $\hat{\chi} = \hat{\psi}\cdot\hat{\varphi} : \hat{X} \longrightarrow \hat{Z}$.

In addition $\hat{\varphi}$, $\hat{\psi}$, $\hat{\chi}$ and i are order isomorphisms, therefore they preserve infima and suprema.

Proof

We obtain (A.58) directly from (A.39) and (A.51).

For (A.59) we note that, in view of (A.57), χ will also be an order isomorphical embedding. Therefore, according to (A.40), $\hat{\varphi}$, $\hat{\psi}$ and $\hat{\chi}$ are order isomorphical embeddings. Furthermore (A.52) implies that i is also an order isomorphical embedding.

Now we proceed to proving (A.59) and (A.60).

Since in (A.59) we deal with three different posets X, Y and Z, it will be convenient to associate indices with the operators $<\,]$, $[\,>$, $(\;)^u$ and $(\;)^\ell$ in (A.5) - (A.6), specifying the posets in which they act. Namely, for the respective operators acting in the poset X, we shall write $<\,]_X$, $[\,>_X$, $(\;)^{u_X}$ or $(\;)^{\ell_X}$, and similarly for the other posets Y and Z.

In view of (A.58), the proof of (A.59) reduces to that of (A.60), for which we note that the following inequality holds for arbitrary posets X, Y, Z and mappings φ and ψ

(A.61) $\hat{\chi}(A) \subseteq \hat{\psi}(\hat{\varphi}(A))$, $A \subseteq X$.

Indeed (A.37) implies that, for $A \subseteq X$, we have

$$\text{(A.62)} \quad \begin{aligned} \hat{\chi}(A) &= (\psi(\varphi(A)))^{u_Z \ell_Z} \\ \hat{\psi}(\hat{\varphi}(A)) &= (\psi((\varphi(A))^{u_Y \ell_Y}))^{u_Z \ell_Z} \end{aligned}$$

thus (A.61) will follow from (A.14) and (A.15).

Further we note that the surjectivity of $\hat{\chi}$ together with the fact that $\hat{\chi}$ is an order isomorphical embedding means that

(A.63) $\hat{\chi} : \hat{X} \longrightarrow \hat{Z}$ is an order isomorphism

which operates according to

$$\text{(A.64)} \quad \hat{X} \ni A \longmapsto \hat{\chi}(A) = \langle \sup_Z \chi(A)]_Z \in \hat{Z} \ .$$

Indeed (A.37) gives for $A \in \hat{X}$ the relation

$$\hat{\chi}(A) = \sup_{\hat{Z}} \{\langle \chi(x)]_Z | x \in A\} \ .$$

On the other hand, (A.24) and (A.55) give

$$\sup_{\hat{Z}} \{\langle \chi(x)]_Z | x \in A\} = \langle \sup_Z \chi(A)]_Z$$

since γ preserves infima and suprema, and therefore the proof of (A.64) is completed.

We prove now (A.60). In view of (A.61) and (A.62) , it suffices to show that

$$\text{(A.65)} \quad \begin{aligned} &\forall \ A \in \hat{X} : \\ &(\psi((\varphi(A))^{u_Y \ell_Y}))^{u_Z \ell_Z} \subseteq (\psi(\varphi(A)))^{u_Z \ell_Z} \end{aligned}$$

which, according to (A.17), is obviously equivalent with

$$\text{(A.66)} \qquad \begin{array}{l} \forall \ \ A \in \hat{X} : \\ \psi((\varphi(A))^{u_Y \ell_Y}) \subseteq (\psi(\varphi(A)))^{u_Z \ell_Z} . \end{array}$$

Therefore, let us take

$$\text{(A.67)} \qquad z \in \psi((\varphi(A))^{u_Y \ell_Y})$$

then

$$\text{(A.68)} \qquad z = \psi(y)$$

for a certain $y \in Y$ which satisfies

$$\text{(A.69)} \qquad \forall \ \ y' \in (\varphi(A))^{u_Y} \ : \ y \leq y' .$$

But $\hat{\chi}$ is surjective, thus in view of (A.62) and (A.64) it follows that

$$\text{(A.70)} \qquad \begin{array}{l} \exists \ \ A_1 \in \hat{X} : \\ (\psi(\varphi(A_1)))^{u_Z \ell_Z} = \langle z]_Z \end{array}$$

and then

$$\text{(A.71)} \qquad z = \sup_Z \psi(\varphi(A_1)) .$$

But (A.71) implies

$$\psi(\varphi(A_1)) \subseteq \langle z]_Z$$

which together with (A.68) and the fact that ψ is an order isomorphical embedding, will give

$$\text{(A.72)} \qquad \varphi(A_1) \subseteq \langle y]_Y .$$

Now (A.72) and (A.69) yield

$$\text{(A.73)} \qquad \varphi(A_1) \subseteq (\varphi(A))^{u_Y \ell_Y} .$$

But we clearly have

$$\varphi(A^{u_X}) \subseteq (\varphi(A))^{u_Y}$$

hence (A.14) gives

$$(\varphi(A))^{u_Y \ell_Y} \subseteq (\varphi(A^{u_X}))^{\ell_Y}$$

and then (A.73) results in

$$\varphi(A_1) \subseteq (\varphi(A^{u_X}))^{\ell_Y}$$

in other words

$$\forall\ x_1 \in A_1,\ \ x \in A^{u_X} \ :\ \varphi(x_1) \leq \varphi(x)$$

which in view of the fact that φ is an order isomorphical embedding, will yield

$$\forall\ x_1 \in A_1,\ \ x \in A^{u_X} \ :\ x_1 \leq x$$

or equivalently

$$\text{(A.74)} \qquad A_1 \subseteq A^{u_X \ell_X} = A .$$

Finally, (A.62), (A.64), (A.70) and (A.74) together with the fact that $\hat{\chi}$ is an order isomorphical embedding, will result in

$$z \in \langle z]_Z = \hat{\chi}(A_1) \subseteq \hat{\chi}(A) = (\psi(\varphi(A)))^{u_Z \ell_Z}$$

therefore the proof of (A.66), and hence of (A.65) and (A.60) are completed.

Concluding the proof we note that (A.60) and the surjectivity of $\hat{\chi}$ imply the surjectivity of $\hat{\varphi}$ and $\hat{\psi}$, and therefore, owing to (A.55), the surjectivity of i as well.

Now, since $\hat{\varphi}$, $\hat{\psi}$, $\hat{\chi}$ and i are order isomorphical embeddings, their surjectivity implies that they are in fact order isomorphisms. In particular, they preserve infima and suprema. ■

Let us now consider the issue of the order completion of an arbitrary Cartesian product of posets.

Suppose $(X_\lambda, \leq_\lambda)$, with $\lambda \in \Lambda$, are posets which are without maximum or minimum elements. Let us define the Cartesian product of posets by

$$\text{(A.75)} \qquad \prod_{\lambda \in \Lambda} (X_\lambda, \leq_\lambda) = (X, \leq)$$

where

$$\text{(A.76)} \qquad X = \prod_{\lambda \in \Lambda} X_\lambda$$

and the partial order $\leq$ on X is given by

$$\text{(A.77)} \qquad (x_\lambda \mid \lambda \in \Lambda) \leq (y_\lambda \mid \lambda \in \Lambda) \Longleftrightarrow \begin{bmatrix} \forall\ \lambda \in \Lambda : \\ x_\lambda \leq_\lambda y_\lambda \end{bmatrix} .$$

It is obvious that the poset $(X, \leq)$ will as well be without maximum or minimum elements.

Proposition A.2

The mapping

$$\text{(A.78)}\qquad \begin{array}{ccc} \prod_{\lambda\in\Lambda} \hat{X}^*_\lambda & \xrightarrow{\quad i \quad} & \hat{X}^* \\ (A_\lambda \mid \lambda \in \Lambda) & \longmapsto & \prod_{\lambda\in\Lambda} A_\lambda \end{array}$$

is an order isomorphism. In other words

$$\text{(A.79)}\qquad \Big[\prod_{\lambda\in\Lambda} (X_\lambda, \leq_\lambda)\Big]^{\hat{}\,*} = \prod_{\lambda\in\Lambda} (X_\lambda, \leq_\lambda)^{\hat{}\,*}$$

which means that, under the restrictions in (A.79), the order completion and Cartesian products *commute* with each other.

Here we used the notation, see (7.26)

$$\text{(A.80)}\qquad \hat{X}^* = \hat{X}\setminus\{\min X, \max X\} = \hat{X}\setminus\{\phi, X\}$$

with the last equality following from (A.13) and (A.22).

Proof

Let us take arbitrary subsets A_λ, with $\phi \underset{\neq}{\subset} A_\lambda \underset{\neq}{\subset} X_\lambda$, for $\lambda \in \Lambda$, and define

$$\text{(A.81)}\qquad A = \prod_{\lambda\in\Lambda} A_\lambda \subsetneq X\ .$$

We shall show that

$$\text{(A.82)}\qquad A^u = \prod_{\lambda\in\Lambda} A^u_\lambda$$

$$\text{(A.83)}\qquad A^\ell = \prod_{\lambda\in\Lambda} A^\ell_\lambda\ .$$

Indeed, assume $x = (x_\lambda \mid \lambda \in \Lambda) \in X$, then (A.77) and (A.6) give

$$x \in A^u \Longleftrightarrow \left[\begin{array}{l} \forall\ y = (y_\lambda | \lambda \in \Lambda) \in A : \\ \forall\ \lambda \in \Lambda : \\ \quad y_\lambda \leq_\lambda x_\lambda \end{array}\right]$$

which in view of (A.81) means that

$$x \in A^u \Longleftrightarrow \left[\begin{array}{l} \forall\ \lambda \in \Lambda : \\ \quad x_\lambda \in A_\lambda^u \end{array}\right]$$

and the proof of (A.82) is completed. The relation (A.83) follows in a similar way.

Now let us particularize the above by taking

(A.84) $\quad A_\lambda \in \hat{X}_\lambda^*, \quad \lambda \in \Lambda$

In view of (A.11), in order to prove that the mapping

(A.85) $$\begin{array}{ccc} \prod_{\lambda \in \Lambda} \hat{X}_\lambda^* & \xrightarrow{\quad i \quad} & \hat{X}^* \\ (A_\lambda | \lambda \in \Lambda) & \longmapsto & \prod_{\lambda \in \Lambda} A_\lambda \end{array}$$

is well defined, it suffices to show the inclusion

(A.86) $\quad A^{u\ell} \subseteq A$.

We note that (A.80), (A.84) give

$$\phi \underset{\neq}{\subset} A_\lambda \underset{\neq}{\subset} X_\lambda, \quad \lambda \in \Lambda$$

hence (A.20) applied to each A_λ, with $\lambda \in \Lambda$, results in

$$\phi \underset{\neq}{\subset} \langle a_\lambda] \subseteq A_\lambda \subseteq \langle b_\lambda] \underset{\neq}{\subset} X_\lambda, \quad \lambda \in \Lambda$$

for suitable $a_\lambda, b_\lambda \in X_\lambda$, with $\lambda \in \Lambda$.

Therefore (A.8), (A.9) imply

$$\phi \underset{\neq}{\subset} A_\lambda^u \underset{\neq}{\subset} X_\lambda, \quad \lambda \in \Lambda$$

But then according to (A.82) and (A.83) we obtain

$$A^{u\ell} = (A^u)^\ell = \Big[\big(\prod_{\lambda\in\Lambda} A_\lambda\big)^u\Big]^\ell = \Big[\prod_{\lambda\in\Lambda} A_\lambda^u\Big]^\ell = \prod_{\lambda\in\Lambda} A_\lambda^{u\ell} = \prod_{\lambda\in\Lambda} A_\lambda = A$$

since $A_\lambda^{u\ell} = A_\lambda$, with $\lambda \in \Lambda$, and the proof of (A.86) is completed.

Further we note that i in (A.85) is surjective. Indeed, let $A \in \hat{X}^*$ and let us consider its projections on each X_λ, with $\lambda \in \Lambda$, namely

(A.87) $\quad A_\lambda = \underset{\lambda}{pr}\, A \subseteq X_\lambda$

We note that $A \in \hat{X}^*$ together with (A.80), (A.20) and (A.87) imply

(A.88) $\quad \phi \underset{\neq}{\subset} \langle a_\lambda] \subseteq A_\lambda \subseteq \langle b_\lambda] \underset{\neq}{\subset} X_\lambda, \quad \lambda \in \Lambda$

We prove that

(A.89) $\quad A = \prod_{\lambda\in\Lambda} A_\lambda$.

Indeed, first we note that given $x = (x_\lambda \mid \lambda \in \Lambda) \in \prod_{\lambda\in\Lambda} X_\lambda$, then

$$\langle x] = \prod_{\lambda\in\Lambda} \langle x_\lambda]$$

which follows easily from (A.77). Now applying (A.23), we obtain

$$A = \inf_{\hat{X}} \left\{ \langle x] \;\middle|\; \begin{array}{l} x = (x_\lambda | \lambda \in \Lambda) \in X \\ A \subseteq \langle x] \end{array} \right\} =$$

(A.90)

$$= \inf_{\hat{X}} \left\{ \prod_{\lambda \in \Lambda} \langle x_\lambda] \;\middle|\; \begin{array}{l} (x_\lambda | \lambda \in \Lambda) \in \prod_{\lambda \in \Lambda} X_\lambda \\ A \subseteq \prod_{\lambda \in \Lambda} \langle x_\lambda] \end{array} \right\}$$

But obviously

$$\text{(A.91)} \qquad A \subseteq \prod_{\lambda \in \Lambda} \langle x_\lambda] \Longleftrightarrow \left[\begin{array}{l} \forall \lambda \in \Lambda : \\ \mathrm{pr}_\lambda A \subseteq \langle x_\lambda] \end{array} \right] .$$

Now we note that in (A.89), the inclusion

$$A \subseteq \prod_{\lambda \in \Lambda} A_\lambda$$

is trivially true. Let us denote

$$B = \prod_{\lambda \in \Lambda} A_\lambda$$

then obviously

$$\mathrm{pr}_\lambda B = A_\lambda = \mathrm{pr}_\lambda A, \quad \lambda \in \Lambda$$

therefore (A.91) gives

$$\begin{array}{l} \forall (x_\lambda | \lambda \in \Lambda) \in \prod_{\lambda \in \Lambda} X_\lambda : \\ \quad A \subseteq \prod_{\lambda \in \Lambda} \langle x_\lambda] \Longleftrightarrow B \subseteq \prod_{\lambda \in \Lambda} \langle x_\lambda] \end{array} .$$

In this way (A.90) results in

$$A = B$$

and the proof of (A.89) is completed.

Now it only remains to show that, see (A.87), if $A \in \hat{X}^*$ then

$$(A.92) \qquad A_\lambda \in \hat{X}_\lambda, \quad \lambda \in \Lambda$$

For that purpose we note that (A.88), (A.8) and (A.9) yield

$$\phi \underset{\neq}{\subset} A_\lambda^u \underset{\neq}{\subset} X_\lambda, \quad \lambda \in X_\lambda$$

In this way, we can apply (A.81) - (A.83) to A in (A.89) and obtain

$$\prod_{\lambda\in\Lambda} A_\lambda = A = A^{u\ell} = (A^u)^\ell = \left[\left(\prod_{\lambda\in\Lambda} A_\lambda\right)^u\right]^\ell = \left[\prod_{\lambda\in\Lambda} A_\lambda^u\right]^\ell = \prod_{\lambda\in\Lambda} A_\lambda^{u\ell}$$

which obviously gives

$$A_\lambda^{u\ell} = A_\lambda, \quad \lambda \in \Lambda$$

and therefore (A.92). ■

One more result we shall need is in a way a converse of Proposition A.1. Indeed, given two posets X, Y and a mapping $\varphi : X \longrightarrow Y$, then (A.40) offers a condition for turning cuts in X into cuts in Y through that mapping φ. This time, however, we are interested in conditions which will turn cuts in Y into cuts in X through the inverse mapping φ^{-1}.

First we need a definition. A subset A of a poset X is called *order dense*, if and only if

$$(A.93) \qquad \begin{array}{l} \forall\ x \in X : \\ \quad x = \sup_X \{a | a \in A,\ a \leq x\} = \inf_X \{a | a \in A,\ x \leq a\} \end{array}$$

In view of (A.23), it follows that any poset X without minimum or maximum is order dense in its MacNeille order completion $\hat{X}$.

Proposition A.3

Suppose that

$$\varphi : X \longrightarrow Y$$

is an order isomorphical embedding. Further, suppose that $\varphi(X)$ is order dense in Y and Y is order complete.

Then

$$\text{(A.94)} \qquad \begin{aligned} &\forall\, y \in Y : \\ &\quad \varphi^{-1}(<y]) \ \text{ is a cut in } X \end{aligned}$$

Proof

Let us take $y \in Y$ and denote

$$A = \varphi^{-1}(<y]) = \{x \in X \,|\, \varphi(x) \leq y\}$$

Then obviously

$$A^u = \left\{x' \in X \;\middle|\; \begin{aligned} &\forall\, x \in X : \\ &\quad \varphi(x) \leq y \Longrightarrow x \leq x' \end{aligned} \right\}$$

However

$$x \leq x' \Longleftrightarrow \varphi(x) \leq \varphi(x')$$

since φ is an order isomorphical embedding. Hence

$$A^u = \left\{x' \in X \;\middle|\; \begin{aligned} &\forall\, x \in X : \\ &\quad \varphi(x) \leq y \Longrightarrow \varphi(x) \leq \varphi(x') \end{aligned} \right\}$$

or

$$A^u = \{x' \in X \mid \sup_{Y} \, (\varphi(X) \cap <y]) \leq \varphi(x')\}$$

But

$$\sup_Y (\varphi(X) \cap <y]) = y$$

since $\varphi(X)$ is order dense in Y. In this way we obtain

$$A^u = \{x' \in X \mid \varphi(x') \in [y>\} = \varphi^{-1}([y>)$$

Similarly, one can prove that

$$A^{u\ell} = \varphi^{-1}(<y])$$

and the proof of (A.94) is completed. ∎

In the proof of Proposition A.3 above, we have actually proved the following more detailed result

Proposition A.4

Under the conditions in Proposition A.3, we have

$$\text{(A.95)} \quad \begin{array}{l} \forall\ y \in Y : \\ \quad \varphi^{-1}(<y])^u = \varphi^{-1}([y>), \quad (\varphi^{-1}([y>))^\ell = \varphi^{-1}(<y]) \end{array}$$ ∎

We end this Appendix with several examples illustrating some of the basic facts mentioned above.

Let $X = Y$ and $\varphi : X \longrightarrow Y$ be the identity mapping. On X we consider the trivial partial order $\leq$ given by the equality $=$, while on Y the partial order $\leq$ is such that

$$\exists\ y_1, y_2 \in Y \ : \ y_1 \leq y_2 \ \text{ and } \ y_1 \neq y_2 \ .$$

Then clearly φ is not an order isomorphical embedding, nevertheless φ preserves infima and suprema.

Further, let X be the set of all real valued continuous functions on $[0,1]$ and Y be the set of all real valued functions on $[0,1]$. Let $\varphi : X \longrightarrow Y$ be the identity mapping of X into Y. The partial order $\leq$ on X is that induced by the partial order $\leq$ on Y, defined for $f,g \in Y$ as follows

$$f \leq g \Longleftrightarrow (\forall\ x \in [0,1] \ :\ f(x) \leq g(x)) .$$

Then obviously φ is an order isomorphical embedding. However φ does not preserve infima or suprema. Indeed, for $\nu \in \mathbb{N}$, let us define $f_\nu \in X$ by $f_\nu(x) = x^\nu$, with $x \in [0,1]$. Then

$$\inf_X \{f_\nu \mid \nu \in \mathbb{N}\} = g, \quad \inf_Y \{f_\nu \mid \nu \in \mathbb{N}\} = h$$

where $g \in X$, and $g(x) = 0$, with $x \in [0,1]$, while $h \in Y \setminus X$, and

$$h(x) = \begin{vmatrix} 0 & \text{if} & x \in [0,1) \\ 1 & \text{if} & x = 1 \end{vmatrix}$$

therefore $g \neq h$.

Finally, let X be the partially ordered set of three elements

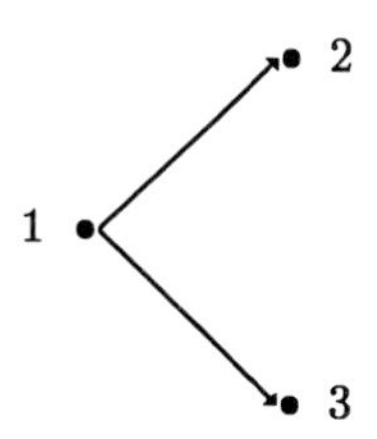

while Y is the partially - in fact, linearly - ordered set of three elements

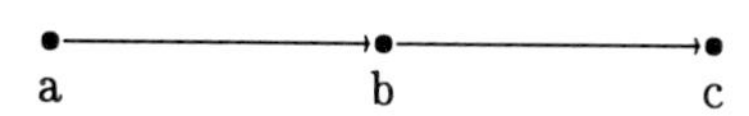

Further, let $\varphi : X \longrightarrow Y$ be defined by $\varphi(1) = a$, $\varphi(2) = b$ and $\varphi(3) = c$. Then φ is clearly bijective and increasing. Nevertheless φ is not an order isomorphical embedding, in particular, it is not an order isomorphism. Indeed, we have $\varphi(2) \underset{\neq}{\leq} \varphi(3)$ in Y, while 2 and 3 are not comparable in X.

Finally, let us indicate why the mapping in (A.78) *cannot* be extended to an injective mapping

$$\prod_{\lambda \in \Lambda} \hat{X}_\lambda \longrightarrow \hat{X}$$

Let us take $\Lambda = \{1,2\}$. If $A_1 \in \hat{X}_1$ and $\phi \underset{\neq}{\subset} A_1 \underset{\neq}{\subset} X_1$, while $A_2 = X_2 \in \hat{X}_2$, then clearly

$$A_1 \times A_2 \notin (X_1 \times X_2)^\wedge = \hat{X}$$

since in view of (A.20), $A_1 \times A_2$ is not bounded above in $X = X_1 \times X_2$. On the other hand, if $A_1 \in \hat{X}_1$ and $\phi \underset{\neq}{\subset} A_1 \underset{\neq}{\subset} X_1$, while $A_2 = \phi \in \hat{X}_2$, then

$$(A_1, A_2) \neq (\phi, \phi)$$

nevertheless

$$A_1 \times A_2 = A_1 \times \phi = \phi \in (X_1 \times X_2)^\wedge = \hat{X}$$

REFERENCES

Abraham, R., Marsden, J.E., Ratiu, T.: Manifolds, Tensor Analysis and Applications. Springer, New York, 1988

Aragona, J., Biagioni, H.A.: An intrinsic definition of the Colombeau algebra of generalized functions. Analysis Math. 17 (1991), 75-132

Baikov, V.A., Gazizov, R.K., Ibragimov, N. Kh.:
[1] Approximate symmetries. Mat Sbor., vol. 136 (178) (1988), no. 3, 427-441

Baikov, V.A., Gazizov, R.K., Ibragimov, N. Kh.:
[2] Perturbation methods in group analysis. J. Sov. Math., vol. 51, no. 1 (1991), 1450-1490

Biagioni, H.A.: A Nonlinear Theory of Generalized Functions. Lecture Notes in Mathematics, vol. 1421, 1990, Springer, New York

Bluman, G.W., Kumei, S.: Symmetries and Differential Equations. Springer, New York, 1989

Bourbaki, N.: Elements des Mathematiques: Livre III, Topologie Generale. Hermann, Paris, 1955

Brosowski, B.: An application of Korovkin's theorem to certain PDEs. In S. Machado (Ed) Functional Analysis, Holomorphy and Approximation Theory, Lecture Notes in Mathematics, vol. 843, 1981, pp. 150-162, Springer, New York

Browder, F.E. (Ed): Mathematical Developments Arising from Hilbert Problems. Proceedings of Symposia in Pure Mathematics, vol. XXVIII, AMS, Providence, 1976

Clifford, A.: Partially ordered abelian groups. Annals of Maths, 41 (1940), 465-473

Colombeau, J.F.: [1] New Generalized Functions and Multiplication of Distributions. North Holland Mathematics Studies, vol. 84, 1984

Colombeau, J.F.: [2] Elementary Introduction to New Generalized Functions. North Holland Mathematics Studies, vol. 113, 1985

Colombeau, J.F.: [3] Multiplication of Distributions. Springer Lecture Notes in Mathematics, vol. 1532, Heidelberg, 1992

Colombeau, J.F., Heibig, A.: Generalized solutions to Cauchy problems. Preprint 1992, Ecole Normale Supérieure de Lyon.

Colombeau, J.F., Heibig, A., Oberguggenberger, M.: Generalized solutions to partial differential equations of evolution type. Preprint 1991, Ecole Normale Supérieure de Lyon.

Colombeau, J.F., Meril, A.: Generalized functions and multiplications of distributions on C^∞ manifolds. Preprint 1992, Ecole Normale Supérieure Lyon

Damsma, M.: [1] An introduction to ultrafunctions. Mem. 841 (1990) Univ. Twente, Enschede

Damsma, M.: [2] Fourier transformation in an algebra of generalized functions. Mem. 954 (1991) Univ. Twente, Enschede

Damyanov, B.P.: Sheaves of Schwartz distributions. IC/91/272 Int. Centre for Theor. Phys., Trieste

Demengel, F., Rauch, J.: Measure Valued Solutions of Asymptotically Homogeneous Semilinear Hyperbolic Systems in One Space Variable. Proc. Edinburgh Math. Soc., 33 (1990), 443-460

De Rham, G.: Differentiable Manifolds. Springer, New York, 1984

De Roever, J.W., Damsma, M.: An algebra of generalized functions on a C^∞ manifold. Indag. Math. 2 (1991), 341-358

Dilworth, R.P.: The normal completion of the lattice of continuous functions. Trans. AMS, 68 (1950), 427-438

Dyson, F.J.: Missed opportunities. Bull. AMS, vol. 78, no. 5, Sept. 1972, 635-652

Egorov, Yu-V.: On generalized functions and linear differential equations (Russian). Vestnik Moskow. Univ., Ser. I, 1990, no. 2, 92-95

Estrada, R., Kanwal, R.P.: Distributional boundary values of harmonic and analytic functions. J. Math. Anal. Appl., 89 (1982), 262-289

Evans, L.C.: Weak Convergence Methods for Nonlinear PDEs. Conference Board of Mathematical Sciences, no. 74, AMS, Providence, 1990

Fine, N.J., Gillman, L., Lambek, J.: Rings of Quotients of Rings of Functions. McGill University Press, Montreal, 1965

Forster, O.: Analysis 3, Integralrechnung im $\mathbb{R}^n$ mit Anwendungen. Friedr. Vieweg, Braunschweig, Wiesbaden, 1981

Gilbarg, D., Trudinger, N.S.: Elliptic Partial Differential Equations of Second Order. Springer, New York, 1977

Gillman, L., Jerison, M.: Rings of Continuous Functions. Van Nostrand, New York, 1960

Gramchev, T.: [1] Le problème de Cauchy pour des systèmes hyperboliques nonlineaires avec données initiales distributions. C.R. Acad. Bulg, Sci., vol. 45, no. 8 (1992), to appear

Gramchev, T.: [2] Semilinear hyperbolic systems and equations with singular initial data. Mh. Math. 112 (1991), 99-113

Grushin, V.V.: A Certain Example of a Differential Equation Without Solutions. Math. Notes, 10 (1971), 449-501

Hellwig, G.: Partial Differential Equations, 2nd ed., Teubner, Stuttgart, 1977

Hermann, R.: [1] Lie Groups for Physicists. Benjamin, New York, 1966.

Hermann, R.: [2] Geometry Physics and Systems. Marcel Dekker, New York, 1973.

Hermann, R.: [3] Quantum Mechanics and Geometric Analysis on Manifolds. Int. J. of Theoretical Physics, 21, 1982, 803-828

Hermann, R.: [4] Geometric Construction and Properties of Some families of Solutions of Nonlinear Partial Differential Equations. J. Math. Phys., 24, 1983, 510-521

Hermann, R: [5] Geometric and Lie-Theoretic Principles in Pure and Applied Deformation Theory, in: M. Gerstenhaber and M. Hazewinkel (eds.), Deformations of Algebras and Applications, D. Reidel, Dordrecht, 1988, 701-796

Hermann, R. and Hurt, N.: Quantum Statistical Mechanics and Lie Group Harmonic Analysis, Lie Groups: History, Frontiers & Applications, vol. 10, 1979. Math Sci Press, Brookline, Mass

Hilbert, D.: Mathematical Problems. Bull. AMS, vol. 8 (1902), 437-479, reprinted in Browder, F.E. (Ed)

Hörmander, L.: Linear Partial Differential Operators. Springer, New York, 1963

Ibragimov, N. Kh.: Transformation Groups Applied to Mathematical Physics. Nauka, Moscow, 1983, English transl., Reidel, Boston, 1985

Illner, R.: Mild solutions of hyperbolic systems with $\mathcal{L}^1_+$ and $\mathcal{L}^\infty_+$ initial data. Transp. Theory and Stat. Phys., 13 (1984), 431-453

Johnson, D.G.: The completion of an Archimedean f-ring. J. London Math. Soc., 40 (1965), 493-496

Kaneko, A.: Introduction to Hyperfunctions. Kluwer, Dordrecht, 1988

Köthe, G.: [1] Topological Vector Spaces I. Springer, New York, 1969

Köthe, G.: [2] Topological Vector Spaces II. Springer, New York, 1979

Lara Carrero, L.: Stability of shock waves. In: Everitt, W.N., Sleeman, B.D. (Eds): Ordinary and Partial Differential Equations. Springer Lecture Notes in Mathematics, vol. 564, 1976, 316-328

Lewy, H.: An Example of a Smooth Linear Partial Differential Equation Without Solution. Ann. Math., vol. 66, no. 2 (1957), 155-158

Lions, J.L.: Quelque Methodes de Resolution des Problemes aux Limites Nonlineaires. Dunod, Paris, 1969

Luxemburg, W.A.J., Zaanen, A.C.: Riesz Spaces I. North Holland, Amsterdam, 1971

Mack, J.E., Johnson, D.G.: The Dedekind completion of C(X). Pacif. J. Math., vol. 20, no. 2 (1967), 231-243

MacNeille, H.M.: Partially ordered sets. Trans. AMS, 42 (1937), 416-460

Meise, R., Taylor, B.A., Vogt, D.: Characterization of the linear partial differential operators with constant coefficients that admit a continuous right inverse. Ann. Inst. Fourier 40 (1990), 619-655

Mikusinski, J.: Sur la methode de generalisation de Laurent Schwartz et sur la convergence faible. Fund. Math. 35 (1948), 235-239

Nakano, K., Shimogaki, T.: A note on the cut extension of C-spaces. Proc. Japan Acad., 38, 8 (1962), 473-477

Narasimhan, R.: Analysis on Real and Complex Manifolds. Masson & Cie, Paris, 1973

Oberguggenberger, M.: [1] Semilinear wave equations with rough initial data : generalized solutions. In Antosik, P., Kaminski, A. (Eds.) Generalized Functions and Convergence. World Scientific Publishing, London, 1990

Oberguggenberger, M.: [2] Multiplication of Distributions and Applications to PDEs. Pitman Research Notes in Mathematics, vol. 259, Longman, Harlow, 1992

Oberguggenberger, M.: [3] The Carleman system with positive measures as initial data - generalized solutions. Transp. Theory and Stat. Phys., 20 (1991), 177-197

Oberguggenberger, M. & Wang, Y.-G.: Delta waves for semilinear hyperbolic Cauchy problems. Preprint 1992

Olver, P.J.: Applications of Lie Groups to Differential Equations. Springer, New York, 1986

Oxtoby, J.C.: Measure and Category. Springer, New York, 1971

Pervin, W.J.: Foundations of General Topology. Acad. Press, New York, 1964

Platkovski, T., Illner, R.: Discrete velocity models of the Boltzman equation: a survey on the mathematical aspects of the theory. SIAM Review, 30 (1988), 213-255

Rauch, J., Reed, M.C.: Nonlinear superposition and absorption of delta waves in one space dimension. J. Funct. Anal. 73 (1987), 152-178

Reed, M.C.: Propagation of singularieties for nonlinear wave equations in one dimension. Com. Part. Diff. Eq. 3 (1978), 153-199

Rosinger, E.E.: [1] Embedding of the $\mathcal{D}'$ Distributions into Pseudo-topological Algebras. Stud. Cerc. Math., vol. 18, no. 5, 1966, 687-729

Rosinger, E.E.: [2] Pseudotopological Spaces. The embedding of the $\mathcal{D}'$ Distributions into Algebras. Stud. Cerc. Math., vol. 20, no. 4, 1968, 553-582

Rosinger, E.E.: [3] Distributions and Nonlinear Partial Differential Equations. Springer Lectures Notes in Mathematics, vol. 684, 1978

Rosinger, E.E.: [4] Nonlinear Partial Differential Equations, Sequential and Weak Solutions. North Holland Mathematics Studies, vol. 44, 1980

Rosinger, E.E.: [5] Generalized Solutions of Nonlinear Partial Differential Equations. North Holland Mathematics Studies, vol. 146, 1987

Rosinger, E.E.: [6] Nonlinear Partial Differential Equations, An Algebraic View of Generalized Solutions. North Holland Mathematics Studies, vol. 164, 1990

Rosinger, E.E.: [7] Global version of the Cauchy-Kovalevskaia theorem for nonlinear PDEs. Acta Appl. Math., vol. 21, 1990, 331-343

Rosinger, E.E.: [8] Characterization for the solvability of nonlinear PDEs. Trans. AMS, 330 (1992), 203-225

Rosinger, E.E., Walus, E.Y.: Group invariance of generalized solutions obtained through the algebraic method. Nonlinearity (to appear).

Schaeffer, D.G.: A regularity theorem for conservation laws. Adv. Math. 11,3 (1973), 368-386

Seebach, J.A. Jr., Seebach, L.A., Steen, L.A.: What Is a Sheaf? Amer. Math. Month. (1970), 681-703

Smoller, J.: Shock Waves and Reaction Diffusion Equations. Springer, Berlin, 1983.

Szpilrajn, E.: Sur l'extension de l'ordre partiel. Fund. Math. 16 (1930), 386-389

Temam, R.: Infinite Dimensional Dynamical Systems in Mechanics and Physics. Applied Mathematical Sciences, vol. 68, Springer, New York, 1988

Treves, F.: Basic Linear Partial Differential Equations. Acad. Press, New York, 1975

Tutschke, W.: Initial Value Problems in Classes of Generalized Analytic Functions. Springer, Berlin, 1989

Varadarajan, V.S.: Lie Groups, Lie Algebras and their Representations. Prentice Hall, Englewood, 1974

Yang, C.T.: Hilbert's fifth problem and related problems on transformation groups. In F.E. Browder (Ed) Mathematical Developments Arising From Hilbert Problems. Proceedings of Symposia in Pure Mathematics, vol. XXVIII, AMS, Providence, 1976, pp 142-146

Zaanen, A.C.: The universal completion of an Archimedean Riesz space. Indag. Math., 45,4 (1983), 435-441

Zaharov, V.: Functional characterization of absolute and Dedekind completion. Bull. Acad. Polon. Sci., XXIX, 5-6 (1981), 293-297

INDEX